DECENTRALISED CONTROL

NORTH-HOLLAND SYSTEMS AND CONTROL SERIES
VOLUME 1

NORTH-HOLLAND PUBLISHING COMPANY
AMSTERDAM • NEW YORK • OXFORD

DECENTRALISED CONTROL

MADAN G. SINGH
Control Systems Centre
University of Manchester
Institute of Science and Technology
Manchester
U.K.

1931 N·H 1981
PC

NORTH-HOLLAND PUBLISHING COMPANY
AMSTERDAM • NEW YORK • OXFORD

Publishers:

NORTH-HOLLAND PUBLISHING COMPANY
AMSTERDAM • NEW YORK • OXFORD

Sole distributors for the U.S.A. and Canada:

ELSEVIER NORTH-HOLLAND, INC.
52 VANDERBILT AVENUE
NEW YORK, N.Y. 10017

ISBN: 0 444 86198 X

Library of Congress Cataloging in Publication Data

Singh, Madan G.
Decentralised control.

(North-Holland systems and control series ; v. 1)
Includes bibliographical references and indexes.
1. Process control. 2. Electronic data processing--Distributed processing. 3. Microprocessors. I. Title. II. Series.
TS156.8.S57 629.8 81-4035
ISBN 0-444-86198-X (Elesevier North-Holland) AACR2

PRINTED IN THE NETHERLANDS

To
Harkrishan
and
Anne-Marie

ACKNOWLEDGEMENTS

This book represents a unification of the literature of decentralised control. I am grateful to all colleagues who contributed to this literature. In particular, the contributions of Varaiya, Ho, Chu, Wang, Davison, Corfmat, Morse, Athans, Sandell, Chong, Witsenhausen, Kurtaran, Aoki, Yoshikawa, Bismut, Geromel, Bernussou, Hassan, etc. were especially inspiring and I have drawn heavily on their works. Many other research workers have also made contributions to this field and they are cited, where appropriate, in the text.

I have made every attempt to provide as broad a treatment as possible. However, I do apologise in advance to any research workers in this field whose work is not cited in this book. My aim was to provide a unified treatment of the subject and I found it necessary to choose only the most significant works. Clearly, since this is the first book on the subject, what I considered significant may not be universally so considered. Nevertheless, I have tried to make the work as balanced as possible.

I am most grateful to my research students, Mr. Armentano, Mr. Xinogalas, Mr. Chen and Mr. Li, who all provided me with material for this book principally in chapters 5 and 6. I am also grateful to Dr. J.P. Elloy from the E.N.S.M., Nantes, whose work I used in writing Chapter 2. I am especially grateful to Mrs. Liz Tongue for her painstaking and diligent typing of the final text, to Mrs. B. Hooley for doing the index, and to Mr. Grace for doing the diagrams. Any errors that remain are my own.

Madan G. Singh

PREFACE

The advent of cheap microprocessors is revolutionising industrial instrumentation and control. It is now possible to use microprocessor based distributed controllers in different parts of the plant. In such a situation the transfer of information and the interconnection of the microprocessor becomes a problem of crucial importance. This problem has been attacked from two different angles in the last few years. On the one hand, computer scientists have been examining the protocols associated with linking computers. There would appear to be little in the way of a comprehensive theory for doing this at least at the present time. On the other hand, control and systems engineers have been examining the control of large scale systems using distributed computing facilities where the control structures are hierarchical or decentralised. Although the field of hierarchical optimisation and control has received much attention with the publication of many articles and of three books (that of Mesarovic et al. in 1970, the book by the present author in 1977, and the one by the author with Titli in 1978), no attempt has been made to bring together the diverse literature on decentralised control. The present book represents an attempt at filling this gap.

The basic literature on decentralised control can be split up into three parts. First of all there is a significant amount of work that has been done on the stochastic decentralised control problem starting with Witsenhausen's important paper of 1968, which describes non-classical information patterns, and the follow on in the area of dynamic teams principally by Ho, Aoki, Athans, Sandell, etc. The other direction, which has been pursued more or less independently, is on the deterministic decentralised control problem using pole placement techniques principally by Davison and Wang, Corfmat and Morse, etc. Some work has also been done on decentralised controller design techniques. The stochastic design techniques have been developed principally by Athans, Chong, Sandell and Aoki. On the deterministic problem, Hassan and Singh, Geromel and Bernussou, etc. have made some contributions. In this book an attempt is made to unify this work in order to present a comprehensive treatment of decentralised control.

The book comprises 6 chapters. Chapter 1 is an introductory chapter which considers hierarchies and decentralisation and the basic decentralised control problem is introduced.

In Chapter 2 microprocessor hardware and software concepts are given and a model due to Varaiya is developed for distributed computing for large scale systems.

In Chapter 3 non-classical information patterns are introduced and team theory is used as an example to illustrate the fundamental difficulties caused by them. In Chapter 4 the stochastic decentralised control problem is continued by considering delay sharing patterns and the design techniques based on fixed structure controllers.

In Chapter 5 the deterministic decentralised control problem is considered from the point of view of pole placement. Here the fundamental results of Wang and Davison and of Corfmat and Morse are outlined. These results are of particular importance in the <u>analysis</u> of decentralised control systems.

The final chapter (6) is devoted to <u>design</u> techniques of deterministic decentralised control.

In the whole book, many examples are given to illustrate the theoretical concepts. The author believes that this should make the book interesting for use as a graduate textbook as well as for self-study by research workers. It should also, hopefully, stimulate more research in this important area.

Madan G. Singh

Manchester, UK, 1981

TABLE OF CONTENTS

CHAPTER 1

HIERARCHIES, DECENTRALISATION AND LARGE SCALE SYSTEMS

1.1 INTRODUCTION

Until very recently, virtually all systems were designed in a decentralised way. Thus, in a power station, the controls for the boilers were designed independently from those of turbines, condensers, synchronous machines, etc. In signalised traffic flow, each intersection's traffic lights changed only in response to the traffic on its approach roads. In chemical plants, the various processes were controlled independently. Unfortunately, such an approach has many drawbacks, the main ones being

a) Since the independently controlled subsystems are in fact interconnected, there is no guarantee that the overall system will remain stable. The blackouts that occurred in France in 1978 and in New York earlier could be attributed to this lack of stability in the interconnected power system.

b) The other drawback is concerned with optimality. Clearly, one could do better in most cases by doing an integrated design. This becomes especially important as the world's non-renewable resources diminish so that optimal or near optimal operation becomes of vital interest.

An integrated design for systems like power stations, large chemical and other industrial plants, environmental and ecological systems, transportation systems, etc. is extremely difficult. The main reason is the complexity and high dimensionality of these systems. Standard multivariable design techniques run into severe computational difficulties once we exceed about 5 inputs and 5 outputs in the frequency domain analysis, or about 15-20 states in the time domain. To get around these computational problems, the last decade has seen the emergence of new design philosophies and the corresponding mathematical tools [1-5]. The basis of the main new philosophy is to use Decomposition. Thus, we recognise that systems in fact often comprise recognisable subsystems which are interconnected in some way. For example, power generation splits in a natural way into steam raising (boilers), electricity generation (turbo-alternators), the grid, etc. Signalised traffic networks split up into junctions or groups of junctions, etc. Standard techniques of multivariable control are applicable to the decomposed subsystems. However, the overall system comprises subsystems which are interconnected. Thus, any design philosophy must take into account the interactions that exist.

In the hierarchical approach [1,2], the overall system is split into subsystems which are treated independently in parallel. A coordinator is then used to take into account the interactions. This is done via an iterative information exchange between the lower level and the coordinator level. Figure 1.1 schematically shows the main idea.

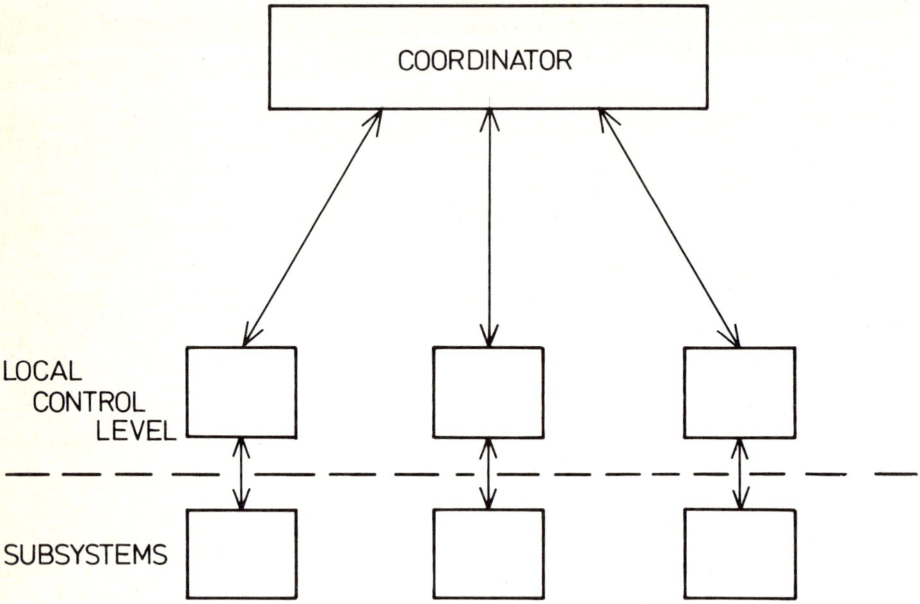

Figure 1.1

Such hierarchies are, of course, quite common in everyday life and particularly in organisations.

From the point of view of information used, hierarchical structures are very similar to centralised structures in that the coordinator can have at least in principle, all the information that the local controllers have. Thus hierarchical structures have what is called a classical information pattern *[6]*. In Chapter 3 we will discuss in detail the notions of classical and non-classical information patterns. Here, we note that, if there had not been a coordination level, then the set of local controllers would only have local information and that would have characterised a non-classical information pattern.

This book deals with decentralised controllers which therefore have non-classical information patterns.

For decentralised control systems, since on the one hand the controllers should take into account the interactions with the other subsystems and on the other, they can only use local information, the classical notions of optimality are no longer meaningful. We will next discuss other optimality criteria for decentralised control.

1.2 OPTIMALITY CRITERIA FOR DECENTRALISED CONTROL *[7]*

The decentralised decision or control problem is characterised by the existence of a number of decision makers whose decisions determine the evolution of the state of a system. If the state of a system is denoted by x and the decision variable for the ith decision maker (or controller) is u_i (i=1,2,...N), then we assume that the evolution of the state is defined by

$$\dot{x} = f(x,u_1,\ldots u_N) \tag{1.2.1}$$

$$t_o \leq t \leq t_f$$

where $x \in R^n$, $u_i \in R^{m_i}$.

The main characteristic of equation (1.2.1) is that no controller can control the whole system on its own. Each controller has a goal which it tries to achieve and the goals of the different controllers may be different. Let us assume that J_i, the goal of the ith controller, is of the form

$$J_i = \int_{t_o}^{t_f} g_i(x,u_1,\ldots u_N)dt \tag{1.2.2}$$

The functions f and g_i, i=1...N will satisfy certain smoothness conditions such that the solution of (1.2.1) exists. To make the problem statement complete, we need to assume an initial and/or final condition for (1.2.1) and specify how t_f, the final time, is defined.

In other chapters of the book, we will extend this formulation somewhat to include noise in (1.2.1). This will enable us to take into account the uncertainties that may exist in the system. Sometimes, instead of a differential equation, we will assume that the system can be described by a difference equation. In that case, the integral in (1.2.2) will be replaced by a summation. In the case where $g_1 = \ldots = g_N$, we have a team decision problem *[8]*. If all the decision makers have the same information about the system, the team decision problem reduces to a classical optimal control problem. If there are only two decision makers and $g_1 \equiv -g_2$, then we have the so-called pursuit-evasion problem. The reason for this is that in pursuit-evasion games, we always have

$$g_1 \equiv -g_2 \equiv 1$$

Problems of this kind are also called two person zero-sum differential games *[9]* because $g_1 + g_2 \equiv 0$.

In order to understand the notion of non-zero sum games, let us consider the problem of two players who both pull a heavy stone through a string. The players can arbitrarily choose the directions in which they pull (2π radians). If one player (say P_1) wants to move the stone as far as possible to the east and the other player (P_2) wants to do the same in the west direction, then the situation is zero-sum. However, if P_2 wants to move the stone as far south as possible then the situation is non-zero sum because the goals are not completely opposite. In the case of three or more decision makers, they cannot all be purely opposite and therefore the theory of many person games and two person non-zero sum games are closely related. In the case of a many-person game, there exists the possibility of coalitions.

Let us now consider the various optimality concepts for decentralised control. We are essentially considering a non-zero sum game situation. The individual controllers (or decision makers) can either cooperate or not cooperate. These lead respectively to the notion of optimality of Pareto and Nash. We note that in the case of a zero-sum game, there is no reason to cooperate and for such games, these concepts are identical.

In the cooperative solution, called the Pareto Solution (after an Italian Railway Engineer), there does not exist a solution which is better for all players. In the rope pulling example that we have considered above where P_1 is heading for the East and P_2 for the South, both players pulling South-East is a pareto solution.

In the non-cooperative situation, one Decision Maker can not unilaterally improve his outcome. Let us suppose that such a solution, also called a Nash solution (after an American economist) is given by $u_1^*, \ldots u_N^*$. Then the definition of a Nash solution is:

$$J_i(u_1^*, \ldots u_N^*) \leqslant J_i(u_1^*, \ldots, u_{i-1}^*, u_i, u_{i+1}^*, \ldots u_N^*) \qquad (1.2.3)$$

$\forall\, u_i$ for $i=1,\ldots N$. Thus, if all the other players stick to their Nash solutions, then P_i can not do better than play u_i^* himself.

Example

As a simple example to illustrate these concepts, let us consider the problem of two decision makers with

$$J_1 = (1 - u_1)^2 + u_2^2$$

and

$$J_2 = (1 - u_2)^2 + u_1^2$$

Then the Nash solution is obtained by solving

$$\frac{\partial J}{\partial u_1} = o \ , \quad \frac{\partial J}{\partial u_2} = o \text{ which yields}$$

$$u_1^* = u_2^* = 1$$

Because of the symmetry of the cost functions, it is obvious that the Nash solution satisfies $u_1 = u_2$. In order to simplify the calculation, let us put $u = u_1 = u_2$ into J_1. If this is then minimised w.r.t. u, then $u = u_1 = u_2 = \frac{1}{2}$ which is clearly different from the Nash solution. It is in fact a Pareto solution.

In the Stackelberg solution, there is a hierarchy amongst the players. In the case of two players, there is a leader and a follower. The leader has to announce his decision ahead of time. The follower can use this information to make up his mind. In the multiple player case, one can visualise either multiple levels (e.g. P_1 is the leader who announces his decision first which will then be known to P_2 and P_3. Subsequently P_2 announces his strategy which will be known to P_3 before P_3 takes his decision) or of more than one leader or more than one follower. The followers can again play Pareto or Nash amongst themselves.

1.2.1 Comments

The important thing to note is that in the decentralised control or decision making situation the classical notion of optimality is no longer meaningful since there is no central agency available for coordination purposes. The decentralised control problem is thus a fundamentally different problem from the centralised one. In Chapter 3, we will see the role information plays in such problems. Here we go on to the hierarchical problem and see how, under certain realistic conditions, it could become a decentralised problem. However, before we discuss that, let us outline some preliminary notions of stability which we will need in our subsequent analysis.

1.3 INTRODUCTION TO STABILITY THEORY *[10]*

Consider a system (mathematical, physical, etc.) which possesses a certain property or is operating at a certain state. Clearly, it is desirable to know how this property or state will be affected by a perturbation of some aspect of the system. Intuitively, for most systems we should like these disturbances to have only a minimal effect on the property in question, and if this is the case, then we shall say it is stable with respect to perturbations of the class being considered.

Let us consider a dynamical system which is in equilibrium and suppose that it is subjected to disturbances which upset this balance (that is, we change the initial state). The problem of stability is then to investigate the resulting behaviour of the system. In particular, we would like to know if the system will continue to operate near the equilibrium (perhaps even returning to it) or if it deviates more and more.

As a simple example, if we have a particle in a vertical gravitational field, with the state of the system described by the position, and the dynamics governed by Newton's second law, then we will call a trajectory of the system stable if small changes in the initial state, x_o, give rise to small deviations from the given trajectory. Figure 1.2 shows the case when x_o is an equilibrium point (if the particle tends to equilibrium as $t \to \infty$, when the stability is asymptotic, as in Fig. 1.2(a)).

Note: We must have some kind of friction between the particle and surface, since otherwise the motion will continue for all time in cases (a) and (b).

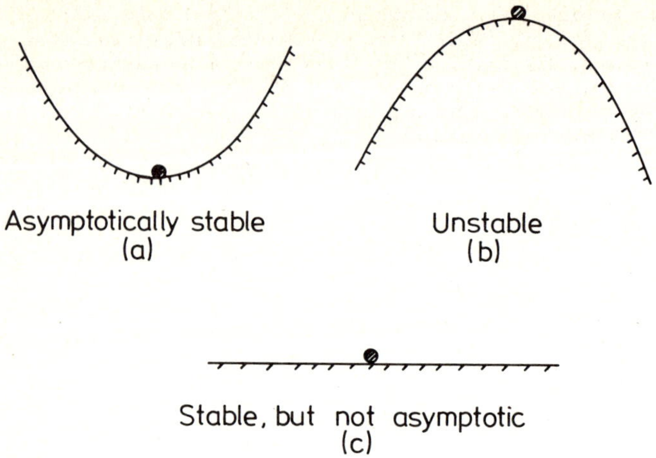

Figure 1.2

We now begin a formal description of this type of stability.

1.3.1 Definition

An equilibrium point $\bar{x} \varepsilon X$ of a dynamical system Σ, with time domain T, unbounded to the right, and a normed state space, $(X, ||.||)$ is said to be stable if $\forall \varepsilon > 0, t_o \varepsilon T, \exists\ \delta = \delta(\varepsilon, t_o) > 0$ such that $\forall x \varepsilon X, t \varepsilon T$ with $t \geqslant t_o$

$$||x-\bar{x}|| < \delta \Rightarrow ||\phi(t;t_o,x)-\bar{x}|| < \varepsilon$$

We may picture this as in Fig. 1.3 or Fig. 1.2(c)

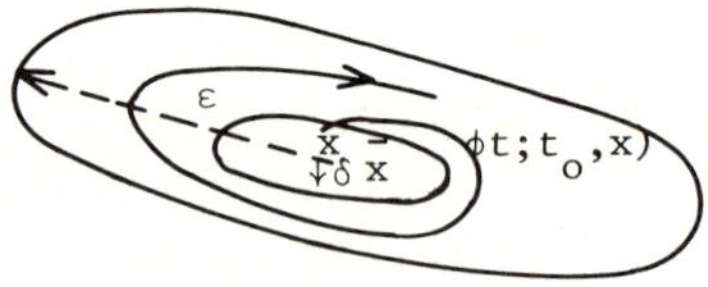

Figure 1.3

It is often required that the system will not only be stable in this sense, but also that, for the initial state sufficiently close to the equilibrium $\bar{x}$, the trajectory $\phi(t;t_o,x)$ becomes closer to $\bar{x}$ as t increases (c.f. Fig. 1.2(a)). We therefore have

Definition

Under the conditions of definition 1.3.1, the equilibrium point $\bar{x}$ is said to be asymptotically stable if it is stable, and in addition, it is attractive i.e. $\forall t_o \varepsilon T, \exists \delta(t_o) > 0$ such that

$$||x-\bar{x}||<\delta(t_o) \Rightarrow \lim_{t\to\infty} ||\phi(t;t_o,x) - \bar{x}|| = 0$$

In this case, the set

$$A(t_o) = \{x\varepsilon X; \lim_{t\to\infty} \phi(t;t_o,x) = \bar{x}\}$$

is called the region of attraction, or the region of asymptotic stability of $\bar{x}$ at time t_o (c.f. Fig. 1.4).

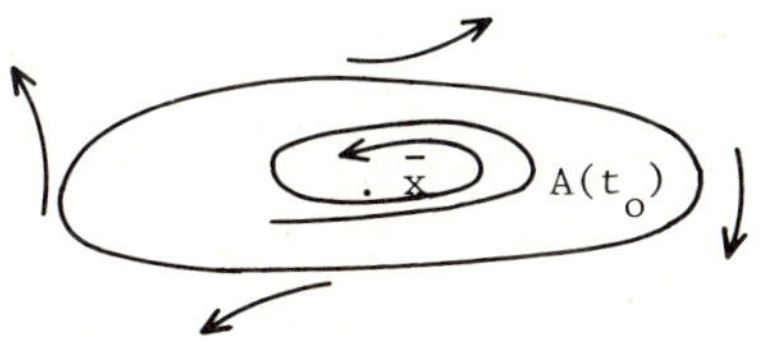

Figure 1.4

Consider now the problem of obtaining stability criteria for non-linear differentiable systems of the form

$$\dot{x}(t) = f(x(t)) \quad x(o) = x_o \ \varepsilon \ R^n \tag{1.3.1}$$

where f is locally Lipschitz on an open domain $G\varepsilon R^n$.

It is often difficult to obtain explicit solutions to this equation. However, in 1892, A.M. Lyapunov developed a method for studying the stability of such systems by considering the function f directly (that is, by examining the vector field defined by (1.3.1)). This technique is known as Lyapunov's direct method and was inspired by the use of the energy function in the analysis of classical mechanical systems, and we shall consider a generalisation of this function.

Before giving a formal description, we first adopt a heuristic approach. Consider, therefore, a time invariant mechanical system having n-degrees of freedom. Its state will be described by n generalised position coordinates, $q_1,\ldots,q_n$ and n generalised momentum coordinates, $p_1,\ldots,p_n$. The kinetic energy, T, of the system is a positive definite quadatic form in p, and from time invariance, the potential energy, W, depends only on the generalised position coordinates $(q_i)^n_{i=1}$. The Hamiltonian function H(q,p) is the total energy of the system

$$H(q,p) = W(q) + T(p) \tag{1.3.2}$$

We derive the following equations of motion from Hamiltons principle

$$\dot{q} = {}^{\partial H}/_{\partial p} \qquad \dot{p} = {}^{-\partial H}/_{\partial q}$$

Thus a state (q_o, p_o) of the system is an equilibrium point if and only if all partial derivatives vanish at this point. At such a state, there are no generalised forces working on the system. For simplicity, assume $(q_o, p_o) = (o,o)$ and $H(o,o) = 0$.

Now suppose that the potential energy $W(q)$ (or equivalently the whole energy $H(q,p)$) has a strict local minimum at $q = 0$. If the system is disturbed from the equilibrium point, so that $q_i(t) \neq o$, $p_i(t) \neq o$ for some $i \varepsilon \{1,\ldots,n\}$, then $W(q(t))$ increases, and from the law of conservation of energy

$$\frac{d}{dt} H(q(t)),p(t)) = 0 \tag{1.3.3}$$

it follows that the kinetic energy must decrease, thus tending to restore the particle to the equilibrium position. Hence, the system is stable.

We have therefore shown that the mechanical system is stable using the function $H(.)$. We now generalise this concept of an energy function in order to study the stability properties of equilibria of non-linear differential equations.

<u>Definition: Lyapunov Function</u>

Suppose that the differential equation (1.3.1) is defined on some domain $G \subset \mathbb{R}^n$ and let G_o be an open connected neighbourhood of 0 $\mathbb{R}^n$ with $\bar{G}_o \subsetneq G$. A continuous function $V(.): \bar{G}_o \rightarrow R$ which is continuously differentiable on G_o is said to be a Lyapunov function for (1.3.1) on G_o if

(i) $V(o) = 0$ and $V(s) > 0 \ \forall s \varepsilon G_o - (o)$

(ii) $\sum_{i=1}^{n} \frac{\partial v(s)}{\partial s_i} f_i(s) \leqslant 0 \quad \forall s \ G_o$

<u>Remarks</u>

(i) The function $\dot{V}(.)$: $G_o \rightarrow \mathbb{R}$ given by

$$\dot{V}(s) = \sum_{i=1}^{n} \frac{\partial v(s)}{\partial s_i} f_i(s) = \langle \text{grad } V(s), f(s) \rangle \tag{1.3.4}$$

is called the derivative of V along the solution of (1.3.1) since the following equation holds for every solution $x(t)$ of (1.3.1).

$$\dot{V}(x(t)) = \sum_{i=1}^{n} \frac{\partial V}{\partial x_i} (x(t)) \quad \Gamma_i(x(t)) = \sum_{i=1}^{n} \frac{\partial V(x(t))}{\partial x_i} \frac{dx_i(t)}{dt} \tag{1.3.5}$$

$$= \frac{d}{dt} V(s(t))$$

(ii) If V is a Lyapunov function for (1.3.1) on G_o, and x(.) is a solution of (1.3.1), then it follows that V(x(t)) does not increase as long as x(t) remains in G_o (cf. Fig. 1.5)

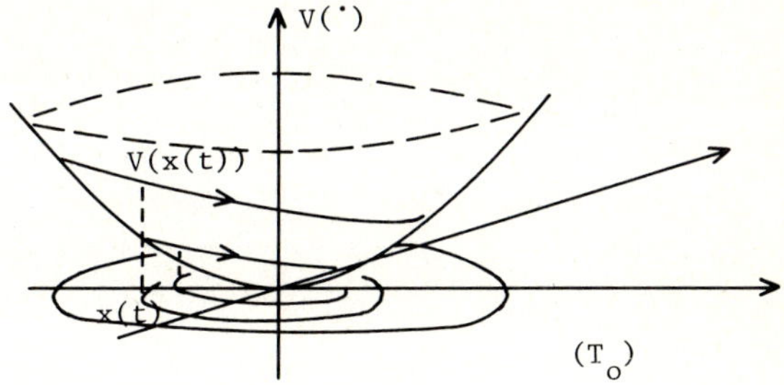

Figure 1.5

It can be shown that the existence of a Lyapunov function guarantees the existence of solutions of (1.3.1) on an unbounded interval $[t_o,\infty]$ for certain initial values x_o.

Proposition

Let $V(.) : \bar{G} \to \mathbb{R}$ be a Lyapunov function for (1.3.1) and suppose that for some $x_o \varepsilon G_o$

(i) $\{x\varepsilon G_o; V(x) \leq V(x_o)\}$ is bounded

(ii) $V(x) > V(x_o)$ for every $x_o \varepsilon \partial G_o$ (the boundary of G_o).

Then the initial value problem

$$\dot{x}(t) = f(x(t)) \quad x(t_o) = x_o, \quad x(t) \varepsilon G_o \tag{1.3.6}$$

has a unique solution on $[t_o,\infty]$ (for each $t_o \varepsilon \mathbb{R}$)

This leads us to Lyapunov's first stability theorem.

Theorem 1.3.1 (First stability theorem of Lyapunov)

Let 0 be an equilibrium point of (1.3.1), i.e. f(o) = 0. If there exists a Lyapunov function V(.) on an open neighbourhood G_o of 0, then (1.3.1) is stable at the origin, i.e. $\forall \varepsilon>o, \exists \delta>o$ such that for every $x_o \varepsilon B = \{x\varepsilon R^n; |x|<\delta\}$ there exists a unique solution $x(.,x_o)$ of (1.3.6) on $\delta[t_o,\infty)$ with $|x(t,x_o)|<\varepsilon \forall t \geq t_o$.

In order to state Lyapunov main theorem, the following definition is required.

Definition

Let $T \subseteq \mathbb{R}$ be a time domain which is unbounded to the right, and consider $x : T \to \mathbb{R}^n$

(i) $\gamma \in \mathbb{R}^n$ is called a (positive) limit point of x(.) if there exists a sequence $(t_n)_{n \in N}$ in T with $\lim_{n\to\infty}(tn) = \infty$ and $\gamma = \lim_{n\to\infty} x(t_n)$

(ii) The (positive) limit set $\Omega(x(.))$ of x(.) is the set of all positive limit points of x(.)

(iii) Consider the differential equation (1.3.1) and assume it has global solutions $x(.;x_o)$ on $G \subseteq \mathbb{R}^n$. A subset, K, of G is called (positively) invariant if

$$x_o \in K \Rightarrow x(t;x_o) \in K \;\forall t \geq 0$$

Note: We may replace the sequence $(t_n)_{n \in N}$ in (i) of this definition for a sequence $(S_n)_{n \in N} \to \leftarrow \infty$ as $n \to \infty$ and we then obtain negative limit set points and invariant sets.

Lemma

Let $T \subseteq \mathbb{R}$ be unbounded to the right, $x(.): T\; \mathbb{R}^n$ be bounded and continuous and $\Omega = \Omega(x(.))$ be the (positive) limit set of x(.). Then

(i) Ω is nonvoid, compact and connected (whenever T is an interval)

(ii) Ω is attractive for x(.), i.e. $\lim_{t\to\infty} d(x(t);\Omega) = 0$

(iii) Ω is a single point if and only if $\lim_{t\to\infty} x(t)$ exists

(iv) Ω is invariant if x(.) is a trajectory such that $\overline{(x(t);t \in T)} \leqslant G$

We may now state

Theorem 1.3.2 (Invariance Principle of La Salle)

Consider the diff. eqn. (1.3.1) and suppose that V(.) is a Lyapunov function on an open neighbourhood G_o. Let x(.) be a solution of (1.3.1) which remains in a compact subset K of G_o. Then if $E = \{x \in K; \dot{V}(x) = 0\}$, $\Omega(x(.)) \subseteq E$ and if M = maximal invariant subset of E, then $x(t) \to M$ as $t \to +\infty$

Proof

Since V(x(t)) is nonincreasing for $t \geq 0$ and is bounded below, $k = \lim_{t\to\infty} V(x(t))$ exists. If $y \in \Omega$, then V(y) = k so also V(x(t;y)) = k, hence $\dot{V}(y) = 0 \forall y \in \Omega$, i.e. $\Omega \subseteq E$.

Theorem 1.3.1 (Lyapunov's Main Stability Theorem)

Let 0 be in equilibrium of (1.3.1), i.e. f(o) = 0 and let V(.) be a Lyapunov function on an open bounded neighbourhood G_o of the origin. If $\dot{V}$ is negative definite on G_o, that is

$$\dot{V}(0) = 0,\; \dot{V}(x) < 0 \;\forall x \in G_o - (0) \tag{1.3.7}$$

Then the equilibrium 0 is asymptotically stable.

Example

Consider a system governed by Duffings Equation

$$\ddot{x}(t) + c\dot{x}(t) + (ax(t) + bx^3(t)) = 0 \qquad a,b,c<0$$

We first reduce this to a pair of first order equations of the form

$$\dot{x}_1(t) = x_2(t)$$

$$\dot{x}_2(t) = -cx_2(t) - (ax_1(t) + bx_1^3(t))$$

the function

$$V(x_1,x_2) = \alpha x_1^2 + \beta x_1^4 + \gamma x_2^2 \qquad (\alpha,\beta,\gamma > 0)$$

has $$\frac{dV}{dt} = 2\alpha x_1\dot{x}_1 + 4\beta x_1^3\dot{x}_1 + 2\gamma x_2\dot{x}_2$$

$$= 2\alpha x_1x_2 + 4\beta x_1^3x_2 + 2\gamma x_2\,(-cx_2(t) - (ax_1(t) + bx_1^3))$$

$$= (2\alpha-2a\gamma)x_1(t)x_2(t) + (4\beta-2b\gamma)x_1^3x_2 - 2\gamma c\, x_2^2$$

If $a\gamma = \alpha$ and $b\gamma = 2\beta$ then

$$\frac{dV}{dt} = -2\gamma c\, x_2^2(t) < 0 \qquad \forall x_2(t) \neq 0 \text{ since } \gamma,c>0.$$

So V defines a Lyapunov function. Hence, by theorem 1.3.1 the origin (which is the only equilibrium point) is stable.

However, the set $E = \{x\varepsilon\mathbb{R}^2;\ \dot{V}(t) = 0\}$ is the x_1 -axis, and clearly the maximal invariant set in E is therefore $\{(0,0)\}$ (This is because $\dot{V}$ does not vanish identically along any solution.) Hence, by theorem 1.3.2 the origin is globally asymptotically stable.

Having discussed some basic stability notions, we now continue with our analysis of decentralised systems by analysing hierarchies and see what happens when they collapse into decentralised control structures.

1.4 STABILITY AND PERFORMANCE OF HIERARCHICAL CONTROLLERS UNDER STRUCTURAL PERTURBATIONS [11,12]

A useful way of looking at decentralised controllers is to consider them to be an extreme case of hierarchical structures under structural perturbations where all the links between the coordinator and the subsystem controllers have been cut. In this section, we will study the effect of loss of information resulting from the breakdown of some of the links within the controller hierarchy. Specifically, we will study the stability of the resulting controller as well as the loss of optimality associated with this new system having incomplete coordination.

Consider the problem

$$\text{Min } J = \tfrac{1}{2}\int_0^\infty (||x(t)||_Q^2 + ||u(t)||_R^2)\,dt \tag{1.4.1}$$

subject to

$$\left.\begin{aligned} \dot{x} &= Ax + Bu + Cz \\ z &= Lx \end{aligned}\right\} \tag{1.4.2}$$

where $||b||_S^2 = b^T Sb$ and A, B, Q, R are block diagonal matrices and L is a full matrix. Such a system comprises N subsystems with the states of the other subsystems being connected as inputs into any subsystem through z.

The problem of minimising (1.4.1) subject to (1.4.2) can be considered as a regulator problem with a special information structure. Under the assumptions that (i) the pair A_i, B_i ; i=1,...,N is completely controllable and (ii) Q_i and R_i are positive semidefinite and positive definite matrices respectively, the optimal solution of this problem can be written as:

$$u(t) = -R^{-1}B^TPx(t) - \xi(t) \tag{1.4.3}$$

$$= -Gx(t) - \xi(t)$$

where P is a block diagonal matrix with each block P_i representing the symmetric positive definite solution of an appropriate Riccati equation

$$A_i^TP_i + P_iA_i - P_iB_iR_i^{-1}B_i^TP_i + Q_i = 0 \tag{1.4.4}$$

It has also been shown *[13]* that

$$\xi(t) = Tx(t) \tag{1.4.5}$$

where T is a full matrix with constant parameters.

From equations (1.4.3) and (1.4.5)

$$u(t) = -Gx(t) - Tx(t)$$

Now the optimal cost for this problem is given by

$$J^* = \tfrac{1}{2}\int_0^\infty ||x(t)||_{Q+W}^2 \quad dt \tag{1.4.6}$$

where $\quad W = (G + T)^T R (G + T)$

This solution is the optimal solution for the global system. Now, it is known *[14]* that if $Q = D D^T$ where Q = diagonal $\{Q_1,Q_2,\ldots,Q_N\}$ such that the pair *[*A_g,D*]* is completely observable (where A_g* is the "A" (or system) matrix

---

* Note that A_g is not the same as the matrix A in equation (1.4.2) since there A is block diagonal, whilst A_g is a full matrix obtained by substituting for z = Lx for z in $\dot{x} = Ax + Bu + Cz$ i.e. $A_g = (A + CL)$.

of the composite system), then the optimal closed loop system is asymptotically stable for a global or hierarchical implementation of the control. The hierarchical implementation here is in the sense of reference [13] i.e. the control u_i for the ith subsystem consists of two parts

$$u_i = -(G_i x_i + \sum_{j=1}^{N} T_{ij} x_j) \qquad (1.4.7)$$

where G_i is obtained by solving the Riccati equation for the subsystem i when interactions between subsystems are ignored, whilst T_{ij} is the ijth block of the matrix T in equation (1.4.5). Thus in the hierarchical implementation $G_i x_i$ is fedback locally to the subsystem i whilst $\sum_{j=1}^{N} T_{ij} x_j$ is supplied by the coordinator. Figure 1.6 shows the implementation of the controller within a hierarchical structure where T acts as a gain matrix for the coordinator whilst the block diagonal G matrix provides local feedback to which must be added the coordinator's input to provide the optimal control.

1.4.1 Stability of the system under certain structural perturbations

Now we will first examine the system in order to investigate the worst perturbed structure from the stability point of view. Then we will give a sufficient condition for this case the satisfaction of which will guarantee the stability of the system under any other perturbation of the communication links between the coordinator and subsystems. It seems intuitively reasonable that the worst perturbed structure is when all the links between the coordinator and subsystems are cut, i.e. under decentralised control. The physical interpretation of this is that, since we have assumed that the pair $\{A_g, D\}$ is completely observable, then the closed loop system is asymptotically stable. This means that, the optimal global gain matrix [G + T] besides insuring the requirements of forcing the system towards the steady state with minimum energy, insures the placement of the poles of the closed loop system inside a certain cone in the left half complex plane. If it is assumed that the system is completely decoupled and that each pair $\{A_i, D_i\}$ where $Q_i = D_i D_i^T$, is completely observable, the block diagonal gain matrix G insures the same requirements. However, with the presence of coupling, some of the poles of the closed loop system with the decentralised gain matrix G may be pushed towards the R.H.S. of the complex plane or at least outside that cone and the role of the matrix T is to push these poles inside the cone.

Since cutting a link means that the particular row of the matrix has all zero elements, that row will not contribute to bringing the appropriate poles from the unstable region of the desired cone. It would therefore appear intuitively reasonable that in the extreme case where all the links are cut, we have a zero T matrix and this is the worst case.

To study the system more precisely let us assume that we have cut some links (say ℓ) between the coordinator and subsystems. Thus we have basically:

(a) ℓ subsystems operating in a suboptimal manner which can be represented by:

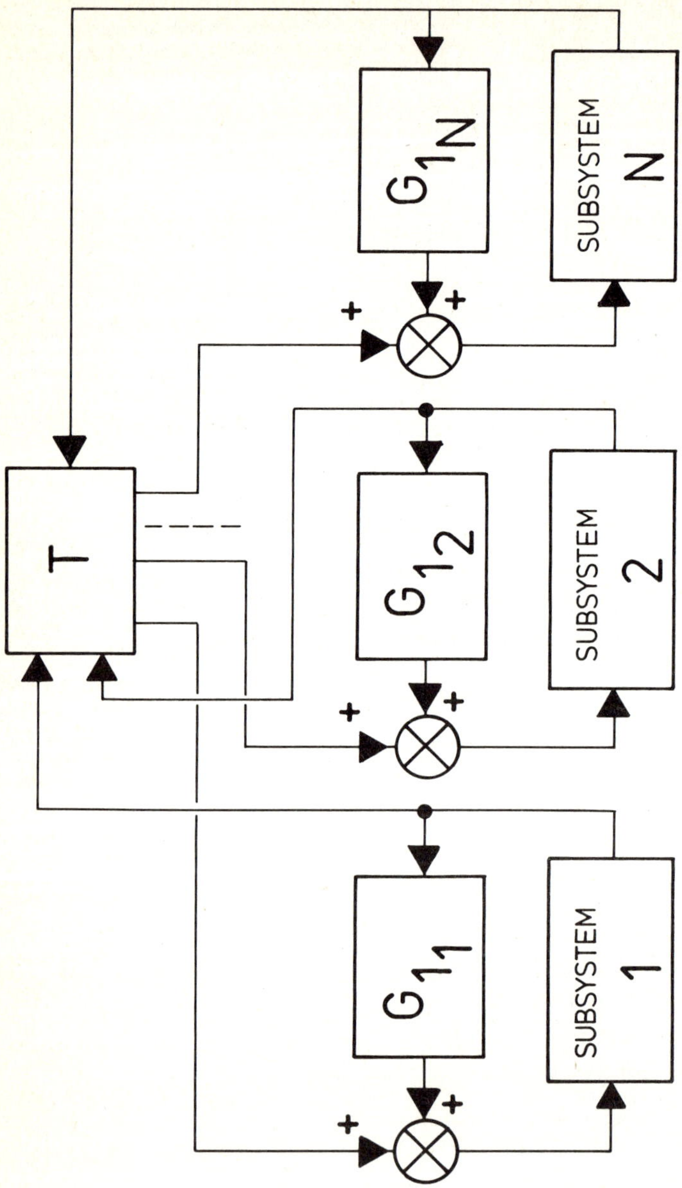

Figure 1.6

$$\dot{x}_p = (A_p - B_p G_p)x_p(t) + C_p \sum_{j=1}^{N} L_{pj}x_j \tag{1.4.8}$$

where $p = 1,2,\ldots, \ell$; $0 \leq p \leq N$

(b) N-ℓ subsystems receiving "optimal" coordination ("optimal" in relation to the original system) and described by

$$\dot{x}_i(t) = (A_i - B_i G_i)x_i(t) + C_i \sum_{j=1}^{N} L_{ij}x_j - B_i \sum_{k=1}^{N} T_{ik}x_k(t) \tag{1.4.9}$$

where $1 \leq i \leq N$; $i \neq p$

Then the composite system can be represented by the equation

$$\dot{x}_{sub}(t) = (A - B\ G)x_{sub}(t) + C\ L\ x_{sub}(t) - BT'\ x_{sub} \tag{1.4.10}$$

where x_{sub} is the suboptimal trajectory resulting from the previously described perturbation,

$$T' = T - T'' = T - D\ T$$

where T" is the part of the matrix T corresponding to the perturbed links and D is a diagonal matrix with diagonal elements 1 and 0, where the zeroes correspond to the links which are not perturbed.

Then equation (1.4.10) can be rewritten as:

$$\dot{x}_{sub} = (A - BG)\ x_{sub}(t) + CL\ x_{sub}(t) - BTx_{sub}(t) + BY''\ x_{sub}(t) \tag{1.4.11}$$

Now, let us consider the following Lyapunov function:

$$V(t) = x_{sub}^T\ S\ x_{sub}(t) \tag{1.4.12}$$

where S is any positive definite symmetric matrix satisfying Lyapunov's second method. For the system to be stable (and since V(t) is positive definite) $\dot{V}(t)$ must be negative semidefinite. Therefore we have:

$$\dot{V}(t) = \dot{x}^T_{sub}\ S\ x_{sub}(t) + x^T_{sub}(t)\ S\ \dot{x}_{sub}(t) \leq 0 \tag{1.4.13}$$

From (1.4.13) in (1.4.14) we have:

$$x^T_{sub}\left[A^TS - G^TB^TS + L^TC^TS - T^TB^TS + T''^TB^TS + SA - SBG + SCL - SBT + SBT''\right] x_{sub} \leq 0$$

$$x^T_{sub}\left[A^TS + SA + L^TC^TS + SCL + T''^TB^TS + SBT''\right]x_{sub} \leq$$

$$x^T_{sub}\left[G^TB^TS + SBG + T^TB^TS + SBT\right]x_{sub} \tag{1.4.14}$$

A sufficient condition for the satisfaction of (1.4.14) is that:

$$2||A^TS|| + 2||L^TC^TS|| + 2||T''^TB^TS|| \leqslant \lambda_{min} (G^TB^TS + SBG + T^TB^TS + SBT)$$

where $\lambda min (\gamma)$ represents the minimal eigenvalue of the matrix γ

since $\quad T''^TB^TS|| \leqslant ||T''^T|| \; ||B^TS||$

then:

$$2||A^TS|| + 2||L^TC^TS|| + 2||T''^T|| \; ||B^TS|| \leqslant \lambda_{min}(G^TB^TS + SBG + T^TB^TS + SBT) \tag{1.4.15}$$

Noting that in the above equation all the terms are constants except T". Since the worst structure is that one which gives the largest values on the L.H.S. of (1.4.15) and since we have:

$$T'' = DT$$

$$||T''|| \leqslant ||D|| \quad ||T|| \qquad\qquad ||D|| = 1$$

$$||T''|| \leqslant ||T||$$

Then the worst condition is given by:

$$2||A^TS|| + 2||L^TC^TS|| + 2||T''^T|| \; ||B^TS|| \leqslant \lambda_{min}(G^TB^TS + SDG + T^TB^TS + SBT) \tag{1.4.16}$$

i.e. there is no case which is worse than the one where all the links between the coordinator and subsystems are cut.

Having discussed this worst condition, let us state the following theorem:

Theorem 1.4.1

Under the worst condition, the system will remain asymptotically stable if the following sufficient condition is satisfied:

$$2||C^TP|| \leqslant \lambda_{min} (Q + G^TRG) \tag{1.4.17}$$

Proof:

Considering now the worst case, we have:

$$\dot{\underline{x}}_{sub} = (A - BG) \; \underline{x}_{sub}(t) + CL \; \underline{x}_{sub}(t) \tag{1.4.18}$$

Choosing the following Lyapunov function:

$$V(t) = x_{sub}^T \; P \; \dot{x}_{sub}$$

where P is the block diagonal symmetric positive definite solution of the Riccati equation. Then we have:

$$\dot{V}(t) = \dot{x}_{sub}^T \; P \; x_{sub} + x_{sub}^T \; P \; \dot{x}_{sub}$$

$$= x_{sub}^T \left[A^TP - G^TB^TP + L^TC^TP + PA - PBG + PCL\right] x_{sub}$$

For the system to be stable $\dot{V}(t)$ must be negative semidefinite, i.e.:

$$x_{sub}^T \left|A^TP - G^TB^TP + L^TC^TP + PA - PBG + PCL\right| x_{sub} \leqslant 0$$

Since $\quad RG = B^TP$, and

$$A^{T''} + PA - PBR^{-1}B^TP + Q = 0$$

we have:

$$x_{sub}^T \left[L^TC^TP + PCL - Q - G^TRG\right] x_{sub} \leqslant 0$$

or

$$x_{sub}^T \left[L^TC^TP + PCL\right] x_{sub} \leqslant x_{sub}^T \left[Q + G^TRG\right] x_{sub} \tag{1.4.19}$$

A sufficient condition for satisfying equation (1.4.19) can be given by:

$$2||L^TC^TP|| \leqslant \lambda_{min}\ (Q + G^TRG) \tag{1.4.20}$$

since $\quad ||L^TC^TP|| \leqslant ||L^T||\ \ ||C^TP||$ and since L can be chosen in equation (1.4.2) such that $||L|| = 1$

Then we have:

$$2||C^TP|| \leqslant \lambda_{min}\ (Q + G^TRG) \tag{1.4.21}$$

which proves the theorem.

Remarks

From the stability condition (1.4.21), it is possible to draw the following qualitative conclusions:

1) If $||C^T||$ is small (i.e. the case of systems with weak coupling) the system will be stable under the perturbations examined here. However, if $||C||$ is not small, and equation (1.4.21) is not satisfied then we can redesign the system by specifying a degree of stability which insures the satisfaction of (1.4.21) and hence the stability of the system under any perturbations.

2) The above result is also valid if one or more of the observed states of the subsystems are not available to the second level controller since in this case the matrix T" will contain columns instead of rows, i.e. T" = TD.

Having studied the stability of the system, we next consider the loss of optimality under structural perturbations, i.e. specifically, what happens when one or more of the links between the coordinator and the subsystems are broken.

1.4.2 Performance Degradation under Certain Structural Perturbations

An expression for the upperbound for the suboptimality will now be developed for stable systems under structural perturbations which can be calculated a priori.

Let us assume again that the links of ℓ subsystems have been broken so that:

$$\xi_r = 0 \qquad\qquad 0 \leqslant r \leqslant N$$

The theorem (1.4.2) below gives a bound on the resulting performance degradation.

Theorem 1.4.2

Composite systems which are asymptotically stable under any structural perturbations, will have a performance degradation ε where:

$$0 \leqslant \varepsilon \leqslant \frac{\mu_M(F_m(t))}{\mu_m(Q+W)} \tag{1.4.22}$$

where $\mu_M(\beta)$ = norm of the matrix β.

$\mu_m(\pi)$ = min eigenvalue of the matrix π

$$F_m(t) = \left[-I + \lambda(t,t_0)\right]^T [Q+W] \left[-I + \lambda(t,t_0)\right]$$

$$+ 2\,(Q + W)\left[-I + \lambda(t,t_0)\right]$$

$$+ \lambda(t,t_0)T^T\, R\, T\, \lambda(t,t_0) - 2\,\lambda^T(t,t_0)\,(G+T)^T RT\,\lambda(t,t_0) \tag{1.4.23}$$

$$\lambda(t,t_0) = \left[I + \Gamma(t,t_0)\,\emptyset(t_0,t)\right]$$

$$= \emptyset'(t,t_0)\,\emptyset(t_0,t) \tag{1.4.24}$$

$$\Gamma(t,t_0) = \int_0^t \emptyset(t,\tau)\, BT''\, \emptyset'(t,\tau)\, d\tau \tag{1.4.25}$$

$$\dot{\emptyset}(t,t_0) = (A - BG + CL - BT)\,\emptyset(t,t_0) \tag{1.4.26}$$

$$\dot{\emptyset}'(t,t_0) = (A - BG + CL)\,\emptyset'(t,t_0) \tag{1.4.27}$$

Proof

Here from equation (1.4.10) we have

$$\dot{x}_{sub}(t) = (A - BG)\,x_{sub}(t) + CL\,x_{sub}(t) - BT\,x_{sub} + BT''\,x_{sub} \tag{1.4.28}$$

and the cost function becomes

$$J_{sub} = \tfrac{1}{2} \int_0^{\infty} (||x_{sub}||_Q^2 + ||u_{sub}||_R^2) dt$$

$$= \tfrac{1}{2} \int_0^{\infty} ||x_{sub}||_Q^2 + x_{sub}^T(t) \; (G + T - T'')^T \; R(G + T - T'')x_{sub} dt$$

$$= \tfrac{1}{2} \int_0^{\infty} (||x_{sub}||_Q^2 + ||x_{sub}||_W^2 - 2||x_{sub}||^2_{(G + T)^T RT''}$$

$$+ ||x_{sub}||^2_{T''^T RT''} \qquad (1.4.29)$$

From equation (1.4.28)

$$x_{sub} = \emptyset(t,t_0)x_{sub}(t_0) + \int_0^t \emptyset(t - \tau) \; BT'' \; x_{sub}(\tau)d\tau$$

$$= x^*(t) + \Gamma(t,t_0) \; x(t_0) \qquad (1.4.30)$$

$$= x^*(t) + \tilde{x}(t) \qquad (1.4.31)$$

where

$$\Gamma(t,t_0) = \int_0^t \emptyset(t,\tau) \; BT'' \; \emptyset'(t,\tau)d\tau$$

$$\dot{\emptyset}(t,t_0) = (A - BG + CL - BT) \; \emptyset(t,t_0)$$

$$\dot{\emptyset}(t,t_0) = (A - BG + CL - BT + BT'') \; \emptyset'(t,t_0)$$

Now since $x^*(t) = \emptyset(t,t_0) \; x(t_0)$

$$x(t_0) = \emptyset(t_0,t) \; x^*(t)$$

Then equation (1.4.31) becomes

$$x_{sub} = x^*(t) + \Gamma(t,t_0) \; \emptyset(t_0,t) \; x^*(t)$$

$$= \lambda(t,t_0) \; x^*(t)$$

where $\lambda(t,t_0) = I + \Gamma(t,t_0) \; \emptyset(t_0,t)$

which is also given by

$$\lambda(t,t_0) = \emptyset'(t,t_0) \; \emptyset(t_0,t)$$

Therefore substituting x_{sub} in equation (1.4.29) by $x_{sub} = x^* + x$

$$J_{sub} = \frac{1}{2}\int_0^\infty ||x^* + \tilde{x}||^2_{G+W} + ||x_{sub}||^2_{T''^T RT''} - 2||x_{sub}||^2_{(G+T)^T RT''} dt$$

$$= \frac{1}{2}\int_0^\infty \{||x^*||_{Q+W} + ||\tilde{x}||^2_{Q+W} + 2\, x^{*T}\ (Q+W)\ \tilde{x}$$

$$+ ||x_{sub}||^2_{T''^T RT''} - 2||x_{sub}||^2_{(G+)^T RT''}\}\ dt$$

Since $x(t) = x_{sub} - x^*$

$$= (\ (t,t_0) - I)\, x^*$$

or

$$J_{sub} = J^* + \int_0^\infty \{||x^*||^2_{(\lambda-I)^T(Q+W)\ (\lambda-I)} + 2||x^*||^2_{(G+W)\ (\lambda-I)}$$

$$+ ||x^*||^2_{{}^T T''^T RT''\lambda} - 2||x^*||^2_{\lambda^T(G+T)^T RT''\lambda}\}dt$$

$$= J^* + \frac{1}{2}\int_0^\infty ||x^*||^2_{P(t)}\ dt \tag{1.4.32}$$

where

$$F(t) = (\lambda(t,t_0)\ -I)^T\ (Q+W)\ (\lambda(t,t_0)-I) + 2(Q+W)\ (\lambda(t,t_0)-I)$$

$$+ \lambda^T(t,t_0)T''^T RT''\ \lambda(t,t_0) - 2\ \lambda^T(G+T)^T RT''\ \lambda(t,t_0) \tag{1.4.33}$$

Let $\mu_M\ (F(t))$ = norm of $F(t)$

$\mu_m\ (Q+W)$ = min eigenvalue of $Q + W$

Then

$$J_{sub} \leqslant J^* + \frac{1}{2}\ \mu_M\ (F(t)) \int_0^\infty ||x^*||^2\ \ dt$$

since

$$\int_0^\infty ||x^*||\ dt \leqslant \frac{1}{\mu_m(Q+W)} \int_0^\infty ||x^*||^2_{Q+W}\ \ dt$$

Then

$$J_{sub} \leqslant J^* + \frac{1}{2}\ \frac{\mu_M(F(t))}{\mu_m(Q+W)} \int_0^\infty ||x^*||^2_{Q+W}\ dt$$

$$J_{sub} \leqslant (1 + \varepsilon\ )J^*$$

$$\varepsilon = \frac{\mu_M(F(t))}{\mu_m(Q+W)}$$

Now, as we can see from equation (1.4.33), F(t) depends on "T" and since its norm will depend on T" and since T" is a part of T then it is clear that the maximum norm of F(t) when we use T", is less or at most equal to that when we use T. Thus the maximum value of ε occurs when T" = T and in that case

$$\emptyset' = (A - BG + CL)\ \emptyset'$$

and

$$F_m(t) = (\lambda - I)^T\ (Q+W)\ (\lambda - I) + 2(Q+W)\ (\lambda - I)$$

$$+ \lambda^T\ T^T\ RT\ \lambda - 2\ \ \lambda^T(G+T)^T\ RT\ \ \lambda$$

or

$$\varepsilon_{max} = \frac{\mu_M(F_m(t))}{\mu_m(Q+W)}$$

Thus the system will be suboptimal with index ε where $0 \leqslant \varepsilon \leqslant \dfrac{\mu_M(F_m(t))}{\mu_m(Q+W)}$ and the theorem is proved.

This theorem enables us to compute a priori an upperbound on the performance degradation resulting from a breakdown of the communication channel.

Next let us consider an example of the application of the stability condition.

1.4.3 Example : Control of a Multimachine System

The system under consideration consists of 11 machines connected together. The mathematical model of an n-machine system consists of a set of non-linear differential equations given by [15] which can, for small disturbances, be linearised around the equilibrium point to give the relation $\dot{x} = Ax + Bu$, where x is the 2 (n-1) state vector of the system and the components of x are the angle and frequency deviations from the nominal values. The matrix A is given in table 1.1 for the 11 machine system considered by Darwish *[5]*. The cost function is

$$\text{Min } J = \tfrac{1}{2} \int_0^{\infty} (\underline{x}^T Q \underline{x} + \underline{u}^T R \underline{u})dt$$

Results

This problem has been previously solved by Hassan and Titli *[16]* using a hierarchical structure and they obtained the local block diagonal gain matrix and the coordination matrix. In this coordination matrix T, all the coordination links were cut and the resulting system was seen to verify the stability condition given by equation (1.4.17) as

$$1.88 < 2.321$$

The uncoordinated and coordinated system were also simulated for the initial conditions

$$x(0)^T = \begin{bmatrix} -0.5 & -2.2357 & -0.30 & 0.3594 & -0.1 & 1.7539 & -0.3 \\ -0.001 & -0.5 & -0.8402 & 0.0 & 1.9104 & -0.5 & 1.1849 \\ -0.5 & -2.02836 & -0.3 & 0.1969 & 0.1 & 1.6257 & \end{bmatrix}$$

Figures 1.7, 1.8, 1.9, 1.10 show the first two states and first two controls for the cases

(a) when the system is optimally coordinated

(b) when the system is completely uncoordinated, and

(c) when only some of the links are cut so that the system is partially coordinated.

It was found that all these cases provide a stable overall system although case (b) was "less" stable than cases (a) and (c). In addition, the costs for these cases were calculated. These were:

case (a) : 1.79266
case (b) : 1.898409
case (c) : 1.850366

These results show that the conditions derived in this section are useful and that they can be invaluable in testing the robustness of hierarchical control structures under certain structural perturbations.

Next we apply some of the ideas developed by Siljak and Sundareshan to the decomposition-coordination framework. In this case, the ideal system against which all comparisons are made is the global optimal control system (as opposed to the decentralised system). Specifically, we study the optimal two level controller of Singh et al *[13]* when the system is subject to structural perturbations of the following two kinds : (a) when there is a failure of a link between the coordinator and the subsystems and (b) when the physical interconnection between the subsystems are cut. For such a system we have developed a stability condition and given a performance bound. Next, we study how we can stabilise the system to degree α if structural perturbations make it initially unstable for either of the two kinds of perturbations. We also develop a performance bound which compares the performance loss under these structural perturbations compared to the globally optimal system.

2	3	4	5	6	7	8	9	10	11	12	13	14	15	16	17	18	19	20
1	0	0	0	0	0	0	0	0	0	0	0	0	0	0	0	0	0	0
0	8.74383	0	5.37437	0	2.62699	0	4.05603	0	-1.64764	0	-2.7157	0	-6.33619	0	-14.37576	0	-12.99014	0
0	0	1	0	0	0	0	0	0	0	0	0	0	0	0	0	0	0	0
0	-140.367	0	9.60086	0	14.016	0	6.681	0	2.67513	0	1.51862	0	-1.99922	0	- 7.02907	0	- 9.53736	0
0	0	0	0	1	0	0	0	0	0	0	0	0	0	0	0	0	0	0
0	10.288	0	-116.061	0	6.21808	0	4.3133	0	0.41594	0	-1.06214	0	-4.6919	0	-11.74903	0	-11.52972	0
0	0	0	0	0	0	1	0	0	0	0	0	0	0	0	0	0	0	0
0	18.07436	0	7.65584	0	-13.972	0	4.075	0	1.7033	0	1.16265	0	-3.055	0	- 7.66032	0	-10.25109	0
0	0	0	0	0	0	0	0	1	0	0	0	0	0	0	0	0	0	0
0	8.60465	0	5.98992	0	4.3135	0	-124.466	0	2.5982	0	-0.0644	0	-1.3065	0	- 9.4617	0	- 9.0164	0
0	0	0	0	0	0	0	0	0	0	1	0	0	0	0	0	0	0	0
0	1.96213	0	1.68471	0	1.2125	0	1.6449	0	-99.694	0	1.9443	0	-0.933	0	- 9.1423	0	- 9.939	0
0	0	0	0	0	0	0	0	0	0	0	0	1	0	0	0	0	0	0
0	10.2119	0	5.69336	0	7.5948	0	5.5155	0	11.7763	0	-166.755	0	5.63471	0	7.2399	0	- 4.4116	0
0	0	0	0	0	0	0	0	0	0	0	0	0	0	1	0	0	0	0
0	-0.31192	0	0.05862	0	-0.397	0	0.4133	0	1.255	0	-0.689	0	-89.399	0	- 6.811	0	- 8.578	0
0	0	0	0	0	0	0	0	0	0	0	0	0	0	0	0	1	0	0
0	7.478	0	3.9759	0	5.6047	0	4.0148	0	6.0777	0	7.286	0	8.009	0	-141.137	0	- 0.465	0
0	0	0	0	0	0	0	0	0	0	0	0	0	0	0	0	0	0	1
0	1.400	0	0.9652	0	0.8061	0	1.8019	0	1.577	0	0.5298	0	1.739	0	- 1.011	0	-110.087	0

TABLE 1.1

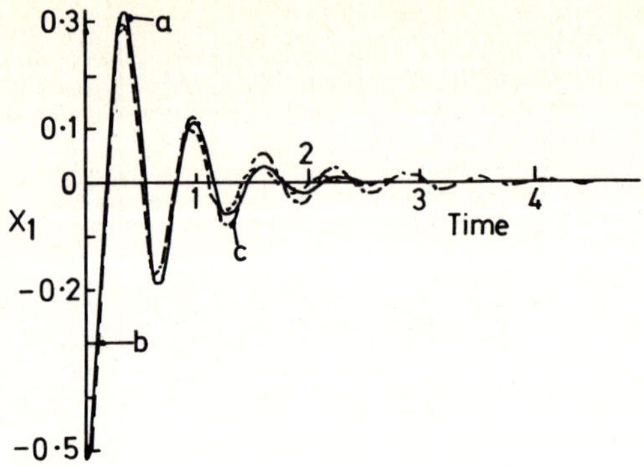

Figure 1.7

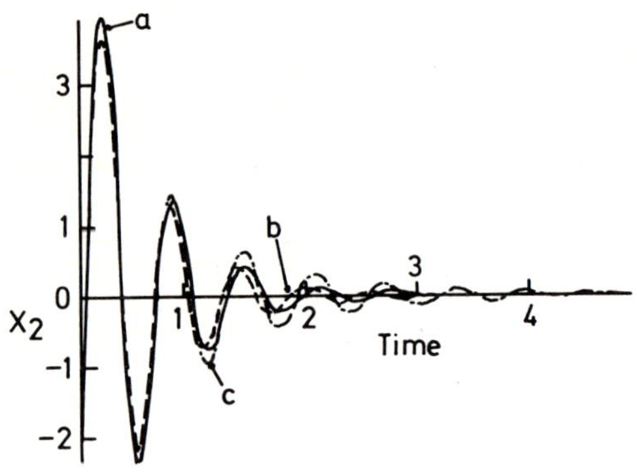

Figure 1.8

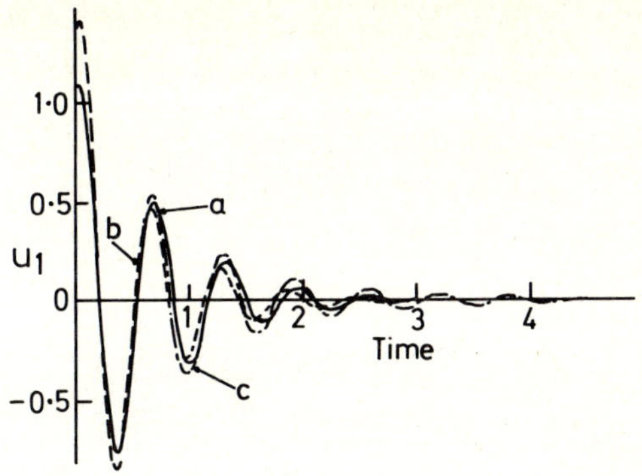

Figure 1.9

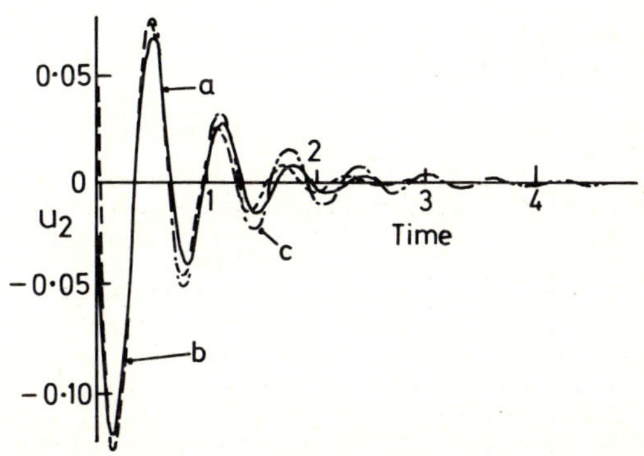

Figure 1.10

1.5 THE CONTROLLER WITH A PRESCRIBED DEGREE OF STABILITY

Consider a system comprising N linear interconnected dynamical subsystems $s_1, s_2, \ldots s_N$, the ith of which is described by:

$$\dot{x}_i = A_i x_i + B_i u_i + \sum_{j=1}^{N} e_{ij} A_{ij} x_j \qquad i = 1,2,\ldots N \tag{1.5.1}$$

where $x_i : R \to R^{n_i}$ is the state vector of the ith subsystem and

$u_i : R \to R^{m_i}$ is its control function.

$A_i \in R^{n_i \times n_i}$ and $B_i \in R^{n_i \times m_i}$ describe the dynamics of the isolated subsystem s_i, $A_{ij} \in R^{n_i \times n_j}$ describes the interconnection from s_j to s_i and $e_{ij} = e_{ij}(t)$ (cf. [17-20]) are continuous functions of time such that $e_{ij} \in [0,1]$ and all t and they are the elements of the interconnection matrix E which defines the connection of the states of one subsystem with those of another.

Now, to design a controller which has a prespecified degree of stability α, we shall use the approach of Anderson and Moore [21] for this case of interconnected subsystems. Thus the problem will be:

$$\text{Min } J = \tfrac{1}{2} \sum_{i=1}^{N} \int_{0}^{\infty} e^{2\alpha t} \left(||x_i||^2_{Q_i} + ||u_i||^2_{R_i} \right) dt \tag{1.5.2}$$

s.t.

$$\dot{x}_i = A_i x_i + B_i u_i + \sum_{j=1}^{N} A_{ij} x_j \tag{1.5.3}$$

where $||\beta||^2_S = \beta^T S \beta$ and e is the exponential function.

In equation (1.5.3) we have assumed for the moment that the system is ideally connected i.e.

$$e_{ij} = 1 \text{ for } i=1,\ldots,N \;;\; j=1,\ldots,N$$

Let: $\hat{x}_i = e^{\alpha t} x_i \;;\; \hat{u}_i = e^{\alpha t} u_i \;;\; \hat{A}_i = (A_i + \alpha I)$

Then using the method given by Singh et al [13] one can show that the optimal control $\underline{\hat{u}}^*_i$ is given by:

$$\underline{\hat{u}}^*_i = - K^1_i \underline{\hat{x}}_i - \sum_{j=1}^{N} K^g_{ij} \hat{x}_j \tag{1.5.4}$$

where $K^1_i = - R_i^{-1} B_i^T P_i$ is the local control given by solving the completely decentralised Riccati equation:

$$\dot{P}_i = A_i^T P_i + P_i A_i - P_i B_i R_i^{-1} B_i^T P_i + Q_i = 0 \tag{1.5.5}$$

$$\text{while } K^g = [K^g_{ij}] = R^{-1} B^T T \tag{1.5.6}$$

where $R = \text{diag } [R_1, R_2, \ldots R_N]$; $B = \text{diag } [B_1, B_2, \ldots B_N]$ and T is a full second level controller matrix which ensures the global optimality of the system s and which can be obtained by the two level computational method given by Singh et al [13].

Using the control law given by equation (1.5.4) and considering again the interconnection functions e_{ij} the state equation of the subsystem s_i can be rewritten as:

$$\dot{\hat{x}}_i = (\hat{A}_i - B_i R_i^{-1} B_i^T P_i)\hat{x}_i + \sum_{j=1}^{N} e_{ij} (A_{ij} - B_i K_{ij}^g)\hat{x}_j \tag{1.5.7}$$

where we have multiplied K_{ij}^g by e_{ij} in order that the second level controller takes into account possible interconnection disturbances.

It is easy to show that the global system in which the ith subsystem is represented by equation (1.5.7) is exponentially stable (i.e. the original system is exponentially stable with degree α) if the following conditions are satisfied:

(a) the pair $(\hat{A},B)$ is completely controllable where A is the global matrix of the system

(b) if Q can be factorised into $Q = DD^T$ where the pair $(\hat{A},B)$ is completely observable.

Next we consider the effect of structural perturbations. First of all we will examine the situation where some of the links between the coordinator and the subsystems are cut so that the subsystems may be operating with limited state feedback. Under these conditions we will examine the stability and performance degradation of the resulting system.

1.5.1 Stability of the System when Communication Links between the Subsystems and the Coordinator are cut

Let us rewrite equation (1.5.1) in the form

$$\dot{x}_i = A_i x_i + B_i u_i + C_i z_i \tag{1.5.8}$$

$$z_i = \sum_{j=1}^{N} L_{ij} x_j \tag{1.5.9}$$

where C_i incorporates the interconnection terms e_{ij}.

Now considering the hierarchical implementation of the control system where we shall assume that ℓ links between the coordinator and the subsystems have been cut, we have:

(a) ℓ-subsystems operating in a suboptimal manner and these can be represented by:

$$\dot{x}_j = (A_j - B_j K_j^1)x_j + C_j \sum_{r=1}^{N} L_{jr} x_r \tag{1.5.10}$$

where $1 \leqslant j \leqslant l$; $0 \leqslant l \leqslant N$

(b) N-ℓ subsystems receiving optimal coordination

$$\dot{x}_i = (A_i - B_iK_i^1)x_i + C_i \sum_{q=1}^{N} L_{iq}x_q - \sum_{k=1}^{N} e_{ij}B_iK_{ik}^g x_k \qquad (1.5.11)$$

$$1 \leqslant i \leqslant N \qquad ; \qquad i \neq j$$

Then the composite system can be represented by:

$$\dot{x}_{sub} = (A - BK^1)x_{sub} + CLx_{sub} - BK^{g'}x_{sub} \qquad (1.5.12)$$

where x_{sub} represents the suboptimal trajectories resulting from this disturbance, $K^{g'}$ is the equivalent perturbed second level controller gain matrix which incorporates the ideal interconnection terms $\bar{e}_{ij}$ where $\bar{e}_{ij}$ = [0, 1]

Using the following Lyapunov functions

$$V = x_{sub}^T P x_{sub} \qquad (1.5.13)$$

where P is the block diagonal Riccati solution resulting from solving equation (1.5.5), it can easily be shown that the resulting system will be asymptotically stable with degree α if the following sufficient condition is satisfied [19,20]

$$2||L^TC^P - K^{g'T}B^TP|| \leqslant \lambda_{min} (Q + PBR^{-1}B^TP) \qquad (1.5.14)$$

Now since $\bar{e}_{ij} \geqslant e_{ij}$ element by element, the above condition also holds for any interconnection matrix E.

However, for all the examples we have tested, the worst case occurs when all coordination links are cut so that K^g = [0] . Then for this case equation (1.5.14) reduces to:

$$2||L^TC^{TP}|| \leqslant \lambda_{min} (Q + PBR^{-1}B^TP) \qquad (1.5.15)$$

Now if L is chosen to be orthogonal, then $||L|| = 1$

Since $\quad ||L^TC^TP|| \leqslant ||L^T||\,||C^TP \;= ||C^TP||$

Therefore we have the following condition which can easily be tested since only subsystem quantities are needed

$$\max_i 2||C_i^TP_i|| \leqslant \min_i \lambda_{min} (Q_i + P_iB_iR_i^{-1}B_i^TP_i) \qquad (1.5.16)$$

Now, if the system is unstable under structural perturbation, then it will be interesting to find the degree α which ensures the stability of the system. This will be given by the following theorem:

<u>Theorem 1.5.1</u>

The degree of stability α that guarantees the exponential stability of the system under structural perturbations is given by

$$\alpha \geq \frac{2 \max_i \lambda_{max}(P_i) \sum_{i=1}^{N} \sum_{j=1}^{N} (\xi_{ij} - ||B_i|| \, ||T_{ij}|| - \min_i \lambda_{min}(w_i)}{2 \min_i \lambda_{min}(P_i)}$$

where $w_i = Q_i + P_i B_i R_i^{-1} B_i^T P_i$

<u>Proof</u>

Consider the Lyapunov function given by:

$$V = \sum_{i=1}^{N} x_{sub_i}^T P_i x_{sub_i} \qquad (1.5.17)$$

where P_i is the solution of a block diagonal Riccati equation given by equation (1.5.5). Since P is positive definite, it is necessary to find the condition under which $\dot{V}$ is negative semidefinite. To do this we have:

$$\dot{V} = \sum_{i=1}^{N} \dot{x}_{sub_i}^T P_i x_{sub_i} + x_{sub_i}^T P_i \dot{x}_{sub_i}$$

Using equation (1.5.12) and noting that P_i is calculated for $\hat{A}_i = A_i + \alpha I$, one can easily get:

$$\dot{V} = \sum_{i=1}^{N} \underline{x}_{sub_i}^T (2\alpha P_i - Q_i - P_i B_i R_i^{-1} B_i^T P_i) \underline{x}_{sub_i} + \sum_{i=1}^{N} 2\underline{x}_{sub_i}^T P_i \sum_{j=1}^{N} H_{ij} \underline{x}_{sub_j}$$

$$+ \sum_{i=1}^{N} 2\underline{x}_{sub_i}^T P_i \sum_{j=1}^{N} B_i K_{g_{ij}} \underline{x}_{sub_j}^T$$

where $\quad H_{ij} = C_i L_{ij} - B_i K_{g_{ij}} \qquad K''_{g_{ij}} = K_{g_{ij}} - K'_{g_{ij}}$

Thus $\dot{V}$ is negative definite if:

$$\sum_{i=1}^{N} 2\underline{x}_{sub_i}^T P_i \sum_{j=1}^{N} H_{ij} \underline{x}_{sub_j} + \sum_{i=1}^{N} 2\underline{x}_{sub_i}^T P_i \sum_{j=1}^{N} B_i K''_{g_{ij}} \underline{x}_{sub_j}^T$$

$$\leq \sum_{i=1}^{N} \underline{x}_{sub_i}^T w_i \underline{x}_{sub_i} + 2\alpha \sum_{i=1}^{N} \underline{x}_{sub_i}^T P_i \underline{x}_{sub_i}$$

where $\quad w_i = Q_i + P_i B_i R_i^{-1} B_i^T P_i$

Since $\quad ||H_{ij} \underline{x}_{sub_j}|| \leq ||H_{ij}|| \, ||\underline{x}_{sub_j}||$

$$||2\underline{x}_{sub_i}^T P_i \sum_{i=1}^{N} H_{ij} x_{sub_j} || \leq \lambda_{max}(P_i) \sum_{j=1}^{N} \xi_{ij} ||\underline{x}_{sub_j}||^2$$

where $\xi_{ij} = ||H_{ij}||$

$$||2\underline{x}^T_{sub_i} K_i \sum_{j=1}^{N} B_i K''_{g_{ij}} \underline{x}_{sub_j}|| \leq 2\lambda_{max}(P_i) \sum_{j=1}^{N} ||B_i|| \, ||K''_{g_{ij}}|| \, ||\underline{x}_{sub_j}||^2$$

$$\leq 2\lambda_{max}(P_i) \sum_{j=1}^{N} ||B_i|| \, ||K_{g_{ij}}|| \, ||\underline{x}_{sub_j}||^2$$

Since $K''_{g_{ij}} = \sum_{k=1}^{N} D_{ik}T_{kj} = D_{ii}T_{ij}$

$$||K''_{g_{ij}}|| \leq ||D_{ii}|| \, ||K_{g_{ij}}|| \quad ; \quad ||D_{ii}|| = 0 \text{ or } 1$$

Hence, one gets finally:

$$2 \max_i \lambda_{max}(P_i) \sum_{i=1}^{N} \sum_{j=1}^{N} (\xi_{ij} + ||B_i|| \, ||K_{g_{ij}}||$$

$$\leq \min_i \lambda_{min}(w_i) + 2\alpha \min_i \lambda_{min}(P_i)$$

where $\lambda_{max}(P_i)$ = maximum eigenvalue of P_i

$\lambda_{min}(P_i)$ = minimum eigenvalue of P_i

$$\alpha \geq \frac{2 \max_i \lambda_{max}(P_i) \sum_{i=1}^{N} \sum_{j=1}^{N} (\xi_{ij} + ||B_i|| \, ||K_{g_{ij}}|| - \min_i \ {}_{min}(w_i)}{2 \min_i \lambda_{min}(P_i)}$$

Now since $\bar{e}_{ij} \geq e_{ij}$ element by element, the above condition also holds for any interconnection matrix E, and the theorem is proved.

Having developed the stability condition as well as the degree of stability α which stabilises the system under structural perturbations of the communication links between the coordinator and subsystems, we shall next develop a bound on the resulting suboptimality of the controller.

1.5.2 Performance Degradation where some Communication Links between the Coordinator and the Subsystems are Cut

Let us assume again that the links of 1 subsystems have been cut for some unknown reason. Then the performance degradation of the composite system is given by theorem 1.5.2.

Theorem 1.5.2

The composite system which is asymptotically stable under structural perturbations is suboptimal with index ε if

$$||K^g|| + \frac{\varepsilon}{\lambda_{max}(R)} ||B^T(P+\tilde{P})|| \leq \left\{ \frac{\varepsilon}{\lambda_{max}(R)} \lambda_{min} [Q + (P+\tilde{P})BR^{-1}B^T(P+\tilde{P})] + \frac{\varepsilon^2}{\lambda^2_{max}(R)} ||B^T(P+\tilde{P})||^2 \right\}^{1/2} \quad (1.5.18)$$

where $\lambda_{max}(\beta)$ = max eigenvalue of β

$\lambda_{min}(\beta)$ = min eigenvalue of β

$\tilde{P}$, $K^{g''}$ will be defined later.

<u>Proof</u>

Let P be the solution of:

$$(\hat{A}^T + L^TC^T - PBR^{-1}B^T)\tilde{P} + \tilde{P}(\hat{A} + CL - BR^{-1}B^TP) - \tilde{P}BR^{-1}B^T\tilde{P} + PCL + L^TC^TP = 0 \quad (1.5.19)$$

Since the global system is optimal under perfect coordination, then the function $V(\hat{x},t) = ||\hat{x}||^2_{P+\tilde{P}}$ satisfies the Hamilton-Jacobi equation (where P is the solution of the decomposed Riccati equation for the decoupled system given by equation (1.5.5)).

Now the total time derivative $\dot{V}(t,x)$ can be calculated along the trajectories of $\hat{x}_{sub}$ of the composite system (1.5.12) but transformed to $\hat{x}_{sub}$ for the fundamental interconnection matrix E.

Thus we get:

$$\begin{aligned}\dot{V}(t,\hat{x}_{sub}) &= \dot{\hat{x}}^T_{sub}(P+\tilde{P})\hat{x}_{sub} + \hat{x}^T_{sub}(P+\tilde{P})\dot{\hat{x}}_{sub} \\ &= \hat{x}^T_{sub}[\hat{A}^T - K^{1T}B^T + L^TC^T - K^{g'T}B^T](P+\tilde{P})\hat{x}_{sub} \\ &\quad + \hat{x}^T_{sub}(P+\tilde{P})[\hat{A} - BK^1 + CL - BK^{g'}]\hat{x}_{sub} \end{aligned} \quad (1.5.20)$$

Let us replace in equation (1.5.20) $K^{g'}$ by $K^g - K^{g''}$ where $K^{g''}$ is the part of the second level controller gain matrix corresponding to the perturbed links and using equations (1.5.5) and (1.5.19) equation (1.5.20) can be rewritten in the following form:

$$\begin{aligned}\dot{V}(t,\hat{x}_{sub}) &= \hat{x}^T_{sub}[-Q - (P+\tilde{P})BR^{-1}B^T(P+\tilde{P}) + K^{g''T}B^T(P+\tilde{P}) \\ &\quad + (P+\tilde{P})BK^{g''}]\hat{x}_{sub}\end{aligned} \quad (1.5.21)$$

Multiplying (1.5.21) by $(1 + \varepsilon)$ and adding to both sides the terms $K^{g''T}RK^{g''}$ one gets:

$$||\hat{x}_{sub}||_F = -(1+\varepsilon)\dot{V}(t,\hat{x}_{sub}) + \hat{x}_{sub}^T [(1+\varepsilon) K^{g''T} R K^{g''} - \varepsilon F] x_{sub} \quad (1.5.22)$$

where $F = Q + (P+\tilde{P}) BR^{-1}B^T (P+\tilde{P}) - K^{g''T}B^T (P+\tilde{P}) - (P+\tilde{P}) BK^{g''T}RK^{g''}$

Integrating (1.5.22) we get

$$J_{sub} = (1+\varepsilon)J_{opt} + \int_0^{\infty} x_{sub}^T [(1+\varepsilon)K^{g''T}RK^{g''} - \varepsilon F]\hat{x}_{sub}\, dt \quad (1.5.23)$$

Thus

$$J_{sub} \leq (1+\varepsilon)J_{opt}$$

if

$$(1+\varepsilon)K^{g''T}RK^{g''} - \varepsilon F \leq 0$$

or

$$(1+\varepsilon)K^{g''T}RK^{g''} - \varepsilon Q - \varepsilon(P+\tilde{P}) BR^{-1}B^T (P+\tilde{P}) + \varepsilon K^{g''T}B^T (P+\tilde{P})$$
$$+ \varepsilon(P+\tilde{P}) BK^{g''} - \varepsilon K^{g''T}RK^{g''} \leq 0$$

$$K^{g''T}RK^{g''} + 2\varepsilon K^{g''T}B^T (P+\tilde{P}) \leq \varepsilon[Q + (P+P) BR^{-1}B^T (P+\tilde{P})]$$

$$||K^{g''}||^2 + \frac{2\varepsilon}{\lambda_{max}(R)} ||K^{g''}||\ ||B^T(P+\tilde{P})|| \leq \frac{\varepsilon}{\lambda_{max}(R)} \lambda_{min}[Q+(P+\tilde{P})BR^{-1}B^T(P+\tilde{P})]$$

$$||K^{g''}|| + \frac{\varepsilon}{\lambda_{max}(R)} ||B^T(P+\tilde{P})|| \leq \left\{ \frac{\varepsilon}{\lambda_{max}(R)} \lambda_{min} [Q+(P+\tilde{P}) BR^{-1}B^T (P+\tilde{P})] + \frac{\varepsilon^2}{\lambda^2_{max}(R)} ||B^T (P+\tilde{P})||^2 \right\}^{1/2} \quad (1.5.24)$$

Now since $K^{g''}$ can be replaced by:

$$K^{g''} = DK^g$$

where D is a diagonal matrix having elements ones and zeroes where the zeroes correspond to the links which are not perturbed. Now since

$$||K^{g''}|| \leq ||D||\ ||K^g||$$

Therefore

$$||K^{g''}|| \leq ||K^g||$$

Thus equation (1.5.24) can be written in the following form:

$$||K^g|| + \frac{\varepsilon}{\lambda_{max}(R)} ||B^T(P+\tilde{P})|| \leqslant \left\{ \frac{\varepsilon}{\lambda_{max}(R)} \lambda_{min} [Q+(P+\tilde{P})BR^{-1}B^T(P+\tilde{P})] + \frac{\varepsilon^2}{\lambda^2_{max}(R)} ||B^T (P+\tilde{P})||^2 \right\}^{1/2}$$

and this proves the theorem.

1.5.3 Stability and Stabilisation of the System under Structural Perturbation between the Subsystems

Here we would like to determine the conditions under which the system will remain asymptotically stable with degree α under any structural perturbation of the subsystem links and to specify α that guarantees these requirements. These are given by the following two theorems.

Theorem 1.5.3

The system represented by equation (1.5.7) with the closed loop gain matrices given by K^l and K^g is connectively asymptotically stable (and hence the original system is connectively asymptotically stable with degree α) if the following condition holds:

$$\sum_{i=1}^{N} \sum_{j=1}^{N} \bar{e}_{ij} \quad \xi_{ij} \leqslant \frac{1}{2} \frac{\min\limits_i \lambda_{min}(w_i)}{\max\limits_i \lambda_{max}(P_i)} \tag{1.5.25}$$

where $w_i = Q_i + P_iB_iR_i^{-1}B_i^TP_i$

$\lambda_m(w_i)$ = minimal eigenvalue of w_i

$\lambda_{max}(P_i)$ = maximal eigenvalue of P_i

$\xi_{ij} = \lambda^{\frac{1}{2}}_{max} (H^T_{ij}H_{ij})$; $H_{ij} = A_{ij} - B_iK^g_{ij}$

This theorem was first proved by Siljak et al [19,20] with a different second level controller. However, the proof for this theorem is exactly the same as that given in [19,20], the only difference being in the numerical value of ξ_{ij} since H_{ij} is different for the two cases.

Theorem 1.5.4.

The degree of stability that guarantees the connective asymptotic stability of the system is given by:

$$\alpha \geqslant \frac{2 \max\limits_i \lambda_{max}(P_i) \sum\limits_{i=1}^{N} \sum\limits_{j=1}^{N} \bar{e}_{ij} \xi_{ij} - \min\limits_i \lambda_{min}(w_i)}{\min \lambda_{min}(P_i)} \tag{1.5.26}$$

Proof

Let us consider the following Lyapunov function:

$$V = \sum_{i=1}^{N} x_i^T P_i x_i \tag{1.5.27}$$

where P_i is the solution of the decomposed Riccati equation given by equation (1.5.5). Since V is positive definite we would like to find the condition for which $\dot{V}$ is negative semidefinite.

$$\dot{V} = \sum_{i=1}^{N} \dot{x}_i^T P_i x_i + x_i^T P_i \dot{x}_i$$

Using equation (1.5.7) after replacing e_{ij} by $\bar{e}_{ij}$ we obtain

$$\dot{V} = \sum \Big\{ [x_i^T (A_i^T - P_i B_i R_i^{-1} B_i^T) x_i + \sum_{j=1}^{N} \bar{e}_{ij}\, x_j^T (A_{ij}^T - K_{ij}^{g\,T} B_i^T)] P_i x_i + x_i^T P_i [(A_i - B_i R_i^{-1} B_i^T P_i) + \sum_{j=1}^{N} \bar{e}_{ij} (A_{ij} - B_i K_{ij}^g) \hat{x}_j] \Big\}$$

Let $H_{ij} = A_{ij} - B_i K_{ij}^g$

then

$$\dot{V} = \sum_{i=1}^{N} \Big\{ x_i^T (A_i^T P_i - P_i B_i R_i^{-1} B_i^T P_i + P_i A_i - P_i B_i R_i^{-1} B_i^T P_i) x_i \Big\} + \sum_{i=1}^{N} \sum_{j=1}^{N} 2\, \bar{e}_{ij}\, x_j^T H_{ij}^T P_i x_i \tag{1.5.28}$$

Substituting from equation (1.5.5) in equation (1.5.27) we have:

$$\dot{V} = \sum_{i=1}^{N} x_i^T (-Q_i - 2 P_i - P_i B_i R_i^{-1} B_i^T P_i) x_i + \sum_{i=1}^{N} \sum_{j=1}^{N} 2\, x_j^T \bar{e}_{ij} H_{ij}^T P_i x_i$$

Let $w_i = Q_i + P_i B_i R_i^{-1} B_i^T P_i$

For $\dot{V}$ to be negative semidefinite we must have:

$$2 \sum_{i=1}^{N} x_i^T P_i \sum_{j=1}^{N} \bar{e}_{ij} H_{ij} x_j \leq \sum_{i=1}^{N} ||x_i||_{w_i}^2 + ||x_i||_{\alpha P_i}^2$$

Since

$$\sum_{i=1}^{N} x_i^T P_i \sum_{j=1}^{N} \bar{e}_{ij} H_{ij} x_j \leq \sum_{i=1}^{N} \lambda_{max}(P_i) \sum_{j=1}^{N} \bar{e}_{ij} \xi_{ij} ||x_j||^2$$

$$\leq \max_i \lambda_{max}(P_i) \sum_{i=1}^{N} \sum_{j=1}^{N} \bar{e}_{ij} \xi_{ij} ||x_j||^2$$

where $\xi_{ij} = \lambda_{max}^{1/2} (H_{ij}^T A_{ij})$

and

$$\sum_{i=1}^{N} ||x_i||_{w_i}^2 \geq \min_i \lambda_{min}(w_i) \sum_{i=1}^{N} ||x_i||^2 \qquad (1.5.29)$$

$$\sum_{i=1}^{N} ||x_i||_{P_i}^2 \geq \min_i \alpha\lambda_{min}(P_i) \sum_{i=1}^{N} ||x_i||^2 \qquad (1.5.30)$$

Therefore

$$2 \max_i \lambda_{max}(P_i) \sum_{i=1}^{N} \sum_{j=1}^{N} \bar{e}_{ij} \xi_{ij} \leq \min_i \lambda_{min}(w_i) + \alpha\min_i \lambda_{min}(P_i)$$

$$\alpha \geq \frac{2 \max_i \lambda_{max}(P_i) \sum_{i=1}^{N} \sum_{j=1}^{N} \bar{e}_{ij} \xi_{ij} - \min_i \lambda_{min}(w_i)}{\min_i \lambda_{min}(P_i)}$$

Since $\bar{e}_{ij} \geq e_{ij}$ element by element the above condition holds for any interconnection matrix.

Finally, we consider the performance degradation under structural perturbation of the links between the subsystems.

1.5.4 Performance Degradation under Structural Perturbation of the Links between the Subsystems

Finally, we derive an expression for the performance degradation index of the system when it is subject to structural perturbations as compared to the performance of the ideal interconnection structure. This is given by theorem 1.5.5.

Theorem 1.5.5

The composite system which is asymptotically stable under structural perturbation is suboptimal with index ε if:

$$||K^g|| + \frac{\varepsilon}{\lambda_{max}(R)} ||(P+\tilde{P})B|| \leq \left\{ \frac{\varepsilon}{\lambda_{max}(R)} \lambda_{min}(w) - \frac{2(1+\varepsilon)}{\lambda_{max}(R)} ||A_0^T(P+\tilde{P})|| + \frac{\varepsilon^2}{\lambda^2_{max}(R)} ||(P+\tilde{P})B||^2 \right\}^{1/2} \quad (1.5.31)$$

where $w = Q + (P+\tilde{P})\ BR^{-1}B^T\ (P+P)$

$A_0 = [A_{ij}]$ and $A_{ij} = 0$ for $i = j$

$\tilde{P}$ is the solution of equation (1.5.19)

$\lambda_{max}(R)$ = maximal eigenvalue of R

$\lambda_{min}(w)$ = minimal eigenvalue of w

We shall omit the proof of this theorem as it involves essentially the same principles used in proving theorem 1.5.2.

It should be noted that our results are somewhat different from those of Siljak et al *[19,20]* since here we compared the performance with the globally optimal one, whereas Siljak's comparisons are with the decentralised case with and without structural perturbations.

Having studied hierarchies and decentralisation, we will now consider a certain number of examples of real life situations where decentralised control is necessary or desirable.

1.6 DECENTRALISED CONTROL APPLICATIONS

1.6.1 Economic applications

Since in socio-economic systems, there are often different decision makers with different information sets, decentralised decision making occurs in a natural way.

Pindyck *[22]* considers a typical example i.e. that of the calculation of optimal stabilisation policies under the assumption that monetary and fiscal control are exercised by different authorities who may have different objectives. He describes the U.S. situation where the monetary decision maker is the Federal Reserve system whereas the fiscal decision makers are the different congressional committees and the President. In Pyndick's analysis *[22]*, each of the two authorities minimise their own quadratic cost functional subject to the constraints of a linear econometric model. The state vector of this model is of 9th order and it includes consumption, residential and non-residential investment, inventory investment, short and long term interest rates, price level, unemployment and money wage rate. The control variables are the following: The monetary decision maker controls the money supply whereas the fiscal decision maker controls exogenous governmental expenditures and a surtax.

The cost functions for the different decision makers are quadratic. They penalise the state and control vectors for deviations from an "ideal" prescribed state evolution. This implies that each Decision Maker has a nominal path in mind which he would like the state to follow, if at all possible. Clearly, the ideal paths for the different decision makers may be different. Similarly, each controller has a nominal control policy for himself and the other players. Deviation from these nominal ideal policies are penalised quadratically which lead to goals of the form

$$J_i = ||x(t) - \xi(T)||^2_{Q_{if}} + \int_o^T ||x - \xi_i||^2_{Q_i}$$

$$+ ||u_i - \eta_i||^2_{R_i})dt$$

where ξ_i and η_i are the ideal paths and policies for the Decision Maker P_i. The Nash deterministic optimal controls have, in this case, the form

$$u_i^*(t) = -R_i^{-1}B_i^T S_i(t)x(t) + \nu(t)$$

where $S_i(t)$ satisfies a Riccati type equation and $\nu(t)$ satisfies a linear differential equation.

In *[22]*, using the model over a time horizon of 20 quarters, over the period 1957-62, the open-loop deterministic Nash policies are calculated. Pyndick shows that

a) It is easier to reduce the unemployment rate than it is to reduce the inflation rate.

b) The fiscal authority has more control over the economy than the monetary authority.

Many other economic applications also exist.

1.6.2 Network flow application *[23]*

Lau et al *[23]* give the example of a dual-mode decentralised scheme for controlling the flow of a commodity through a capacitated network. In the first mode, the local controllers which are located at the nodes of the network exchange some simple messages with their neighbours and determine the local optimum flows. In the second mode, the local controllers adjust their flows to achieve the optimum steady state conditions in finite time.

The network (N,A) is considered where $N = \{1,2,...n\}$ is the set of nodes and $A \subset N \times N$ is the set of directed edges such that $(i,j) \in A$ if and only if there is a directed edge from node i to node j. Node 1 is the source and node n is the sink. Associated with each edge (i,j) is a non-negative integer C_{ij} which is the capacity of that edge i.e. the maximum permissible flow of the commodity through the edge (i,j). A controller is located at each node n and controls the flows at which the commodity is withdrawn from the edges coming into m and transferred onto the edges leaving m.

Lau et al *[23]* describe a simple control scheme based on the Ford-Fulkerson labelling algorithm *[24]* wherein each controller transmits some simple messages to its neighbours and based on the information received, each controller computes the optimal flows associated with the edges incident to itself. The messages transmitted by the controller at node m are taken from a finite alphabet whose size depends only on the number of edges incident to m. The scheme is adaptive in the sense that any change in the capacity of an edge is detected and a new optimum is computed automatically. The authors also present a decentralised control scheme which transfers the system, in a finite time, from any initial state to the optimum steady state

1.6.3 Dynamic File Assignment in a computer network *[25]*

A computer network comprises of a number of computers located at different sites and connected by communication lines. One of the main advantages of having a computer network is the possibility of sharing data files which may be located at one site and accessed by others through the communication network. This can be provided by having a central controller. However, the amount of data flowing to the controller to update its information pattern and the commands sent from the controller back into the network may take up a large fraction of the capacity of the network lines and thereby significantly deteriorate its effective performance in terms of delay and throughput. It is therefore important to be able to design the system such that it is controlled by multiple controllers acting on partial information.

In *[25]*, Segall and Sandell consider the problem of distributed file assignment in a computer network. They investigate solutions for two extreme information patterns, i.e.

(1) where the entire information is exchanged by the controllers but because of the inherent delays in the network, it arrives at the receiving controllers with a one-step delay.

(2) No information is obtained.

For the one-step-delay pattern, they have obtained explicit solutions for the optimal strategy in terms of the current estimates of the subsystems states. For the no-exchange of information case, they have shown how the solution can be obtained in principle but the expressions are too complicated to be calculated explicitly.

1.6.4 Detection with distributed sensors *[26]*

The classical theory of optimal signal processing is based on statistical estimation and hypothesis testing methods. The need for such a theory arose in the optimal detection and tracking of targets using single sensor such as a radar or a sonar. Thus all the sensor signals were assumed to be available in one place for processing.

The situation is, however, more complicated when we consider a distributed sensor network. If it were possible to transmit all sensor signals to some central location with negligible delay, then the classical theory is applicable, at least in principle. However, because of such considerations cost, reliability, survivability, communication bandwidth, compartmentalisation, sensors on platforms under emmission control or even simply the problem of flooding the fusion centre with more information than it can process, there is never a total centralisation of information.

In *[22]*, the authors attempt the development of a detection theory for distributed sensors by considering decentralised hypothesis testing.

Whilst hypothesis testing is a well understood technique, its extension to a decentralised formulation yields behaviour more complex than might be initially expected. In the case of total decentralisation (Fig. 1.11) one would expect each detector to operate independently and base its decision on the familiar likelihood ratio test. While this is true in special cases, Tenney and Sandell *[26]* have shown that it is untrue in general. When the detectors choose their decisions to achieve a system wide optimisation, then they will often use different strategies than in cases where the joint cost associated with their decisions separates into a cost for each.

The problem of constructing decentralised hypothesis testing rules can be viewed in the framework of decentralised optimal control theory. Unlike most decentralised control problems, the hypothesis testing problem can be solved in a relatively straightforward way. This is due principally to the fact that the decisions made do not feed back into the system dynamics, and thus affect the information of other decision makers. For example, Tenney and Sandell *[21]* prove that in the important special case where signals are independent when conditioned on the hypotheses, each detector uses an independent, local likelihood ratio test, but with thresholds determined via a coupled computation. The computation of thresholds decouples only in those special cases where the payoff function separates.

However, even in the case of independent observations, several types of unusual behaviour can occur. The threshold computations can yield locally optimal thresholds which are far from the globally optimal values. Moreover, even when identical sensors and a symmetric cost function are considered, the optimal thresholds need not be equal: the two detectors need to hedge their decisions in non-identical ways.

1.6.5 Power System applications

One of the most important problems associated with the operation of large power systems is the automatic generation control problem. This problem occurs when various operating power system areas are interconnected for mutual help in the sharing of the overall power generation and load requirements of the composite power system. The problem can be solve using a central controller which regulates in particular the frequency and power exchange over the inter-area-tie lines of the interconnected power system. However, a decentralised controller is attractive from a computational point of view. Davison and his co-workers have presented a number of decentralised schemes for solving such problems *[27,28]*.

1.7 CONCLUSIONS

In this chapter, we have given an introduction to hierarchies and decentralisation for large scale systems. We have seen that decentralised control could be viewed as a special case of hierarchical control where all the links between the coordinator and the subsystem controllers have been cut. This leads to a special kind of an information structure called a

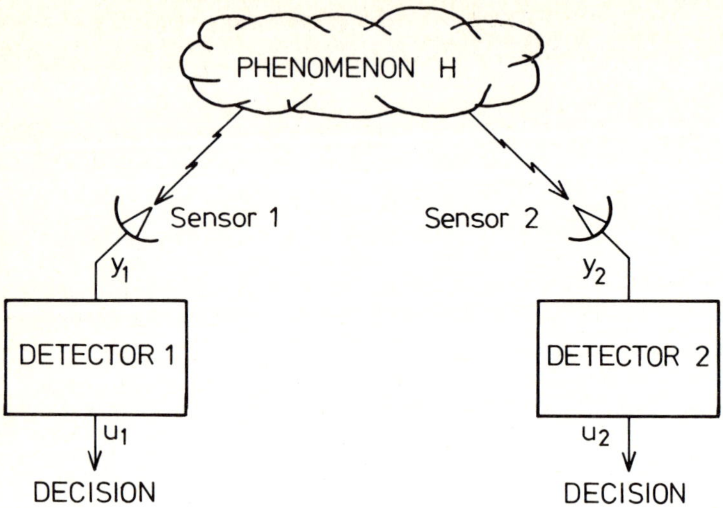

Figure 1.11

non-classical information pattern. For such systems, the classical optimality notions become meaingless and we have to consider other notions of optimality. We have also considered hierarchical structures with limited information as representing a typical practical situation lying between the extremes of ideal hierarchical control and completely decentralised control. Finally, to motivate the development in the rest of this book, we have outlined a number of examples taken from different fields where decentralised control is particularly relevant.

In the rest of this book, we study the current state of the art of decentralised control. However, before starting the analysis, we devote chapter 2 to the hardware and software aspects of microprocessors since the decentralised controllers that we will subsequently discuss will need microprocessors for their implementation. In chapter 3 we begin by considering non-classical information patterns and study team theory as an example. We study in this chapter as well as in chapter 4 aspects of decentralised stochastic control. Chapters 5 and 6 are, on the other hand, devoted to deterministic system stabilisation. In chapter 5, decentralised pole placement is examined whilst in the 6th and final chapter, deterministic design techniques are outlined. In each chapter, we outline the necessary background mathematics before we use it in the context of decentralised systems. The reader will notice that we have done this already in this chapter when we introduced stability theory before considering hierarchies under structural perturbations.

1.8 REFERENCES

1. Singh M.G. "Dynamical hierarchical control", North Holland, Amsterdam, 1977. Revised edition, 1980.

2. Singh M.G. and Titli A. "Systems: decomposition, optimisation and control", Pergamon Press, Oxford, 1978.

3. Singh M.G. and Titli A. (editors) "Handbook of large scale systems engineering applications", North Holland, Amsterdam, 1979.

4. Mesarovic M., Macko D., and Takahara Y. "Theory of hierarchical multi-level systems", Academic Press, New York, 1970.

5. Siljak D. "Large Scale Dynamic Systems", North Holland, New York, 1978.

6. Witsenhausen H. "A counterexample in stochastic optimal control", SIAM J. of Control, 6, 1, 1968.

7. Olsder G.J. "Decentralised control and economic applications", Journal A, Volume 20, 2, pp. 82-91, 1979.

8. Ho. Y.C. and Chu K.C. "Team decision theory and information structures in optimal control problems - Part I", IEEE Trans. AC 17, pp. 15-22, 1972.

9. Isaacs R. "Differential games" Second edition, Wiley, New York, 1978.

10. Prichard A. et al, University of Warwick Lecture notes "Introduction to mathematical systems theory", 1979.

11. Hassan M., Hurteau R., Singh M.G. and Titli A. "Stability, stabilisation and performance of hierarchical controllers subject to structural perturbations Part I", Proc. IEE, September, 1980.

12. Hassan M.F. and Singh M.G. Abid Part II, Proc. IEE, September, 1980.

13. Singh M.G., Hassan M.F. and Titli A. "Multi-level feedback control for interconnected dynamical systems using the prediction principle", IEEE Trans. SMC 6, pp. 223-229, 1976.

14. Sundareshan M. "Large scale discrete systems: a two level optimisation scheme", IJSS 7, 8, pp. 901-909, 1976.

15. Darwish M. and Fantin J. "The application of Lyapunov methods to large scale power systems using decomposition", IJC, 1979.

16. Hassan M.F. and Titli A. "closed loop hierarchical control for a practical large scale system using the prediction principle", Proc. 12th Statistics Conference, Cairo, 1976.

17. Siljak, D.D. "Stability of the large scale systems under structural perturbations", IEEE Trans. SMC, 2, pp. 657-663, 1972.

18. Siljak, D.D. "Connective stability of competitive equilibrium", Automatica, Volume 11, pp. 389-400, 1975.

19. Siljak D.D. and Sundareshan M.K. "Large scale systems: optimality vs reliability", In directions in decentralised control, many pearson optimisation and large scale systems (Ed. Y.C. Ho and S.K. Mitter) Plenum, New York, 1976.

20. Siljak D.D. and Sundareshan M.K. "A multilevel optimisation of large scale dynamics systems", IEEE Trans. AC 21, pp. 79-84, 1976.

21. Anderson B.D. and Moore J.B. "Linear optimal control", Prentice Hall Inc., New Jersey, 1971.

22. Pindyck R. "Optimal economic stabilisation policies under decentralised control and conflicting objectives", IEEE Trans. AC 22 (4), pp. 517-530, 1977.

23. Lau R., Persiano C. and Varaiya P. "Decentralised information and control: a network flow example", IEEE Trans. AC 17, 4, pp. 466-473, 1972.

24. Ford L. and Fulkerson D. "Flows in networks", Princeton University Press, 1962.

25. Segal A. and Sandell N. "Dynamic file assignment in a computer network - Part II: Decentralised Control", IEEE Trans. AC 24, 5, pp. 709-716, 1979.

26. Tenney R. and Sandell N. "Detection with distributed sensors", Unpublished paper, Laboratory for Information and Decision systems, MIT, USA, 1980.

27. Davison E.J. and Tripathi N. "The optimal decentralised control of a large power system: load and frequency control", IEEE Trans AC 23, pp. 312-325, 1978.

28. Davison E.J. and Tripathi N. "Decentralised tuning regulators: an application to solve the load and frequency control problem for a large power system", LSS 1, 1, pp. 3-15, 1980.

CHAPTER 2

DISTRIBUTED COMPUTING AND MICROPROCESSORS

One of the main motivations in implementing decentralised control schemes is their relatively low cost. This low cost arises from the fact that unlike for centralised control, where a lot of computing is required, here the local calculations are not very significant. Therefore, a small microcomputer could be used to perform all the control tasks. In this chapter, we introduce the hardware and software aspects of microprocessors. However, before we go on to do this, we will develop a model for distributed algorithms based on the recent work of Courcoubetis and Variaya *[1]*.

2.1 A MODEL FOR DISTRIBUTED COMPUTING *[1]*

Distributed computing occurs when more than one computing machine is used in a cooperative way to solve a common problem. The different machines operate simultaneously and asynchronously. In order for the machines to be able to cooperate with each other, some coordination is required and this is achieved by an exchange of messages over a communication network.

We have seen in the previous chapter that our interest in using distributed computing has essentially arisen from the need to control a large scale system via several control devices each of which has only partial information of the overall state of the system and it can only manipulate a subset of the control variables. Here also, some degree of coordination may be required and this can be achieved by an information exchange between the controllers. Courcoubetis and Varaiya *[1]* have combined some of the ideas taken from the literature of distributed computing as well as from Large Scale Systems control in order to develop a model for distributed algorithms. Their model emphasises certain aspects of coordination which can be achieved through an exchange of messages.

2.1.1 Contrasting features of sequential and distributed algorithms

We can conceive of a sequential algorithm (SA) as finite sequence of computational steps. Starting from an initial input data u_o, the SA computes u_k at step k=1,2,... K. The calculation of u_k depends only on (a subset of) the previously calculated values $u_o,\ldots u_{k-1}$. The important feature of these algorithms is their sequential or recursive nature. Thus one may model an SA as an (ordered) sequence of functions, say $(f_1,\ldots,f_k)$ where

$$f_k: (u_o,\ldots,u_{k-1}) \to f_k(u_o,\ldots u_{k-1})$$
$$:= u_k$$

This model then defines the function

$$f: u_o \to F(u_o) := f_k(u_o, f_1(u_o), \ldots f_{k-1}(u_o, f_1(u_o), \ldots)).$$

It is reasonable to assume that the $SA(f_1, \ldots f_k)$ was implemented on a single machine which carries out the following task at step k, k=1,...K as shown in Fig. 2.1.

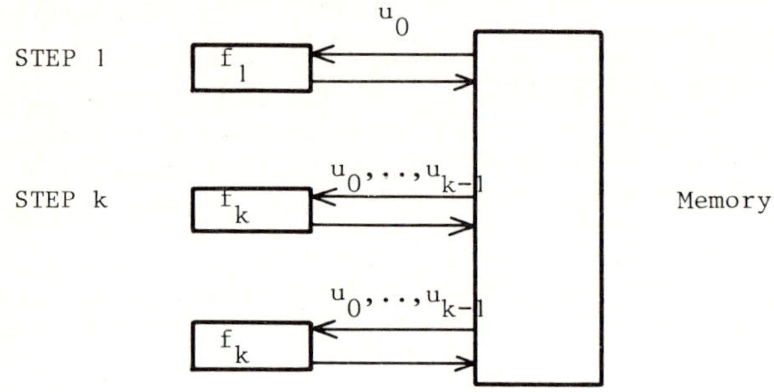

Figure 2.1 Sequential Algorithm

a) Pick out $u_o, \ldots u_{k-1}$ from the memory

b) Calculate $u_k = f_k(u_o, \ldots, u_{k-1})$,

c) Store u_k in memory.

This same task could also be implemented in a <u>distributed</u> way by 1+K machines which communicate with each other. The task of the machine M_o is to send the initial data to the other machines:

Transmit "message" $\langle u_o \rangle$ to $M_1 \ldots M_k$ and for K=1,... K the task of the kth machine (M_k) could be described as:

a) Check if the messages have been received from $M_o, \ldots M_{k-1}$

b) If $\langle u_o \rangle, \ldots \langle u_{k-1} \rangle$ have been received, then calculate

$$u_k = f_k(u_o, \ldots u_{k-1})$$

c) Transmit the message

$$\langle u_k \rangle \text{ to } M_{k+1}, \ldots, M_K.$$

In addition to the K+1 machines, we also require the communication network shown in Fig. 2.2 in order to implement this algorithm.

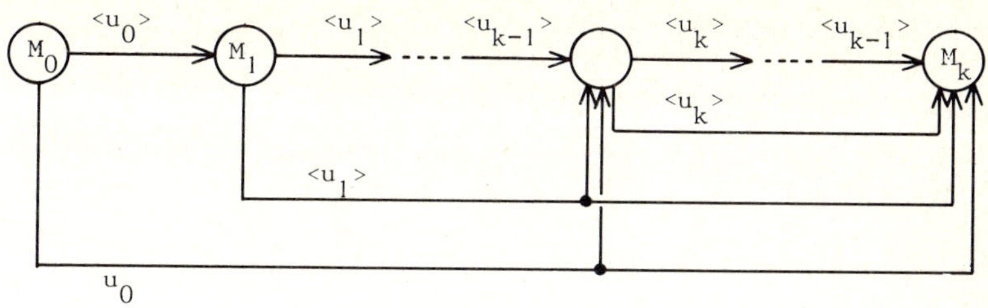

Figure 2.2 An equivalent Distributed Algorithm

Although this DA has been obtained by a straightforward transformation of the SA, it may nevertheless offer two distinct advantages over the SA. First of all, we could imagine that the various M_k are physically separated and that the implementation of f_k requires information which is easily available only at M_k. (This would be the case if the calculation of f_k uses information which is a part of the database only accessible to M_k.) In such a case, it may well be more convenient to transmit $\langle u_o \rangle, .., \langle u_{k-1} \rangle$ to M_k than to transmit the information necessary to implement f_k to a central machine. However, the more important advantage of the DA is the possibility of doing parallel or simultaneous computation. Suppose that f_k in fact only depends on $u_o, \ldots u_{k-2}$ and not on u_{k-1}. Obviously, in that case M_k could calculate u_k simultaneously with M_{k-1} calculating u_{k-1}. The model of Courcoubetis and Varaiya *[1]* explicitly displays this potential for parallel operation.

Another important aspect arises from the fact that if we examine the graph in Fig. 2.2, it has the form of a directed tree i.e. it has no directed loops. This arises from the fact that this graph represents a DA which is an implementation of step k of a SA. This makes it easy to reconstruct the SA of Fig. 2.1 from the DA of Fig. 2.2. However, in general, if we concieve of more complex communication networks, it may not be possible to recover an SA from a DA.

To sum up, a DA is a set of machines $\{M_k\}$ located at the nodes of a directed graph (the communication network). The task of each Machine M_k is:

a) It receives messages from some of its (upstream) neighbours.

b) On the basis of the information contained in the messages, it carries out a computation and

c) It then sends messages to some of its (downstream) neighbours.

We are now in a position to construct a model for these tasks *[1]*.

Let us model the network for a DA as a directed graph G = (N,E) where N = 0,1,...K is the set of nodes and $E \subset N \times N$ is the set of directed edges. G is assumed fixed throughout.

The Machine M_k is located at node k. The sets $A_k := \{i | (i,k) \varepsilon E\}$ and $B_k := \{j/k,j) \varepsilon E\}$ are respectively the upstream and downstream neighbours of k.

Next we define the capabilities of the machine and the network in terms of the following conditions

1) M_k can receive messages from machines in A_k and it can transmit messages to machines in B_k; no other communication is possible. Each message is subject to an unknown but finite delay.

2) Within each machine, the local time is an ordered set and there is no time common to all the machines i.e. the machines operate asynchronously.

These two conditions mean that the information available to a machine at any time consists of the messages received until then together with the order in which they are received. Assume that at some time M_k has received r messages. The information that M_k has available to it can be represented by its observation or log $l_{kr} := (<i_1,y_1>,\ldots,<i_r,y_r>)$ which is the ordered set of sequence of messages received by M_k. Here $<i,y>$ means that M_k has received message y from M_i.

3) At each instant, M_k examines its current log l_{kr} and decides to <u>WAIT</u> or to <u>ACT</u>. This can be described by a function w_k with values in $\{0,1\}$ such that $w_k(l_{kr}) = 0$ or 1 depending on whether M_k decides to wait or act when its log is l_{kr}. An <u>action</u> of M_k comprises two parts, i.e. first M_k chooses a subset of its downstream neighbours $M_k := H_k(l_{kr}) \subset B_k$ and then for each j in H_k it transmits to M_j the message $y_{kj} := h_{kj}(l_{kr})$. We use the notation $h_k := \{h_{kj}, j\varepsilon H_k\}$, $y_k := \{y_{kj}, j\varepsilon H_k\}$.

4) Each machine can act at most once.

Courcoubetis and Varaiya *[1]* point out that condition 4) is not really as restrictive as it appears to be since repeated actions of a single machine could be viewed as the single actions of several copies of the same machine.

For $(k,j)\varepsilon E$ let y_{kj} be the finite set of all possible messages that M_k can send to M_j. These sets define l_{kr}, the set of all possible logs of M_k which contain r messages. Because of condition 4, L_{kr} is empty if $r>|A_k|$. Let $L_k := \bigcup_{r\geq 0} L_{kr}$; here $L_{ko} = \{l_{ko}\}$ and l_{ko} is the empty log. We assume that the sets Y_{kj} and therefore L_k are fixed throughout.

In terms of the above notation, the behaviour of the machine M_k can be described by a function $d_k := (w_k,H_k,h_k)$. It is assumed that $d_k(l_k)$ is defined for all k in L_k.

Conditions (3,4) imply that once M_k has acted, it cannot change its action even if it receives another message. In terms of d_k this restriction can be expressed as the condition

5) If $w_k(l_{kr}) = 1$ and $l_{kr+1} = (l_{kr},<i,y>)$, then $d_k(l_{kr}) = d_k(l_{kr+1})$

With this formalism, we are now in a position to give the main definitions:

<u>Definition 2.1.1</u>

A distributed algorithm is any collection $d := \{d_k, k\varepsilon N\}$, where for each k, d_k satisfies condition 5.

Next, let us see how we can analyse the solutions (if any) which are computed by a DA$\{d_k\}$. Here we have two possibilities. Firstly, we could regard the DA in an abstract sense as simply a system of "simultaneous equations". From this viewpoint, the solution concept is given by the next definition.

Definition 2.1.2

The collection of messages $\{y_k, k\varepsilon N\}$ is a solution of $d = \{d_k = (w_k, H_k, h_k)\}$ if there exists a set of logs $\{l_k, k\varepsilon N\}$ such that for every k:

(a) $w_k(l_k) = 1$

(b) $y_k = \{y_{kj} = h_{kj}(l_k), j\varepsilon H_k(l_k)\}$

(c) $\langle i,y\rangle \ \varepsilon |l_k|$ only if $k\varepsilon H_i(l_i)$ and $y = h_{iR}(l_i)$. Here $|l_k|$ denotes the unordered set of messages in the log k.

In order to clarify this definition further, we note that two types of events are relevant when describing the operation of the machine M_k i.e. the reception of a message (let us call this an R-event) and the execution of an action which is simply the transmission of messages (call it a T-event). By condition (2) above, the events relative to H_k can be ordered in their own local time so that it becomes meaningful to say of two such events E_1^k, E_2^k that E_1^k occurs before E_2^k. We could write this as $E_1^k < E_2^k$. By condition (4) above, at most one of the E^k is a T-event and by condition (3), this can occur only after a sequence of R-events such that for the corresponding log l_k, $w_k(l_k) = 1$ (cf. Fig. 2.3). All of this, of course, concerns events for a single machine. To obtain a solution, the Definition 2.1.2 relates events relevant to different machines by requiring that only messages which are transmitted can be received.

To illustrate this, let us consider an example.

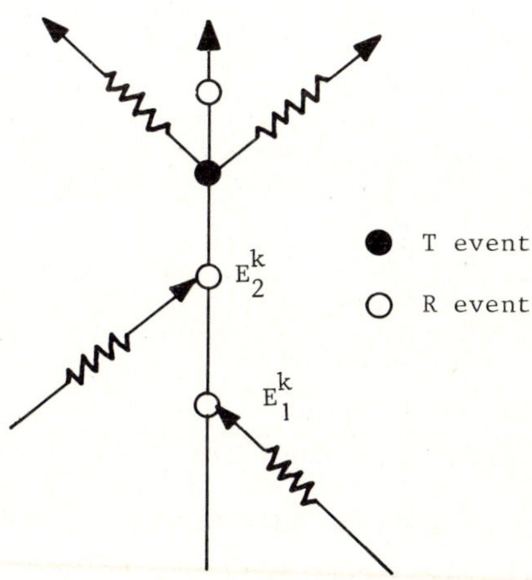

Figure 2.3 Events in M_k

Example

Let us consider a DA with 2 machines M_o, M_1 and let us assume that $Y_{01} = Y_{10} = Y$; assume that each machine waits for one message before it acts so that $w_k(1_{ko}) = 0$, $w_k(1_{k1}) = 1$. Suppose that when it acts, the machine k transmits the message which it has in its log i.e. $h_{kj}(1_{k1}) = 1_{k1}$. We can see that (y_1, y_2) is a solution if and only if $y_1 = y_2 = y$. In Fig. 2.4 E^k_1 denotes the reception of the message y and E^k_2 the subsequent transmission so that

$$E^o_1 < E^o_2 \text{ and } E^1_1 < E^1_2$$

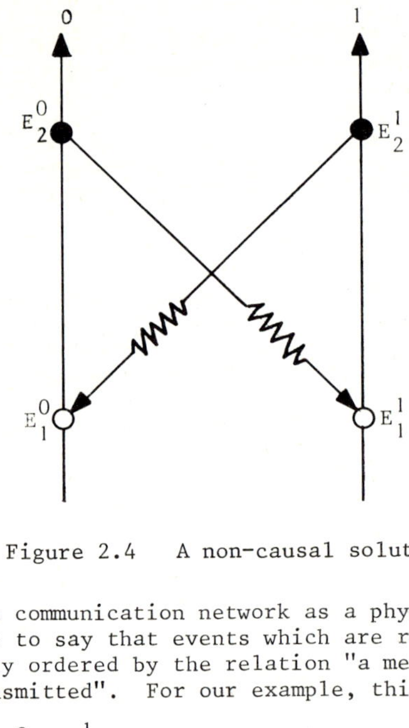

Figure 2.4 A non-causal solution

Now if we imagine the communication network as a physical implementation then it is reasonable to say that events which are relevant to different machines are partially ordered by the relation "a message can be received only after it is transmitted". For our example, this is equivalent to:

$$E^o_2 < E^1_1 \text{ and } E^o_1 < E^1_2$$

However, combining these two relations gives us $E^o_1 < E^o_1$ and $E^o_2 < E^o_2$ from which we conclude that (y,y) is <u>not</u> physically realisable.

This example suggests the following definition:

Definition 2.1.3

The collection $y = \{y_k, k\varepsilon N\}$ is a <u>causal solution</u> if there is a set of logs $\{1_k\}$ satisfying (a) - (c) above and in addition there is a permutation or ordering $p = \{i_o, \ldots i_k\}$ of N such that for each k

(d) $\langle i,y\rangle \varepsilon |1_k|$ only if i precedes k in the ordering p.

To clarify the above ideas, let us consider a solution and the events associated with it. It is easy to see that the solution is <u>CAUSAL</u> only if the partial order induced by the communication network and that induced by each machine are compatible. The above example has no causal solution. However, a DA can have many causal solutions. Also, the same causal solution may be produced by machines acting in different orders. In fact the number of such orders or perturbations is a good index of the degree of parallel operation possible. For instance, in the extreme case of a completely decoupled algorithm $\{(w_k,H_k,h_k)\}$ with $w_k \equiv 1$ and $h_k \equiv \bar{y}_k$, the solution $\{y_k\}$ can be produced by any ordering.

2.1.2 Fixed message structure algorithms

A DA is said to have a fixed message structure if for each k in N,

(i) The subset $H_k(l_k)$ of downstream neighbours to which M_k transmits is fixed independently of the log l_k and

(ii) The wait decision w_k and the action h_k, depend only on the set $|l_k|$ of messages received and not on the order in which they are received.

Most algorithms which are proposed in the literature have a fixed message structure and this makes their analysis much simpler. To see this we note that since H_k is fixed, we can regard $y_k = \{y_{kj}, j\varepsilon H_k\}$ and $y = \{y_k, k\varepsilon N\}$ as fixed dimension vectors. The set of messages that k receives can be expressed as $|l_k| = g_k(y) := \{y_{ik}, k\varepsilon H_i\}$. Obviously $\{H_k\}$ and $\{g_k\}$ are equivalent descriptions of the fixed message structure.

Courcoubetis and Varaiya *[1]* have used their model also to analyse the algorithm of Merlin and Segal *[3]*. Although their model is a tentative one, it does appear to be useful in analysing distributed computing situations.

In the rest of this chapter, we will concentrate on a single one of the computing machines in order to describe the current state of the technology. Thus the rest of the chapter is devoted to microprocessor hardware and software. We begin with notions of hardware.

2.2 MICROCOMPUTER HARDWARE

A fundamental component which is the basis of all computing machines is the memory. The central memory of all computers holds both the programme and the data.

2.2.1 Organisation of the Memory

The memory could be considered to comprise a set of cells, each one of which could store a binary digit (bit) in the form of a voltage which has a value of zero or five volts. Each cell could only take up two stable states which we denote by zero or one. A memory usually contains a large number of such cells; a selection and waiting mechanism allows us to load a 0 or 1 into any of these cells; a read mechanism allows us to know the value of any bit in the memory. It is for this reason that for all memory units there is a corresponding "address" register, a "word" register, a control and read link and a link for the control of the write as shown in Fig. 2.5.

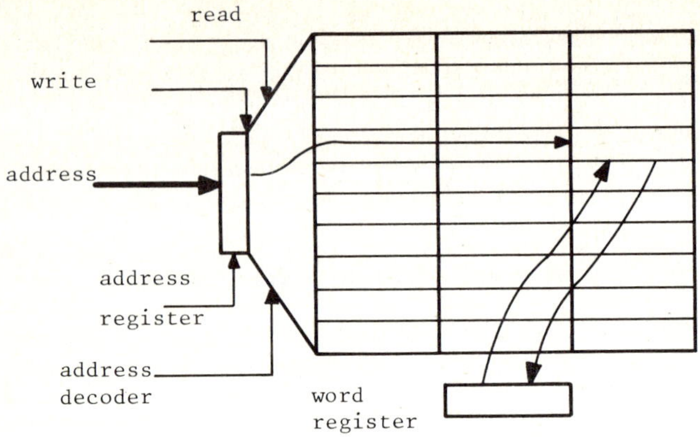

Figure 2.5

The address register will contain the number of the bit which we wish to read or write (we call the number of a bit its "address"); the word register will contain the value that we wish to give to a bit in the case of a write or the value the bit has, in the case of a read. Thus, in order to write the value 1 in bit number 7, the programme needs to put the value 7 in the address register, the value 1 in the word register and give a write order. In the same way, in order to read the value of the bit at address 8, the programmer needs to put the value 8 in the address register and give a write order; he/she will thus find the value of the bit in the word register. All these operations need to be ordered explicitly by the programmer.

The organisation of the memory that we have just described is, in fact, elementary since each bit can only contain the values 0 or 1. In order to represent bigger numbers (for example a numerical value), we group the bits into words of fixed size. The word length (i.e. the number of bits it contains) is a fundamental characteristic of the memory and it reflects the power of the computer. Thus, microcomputers have word lengths of 4, 8 or 16 bits; the mini-computers: 16 or 32 bits; big computers: 64 or even 128 bits. This classification is relatively fluid, however, since 16 and 32 bit microcomputers are already on the market and 64 and 128 bit ones will be available soon.

Comment

In fact, the access to the memory cell is not done on the level of the bit but on the level of the word; the address register will not contain thus the address of a bit but the address of a word and the word register will contain the value that is read or which is to be written into a complete word of the memory.

The number of words contained in a memory defines its capacity. It is measured in "K" ($1\ K = 2^{10} = 1024$ words). Thus, in a memory of 65K there are 65536 words. We call an octet all combinations of 8 bits.

2.2.2 Representation of Information in the Memory

Since a memory word only contains zeros and ones, the representation of all numerical values higher than one must be coded. This coding is based on the following principle: if N is an integer to be represented in the memory, we could write it as:

$$N = a_n x 2^n + a_{n-1} x 2^{n-1} + \dots a_2 x 2^2 + a_1 x 2 a_o$$

where a_i is 1 or 0.

We will denote such a decomposition by

$$N = (a_n a_{n-1} \dots a_2 a_1 a_o)_2$$

The position i occupied by the symbol a_i is called the weight given to the symbol.

Example: $193 = 1x2^7 + 1x2^6 + 1 = (11000001)_2$

We note that this binary representation which is directed by the technological structure of the memory leads to complicated coding for high numbers. It is for this reason that we usually prefer a more compact representation, called Hexadecimal, which consists of writing:

$$N = b_m \times 16^m + b_{m-1} \times 16^{m-1} + \dots + b_2 \times 16^2 + b_1 \times 16 + b_o$$

This representation thus consists of grouping the bits of a memory word 4 x 4 and representing the value of each group by a unique symbol. As with 4 bits we could code 16 different values from $(0000)_2$ to $(1111)_2$; the hexadecimal representation uses 16 symbols as shown in the table below.

$a_3a_2a_1a_o$	decimal	b_o
0 0 0 0		0
0 0 0 1		1
0 0 1 0		2
0 0 1 1		2
0 1 0 0		4
0 1 0 1		5
0 1 1 0		6
0 1 1 1		7
1 0 0 0		8
1 0 0 1		9

$a_3a_2a_1a_o$	decimal	b_o
1 0 1 0	10	A
1 0 1 1	11	B
1 1 0 0	12	C
1 1 0 1	13	D
1 1 1 0	14	E
1 1 1 1	15	F

Thus $b_i = 0,1,2,\dots,9,A,B,\dots,E,F$ Example: $193 = C \times 16^1 + 1 = (C1)_{16}$

2.2.3 Representation of Signed Integers

In the representation of an integer in an M bit word we use the bit with the highest weight in order to represent the sign. Thus, a value of 0 for this bit will indicate a positive number and a value of 1 a negative number. There remains thus N-1 bits in the word for representing the numerical value of the number, and this allows us to code 2^{M-1} values. For example, in a 4 bit word length machine 7 will be the biggest positive integer that could be coded ($7 = (0111)_2$). In order to represent negative integers, the simplest procedure would be to maintain the same coding for the numerical value and simply change the bit for the sign. Thus, in the previous example, $7 = (1111)_2$. However, if we adopt this solution, the central processing unit (CPU) of the computer will need to have "addition" and "subtraction" done by two distinct hardware operators. It is for this reason that all existing machines use a different representation for negative numbers. This representation is based on the following principle. If A and B are two positive numbers, then A-B could be written as:

$$A-B = [A + (100000)_2 - B] = A + [(100000)_2 - B]$$

because A and $A+(10000)_2$ have the same representation in a 4 bit word. Thus, if we represent the negative number -B by its complement of $(10000)_2$ the subtraction operation becomes the following addition operation:

$$A-B = A + [\text{complement of } (10000)_2 \text{ of } B]$$

Such a representation of negative numbers is called 2's complement.

2.3 FIXED POINT REPRESENTATION

In order to represent positive fractions less than 1 we decompose this number in negative powers of 2,

$$N = C_1 \times 2^{-1} + C_2 \times 2^{-2} + \ldots + C_p \times 2^{-p}$$

where $C_i = 0$ or 1.

The first bit of the word (i.e. the one with the highest weight) represents the sign of the number and the decimal point, which is not explicitly represented, is implicitly placed after the sign bit. Thus:

$$(0\overline{1\!\downarrow\!01}) = +^{\downarrow}(1\text{x}2^{-1} + 0\text{x}2^{-2} + 1\text{x}2^{-3}) = 0.5 + 0.125 = 0.625$$

In this method of coding, the smallest positive fraction that it is possible to code in an M bit word is thus 2^{-M+1}. For example, for a 16 bit word this value will be 0.000030517.

In order to represent negative fractions, we use as before the technique of 2's complement.

<u>Example</u>: $-0.625 = (1111)_2 - (0101)_2 + (0001)_2 = (1011)_2$

2.3.1 Floating Point Representation

This means the representation of a numerical value in the form:

$$N = M \times 2^E$$

where M is a signed fixed point number and it is called the mantissa, while E is a signed integer called exponent.

There exist many representations of a number in floating point; we will call the normalised floating point representation the one where the first bit of the mantissa is significant. In microcomputers where the words are often 8 bits long, the representation of a floating point number usually occupies 4 words, i.e. 32 bits. The first bit contains the sign of the mantissa as well as the exponent with its sign; the 3 other words contain the value of the mantissa in fixed point. Such a coding allows us to represent numerical values that lie between $2^{-64} = 5.4 \times 10^{-20}$ up to $2^{63} = 9.2 \times 10^{18}$.

Such a representation of numbers is necessary in scientific calculations, and particularly in the control of processes. Indeed, even if the measured signals that have been taken from a process and the control signals that have been applied are generally represented in fixed point (for example a signal varying from 0 to 10 volts being represented by binary values from 0 to $(FF)_{16}$), the intermediate calculations which are used to calculate the control signal often needs to be done in floating point.

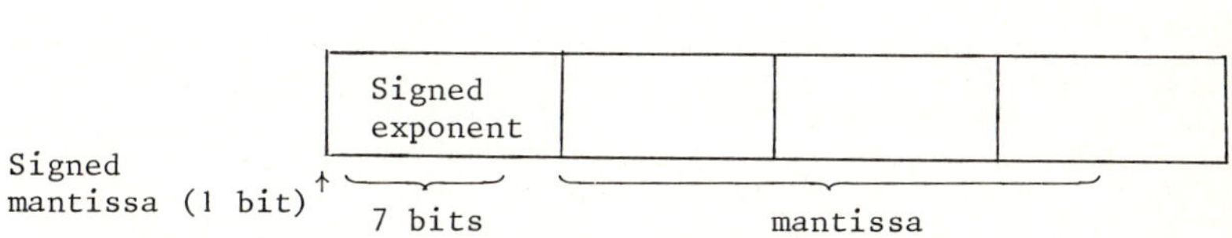

Figure 2.6 Signed exponent

2.3.2 Representation of Alpha Numeric Characters

When we connect a microprocessor to a peripheral like a teletypewriter or a line printer, the informations exchanged are alphanumeric characters, i.e. letters of the alphabet plus typographical signs. These typographical symbols need to be put into the memory and for this purpose they should be coded in some special form of representation of 0's and 1's. There exist at present 2 types of coding that are used. The most frequently used one, especially on mini and microcomputers is the ASCII (American Standard Code for Information Interchange). It is in this way that the typing of the number 0 on the Teletype keyboard causes the transmission of an octet containing the value $(BO)_{16}$. It will be necessary for the programmer to decode this value in order to replace it by the value zero in the memory if he/she wishes to use it in the calculations. Inversely, before writing on the printer the value of a result, the programmer needs to code each number into this numerical value before sending it to the printer.

Next, we consider the technology of memories.

2.4 Technology of Memories

The general characteristics of memories are:

(i) Their volatility: The information stored in memories is said to be volatile if there is a risk of it being changed in the case of a power failure.

(ii) Live memory: A memory is said to be live if we can read and modify its contents using a programme.

(iii) Dead memory: In this case we can read the memory contents using a programme but we cannot modify it.

(iv) Destructibility: We call a memory destructible if the read operation destroys its contents.

(v) Access time: This is the time it takes to obtain a piece of information after a read request.

(vi) Cycle time: This is the duration of separation of 2 successive calls to the memory for a read or a write.

Ferrite Memories

In such memories, each memory cell is a miniscule ferrite ring with conducting wires. According to the direction in which the current is sent, the ring is magnetised in one direction or the other, thus registering a 1 or a 0. In order to read it, we send a reverse current through a wire; the stored information shifts the magnetisation which causes an induced current in a particular conductor, i.e. the read conductor. Such a memory is thus destructible. It is for this reason that such memories are always attached to a unit which rewrites after the read. It is a non-volatile memory which is its main strength. However, such memories are slower than semi-conductor memories.

Dynamical RAM

The RAM (Random Access Memories) are live semi-conductor memories which store information, i.e. the value of a 0 or 1 of a bit, in capacitors. The basic cell of a dynamic memory is shown in Fig. 2.7

The information is stored in the capacitor C_{STG}.

Over a period of time, the capacitors discharge so that after a few milliseconds the stored information becomes erronious. It is therefore necessary to periodically recharge the capacitor to its logically correct value; this is called refreshing of the memory. It is for this reason that a RAM memory unit has attached to it a special input plug through which a periodic signal (every 2 milliseconds) is sent to excite an intermediate circuit which refreshes the cells.

RAM memories are volatile but non-destructible.

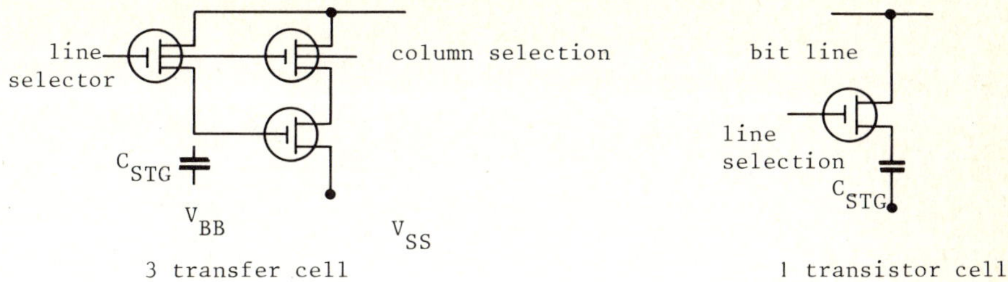

Figure 2.7

Static RAMs

These are also live semi-conductor memories. However, the memory element is in an electronic bistable with 6 or 4 transistors. In the latter case, the transistors T2 and T5 of Fig. 2.8 are replaced by resistors. In this diagram the value of the bit is stored either in T3 or in T4.

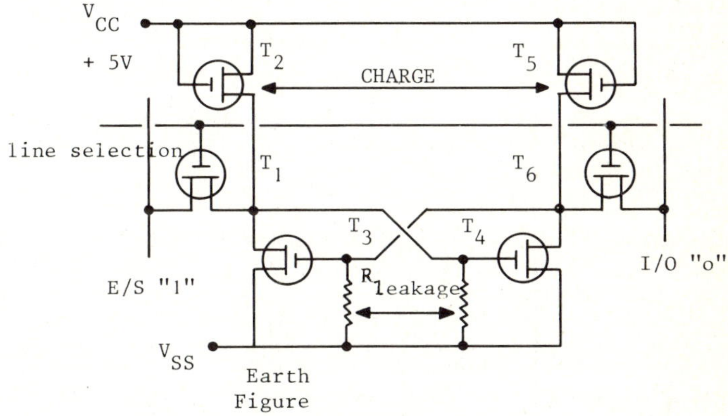

Figure 2.8

Static RAMs are thus more complex than dynamic RAMs. They are nevertheless desirable since:

(a) they do not require refreshing

(b) they are less sensitive to noise, and

(c) they work at high speeds with a single power supply voltage.

In the same way as dynamic memories, static RAMs are volatile though non-destructible.

ROM

ROMs (Read Only Memories) are dead memories which have been programmed once and for all at the time of construction by the manufacturer. Such a memory can be constructed very easily as shown in Fig. 2.9, which shows a 4 word 4 bit ROM.

In the case presented in Fig. 2.9 the contents of the memory are:

Address	Contents
00	1010
01	1100
10	0100
11	0001

Since the making of a "mask" for the manufacture of a ROM is quite expensive only the need to have many thousand identical memories allows one to amortise the price. Such a memory is therefore reserved for the storing of utility programmes for apparatuses made in a large scale way.

PROM (Programmable Read Only Memory)

For the case where it is desirable to fix the programmes in small scale production, the user has the possibility of utilising another type of dead memory. In these, the user may electrically write the desired programme himself. These memories are called Programmable Read Only Memories. Such memories are provided blank by the manufacturers. There exist many kinds of PROM. For example in the case of fusible PROM, the writing of a piece of information is done by melting the fuses. This melting which allows one to establish the value 1 at a given point of the memory is obtained by sending a high current in the corresponding fuse.

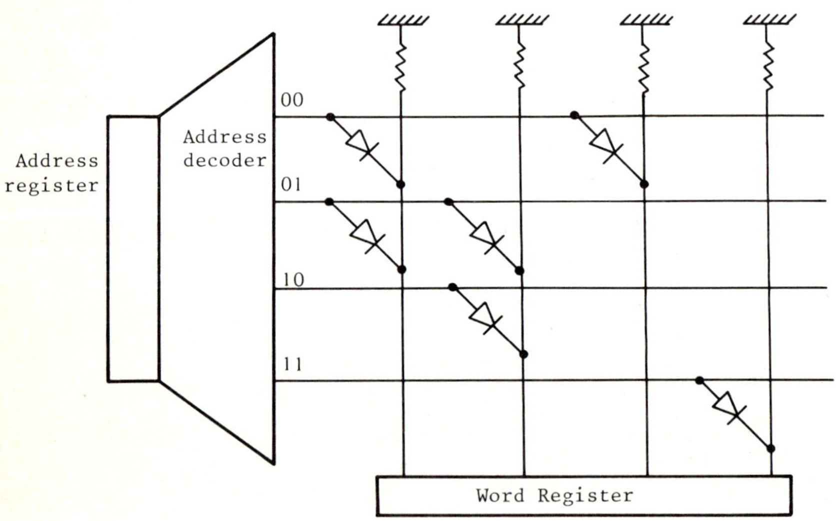

Figure 2.9

Another type of PROM uses the establishing of a connection. The putting of a zero at a particular point in the memory is obtained by creating a link which did not initially exist (whilst in the previous PROM that we have considered, a value of 1 is stored by cutting a connection). For this purpose, we apply at the appropriate point in the memory an impulse that is sufficiently strong for obtaining the breaking of the junction corresponding to a diode polarised in the reverse direction.

EPROM Memories

The memory point of an EPROM is made by a MOS transistor which ensures the passage or otherwise of the current depending on whether this transistor is conducting or is blocked. These EPROM memories are put in a little box in ceramic with a transparent quartz window. Shining ultraviolet light through this window causes the neutralisation of the charges captured on the floating grills of its transistors and this causes the rubbing out of the memory contents.

We note that the new generation of bubble memories would appear to be very efficient indeed. However, they are still in an experimental state.

2.5 ROLE OF THE PROCESSOR

We have previously seen how we can code and store data in the memory. However, the memory is not only used for storing data, it also contains the programme which is executed by the CPU. As in the case of any information that is stored in the memory, a programme needs to be coded. Each processor has its own instruction set. It is therefore necessary to adapt to each processor the coding of the programme that we want it to execute. Although these codings are different, depending on the type of processor used, they all respect a certain number of common characteristics. This is because the Hardware Architecture of all processors are very similar. All processors possess an instruction register I, an ordinal counter P, a sequencer, general purpose registers and an arithmetic and logic unit. The execution of all programmes therefore use these basic components.

2.5.1 Instruction Register and Sequencer

Suppose we wish to execute a programme which subtracts the value 1 from a piece of data D (for example D = 168). Assume that in the specification of the 8 bit processor used, there exists a coded instruction $(07)_{16}$ on an octet which achieves the following function:

"The contents of the following octet is decremented by 1". Since, conceptually, all processors have in their memory the programme that it is to execute as well as the data to be manipulated by this programme, it is therefore necessary to put into the memory this instruction and the data D before starting its execution. Let us assume that this instruction and the data are placed in two consecutive addresses in the memory, for example 10 and 11; it is then possible to start the execution of this programme, as shown in Figure 2.10. Since the memory is only a unit for storing information, we will complete our computer by new units which will enable us to execute our programme.

The first of these units is the instruction register. The function of this register is to receive all instructions to be executed and to proceed to interpret them using an associated logic module called the sequencer. If the instruction register contains at a given moment an executable instruction, its interpretation will cause the transmission of electronic signals which will trigger the desired execution by the processor. The problem that remains is to send to the instruction register the instruction currently in the memory that we wish to execute. This is the role of the ordinal counter P. The ordinal counter is a special register of the processor which contains at any instant the address of the instruction to be executed. Thus, to execute our example programme, the user needs to load the ordinal counter to the address 10, this can be done by incorporating a special push button (RUN or EXEC) for the instruction contained in the address 10 to be extracted from the memory, placed in the instruction register and executed (Fig. 2.11).

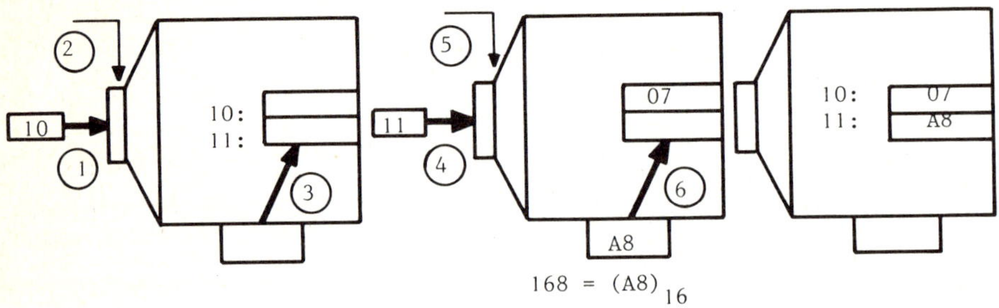

Figure 2.10

The example that we have just given is of a programme which only has one instruction. In reality, all applications require programmes with hundreds or even thousands of instructions. It is thus unrealistic to repeat such a procedure in order to execute each instruction. It is for this reason that the execution of each instruction is followed automatically by the incrementation of the ordinal counter P. This increment is not always the same. It depends on the instruction that has just been executed. In our programme, the execution of instruction 7 will be automatically followed by the incrementation of the value in P by 2 which will thus become 12. If, therefore, the following instruction in the programme is put into this address 12, this instruction will be executed automatically without intervention of the user. In our example, the incrementation by 2 of the value P is justified by the fact that in the address 11 is placed a piece of data and not an instruction; this is in fact a special case in that most instructions only cause the incrementation of the ordinal counter P by 1. This is the reason why all programme instructions are usually placed in consecutive memory addresses and this causes the execution of the entire programme once the value of the ordinal counter is initialised to the first instruction.

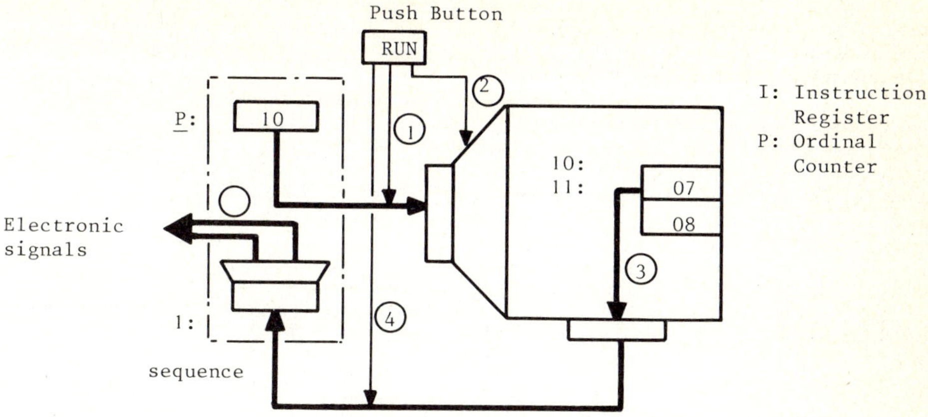

Figure 2.11

2.5.2 The Arithmetic and Logic Unit (ALU) and the General Registers

In a previous example we have seen how an instruction is decoded and how the sequencer automatically generates the signal ordering the execution of this instruction. These signals are, in fact, sent to the ALU whose function is the real execution of the instruction (in the case of our example the subtraction of 1 from 168).

This unit is the part of the processor having the circuits capable of doing the treatment required by the instructions. It consists essentially of the following elements:

(a) The arithmetic and logic calculations: This is a complex circuit capable of doing all the calculation instructions which form a part of the instruction set of the processor. These operations usually are: addition, subtraction, comparison of integers or fixed point numbers, complementation logic functions AND, OR, NAND, etc., left or right shift by one or more bits of information.

(b) The general registers: These registers serve to temporarily store the operands presented to the ALU and the results of the operation of this unit after the execution of an instruction. These registers also receive data from the external environment (peripherals connected to the computer); they also contain data to be sent outwards.

(c) The state register: This is a multi-bit register (8 to 16) associated with the ALU which provides certain indications with respect to the last operation that has been performed. It is possible to test using a programme each of these bits individually using special instructions. These indicators contain the following information: the carry for an addition or subtraction operation, overflow, the equality of two numbers, the sign of a result, etc. Figure 2.12 shows the use of the ALU in the course of executing our example. We note especially a register marked A and called Accumulator, which we find in many processors. It is this register which contains the result produced by the ALU. In computers possessing an accumulator, there exist many instructions which allow one to manipulate its contents. It is usually the register that is most often used by the programmer.

Another frequently used general register is the index register X. We will see how it is used when we examine memory addressing.

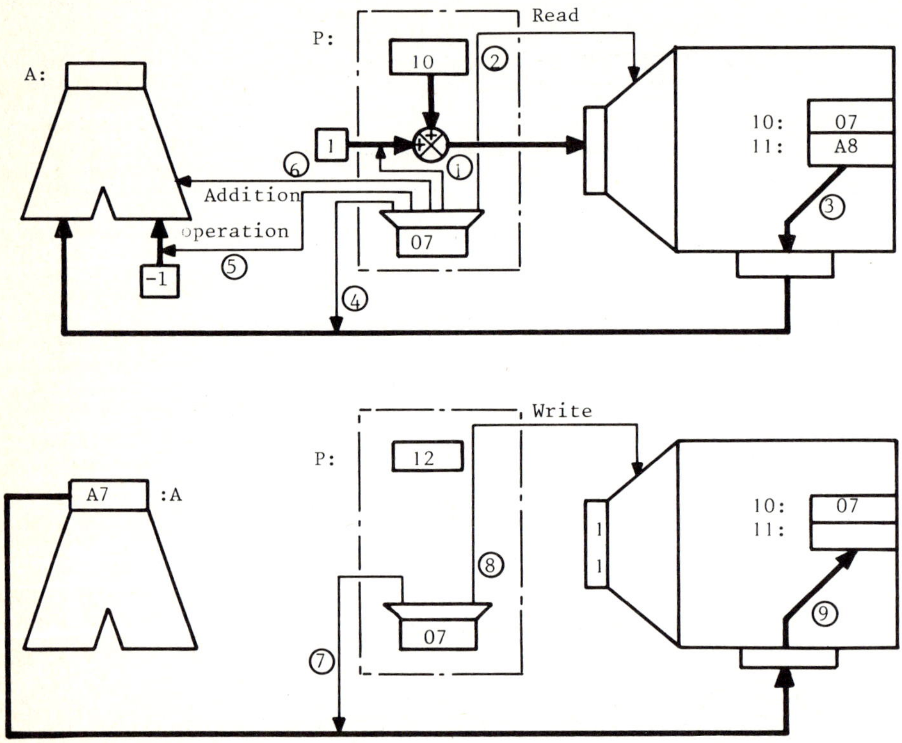

Figure 2.12 Instruction Set and Memory Addressing

2.6 THE INSTRUCTION SET

The set of instructions that are executable by a processor often allows one to write the same programme in different ways. Thus, we could write the programme for the previous example (subtraction of 1 from 168) using three instructions:

1) The instruction coded $(02)_{16}$ achieves the loading into the accumulator of the contents of the memory cell whose address is put into the octet that follows the instruction.

2) The instruction coded $(104)_{16}$ which decrements the accumulator by 1.

3) The instruction coded $(01)_{16}$ which causes the putting of the contents of the accumulator into the memory cell whose address is placed in the octet which follows this instruction.

On using these three instructions, the numerical value of D could be put anywhere in the memory and not necessarily immediately following an instruction, as was the case previously. In this way, by putting the value of D at the address 20, the programme can be written as:

Addres	Memory contents	Comments
10	02	A ← (20)
11	20	
12	04	A ← (A)-1
13	01	20 ← (A)
14	20	
----	-----	
20	A8	Data D

In the notation used in the comments, (XX) means the contents of the memory address XX.

This example of instruction writing highlights two types of special instructions: <u>memory reference instructions</u> whose operand is not a numerical value but the address of a memory cell where we can find the numerical value and the <u>instructions on the register</u> (instruction $(04)_{16}$) which directly manipulate the contents of one or more registers and which do not have an operand.

The programme that we have just described, as well as its previous version, is incomplete. As we have seen, a processor systematically executes the contents of the memory cells sequentially (P being incremented by one or more depending on the instruction). When the processor executes the last instruction of a programme, it is necessary to indicate to it that the programme has terminated. This is the objective of a specific instruction coded $(0F)_{16}$. In the previous example, it will be necessary to place this instruction at the address 15. The execution of this instruction will cause the stopping of the processor.

In order to introduce a new type of instruction, we will code a programme which causes the decrementation of the data D up to the point where it has the value zero. Since D = 168, it is obviously possible to do the above by repeating 168 times the 5 consecutive instructions of the programme that we have just described. Such a procedure is too cumbersome since the programme will occupy a large slice of memory and this is only desirable at the value 168 of the data D. To get around this, the instruction sets of all computers have special <u>JUMP</u> instructions whose function is to modify the value of the ordinal counter P.

One of these instructions coded $(05)_{16}$ will cause the loading of the ordinal counter by the value contained in the octet which follows this instruction; this value should evidently be the address of an instruction, since this is the only type of information that P can contain. Another of these instructions which is coded as $(08)_{16}$ will do the same thing as $(06)_{16}$ except for the fact that it will only do this provided the contents of the accumulator are zero. This is called a <u>conditional Jump</u> instruction.

Thus the programme for our example will be:

Address	Contents	Comments
10	02	$A \leftarrow (20)$
11	20	
12	08	$P \leftarrow 19$ if $A = 0$
13	19	
14	04	$A \leftarrow (A) - 1$
15	01	
16	20	$20 \leftarrow (A)$
17	05	$P \leftarrow 10$
18	10	
19	OF	Stop

Aside from the instructions that have already been mentioned, there exist input-output instructions which allow us to put into a register data coming in from outside or to send it out as well as test instructions for the state register, etc.

2.6.1 Memory Addressing

We have already mentioned that there exist instructions which refer to the memory i.e. instructions which have the memory address of the operand and not its value. In an 8 bit computer, often this address is coded on a single octet. However, with 8 bits, only 255 values can be coded whereas many microcomputers have memory extendable to 64K. It is for this reason that the address is sometimes manipulated before the real address of the appropriate operand is fixed. This type of manipulation is called memory addressing and since there exist many possible modes of addressing, its coding is integrated into that of the instruction. We therefore say that a memory reference instruction has three parts: an operational code which specifies the function to be executed, the address condition which specifies the manipulation to be done on the address, and the address of the operand itself. The different memory addressing generally available on all machines are:

(1) Direct or absolute addressing: In this case the specified address is the real address of the operand as shown in Fig. 2.13.

(2) Indirect addressing: In this case the specified address denotes a memory cell where the operands address exists (Fig. 2.14)

(3) Relative addressing: In this case the address of the operand is obtained by adding to the specified address in the instruction, the value of the ordinal counter. This specified address therefore represents a displacement with respect to P which could be positive or negative (Fig. 2.15).

Index Addressing

In this case the address of the operand is obtained by adding to the address specified in the instruction the value of the index register X. This type of addressing is very often used in order to sequentially access elements of a data table. In that case the address specified in the instruction is the address of the first element of the table; X contains the number of the element in this table that we wish to manipulate (Fig. 2.16).

Most machines accept certain combinations of the different methods of addressing given here, for example, relative indirect addressing, indirect index addressing, etc.

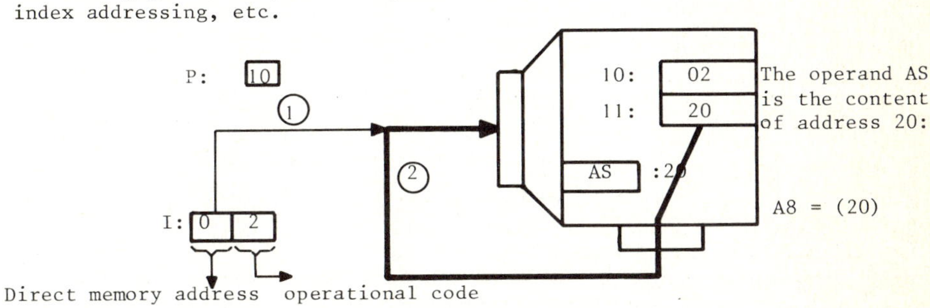

Figure 2.13

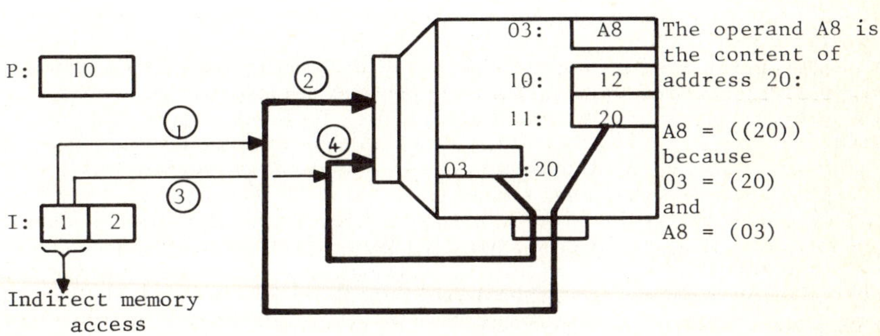

Figure 2.14

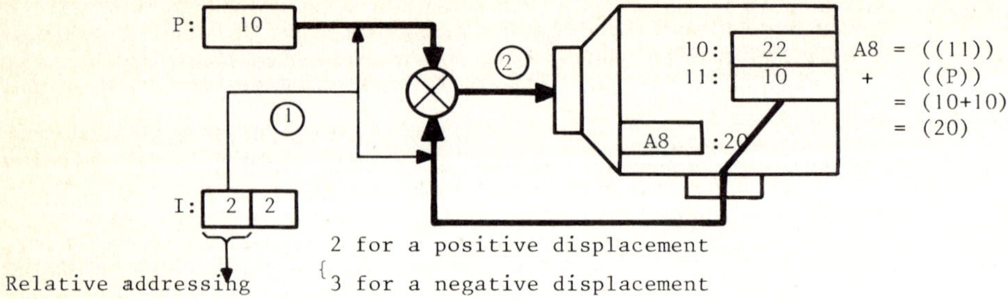

Figure 2.15

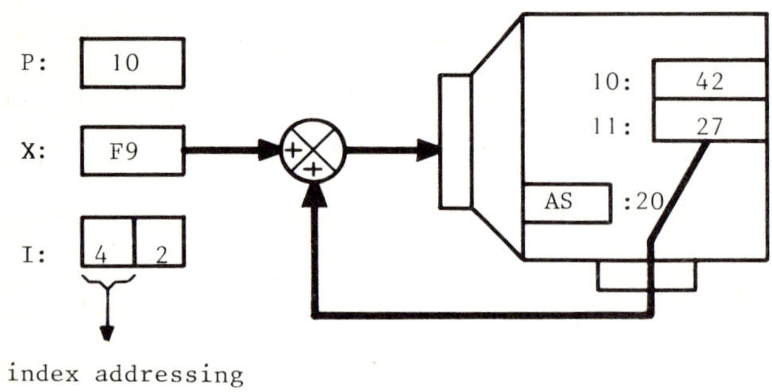

$(X) - (F9)_{16} = -7$ AS = ((11)) + (X))
= (27 -7) = (2)

Figure 2.16

2.7 INPUT-OUTPUT INTERACTION

If we should wish to connect an input/output peripheral to a computer, it is to enable the calculator to communicate with the exterior, either by receiving information (from a keyboard, a measurement, an external bulk memory, etc.), or by sending information (to a printer), an actuator, a bulk memory, etc.). In the case of an output, for example, the information provided by the computer is usually the result of a calculation made by the arithmetic logic unit and this information is available either in one of the processing unit's general registers, or in the memory bank. One could, then, consider connecting all the input/output peripherals directly, either to the processing unit or to the memory bank. Unfortunately this has one obvious drawback - it would be necessary to make as many access plugs on the microprocessor or on the memory bank as there were peripherals. Moreover, the number of peripherals connected can be high since a complicated device of the teletype or disc unit kind can be considered as a peripheral just as much as a simple logic line recording a state on a process. Therefore, it quickly becomes impossible to create as many access routes to the microprocessor as there are connectable peripherals. This is why, in order to connect a peripheral to a computer, the structure of the control unit itself is used. Indeed, the construction of the central unit is hinged around three principle modules: the memory bank, the sequencer and the processing unit.

Three types of information circulates between these three modules: data (exchanged between the memory and the arithmetic logic unit), addresses (sent to the address register of the memory bank) and control signals (released by the sequencer). These three types of information circulate on three paths called "buses". They are the data bus, the address bus, and the control bus. So, instead of connecting a peripheral directly with the processing unit, it will be linked up with these three "buses" which communicate with all parts of the central unit. The peripheral will then use these "buses" in the following way:

- on the data 'bus' the data exchanged between the microprocessor and the peripheral will be conveyed. This is a two-directional 'bus' comprising a number of conductors corresponding to the size of the processor words (4, 8 or 16 bits).

- the address 'bus' makes it possible to identify the peripheral concerned.

- on the control bus will circulate the information which will manage and synchronise the exchanges of information between microprocessor and peripherals.

This is why a peripheral will always be connected to the computer through an interface whose function will be to guarantee the interpretation of the signals released on these 'buses' and to conduct the exchanges between the peripheral and these buses (Fig. 2.17)

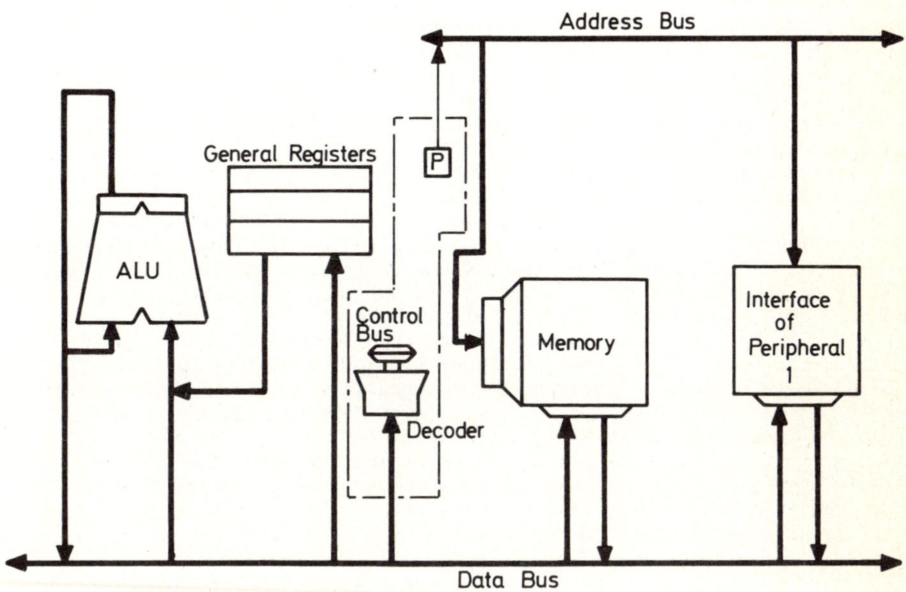

Figure 2.17

2.7.1 Exchange in Programmed Mode

Input/output transfers take place at very precise moments in the user's programme. The exchange is made at the moment determined by the execution of the input/output operation which concerns the peripheral. The first thing to be done in order to make such an exchange is to give the calculator the role of master and to the peripheral that of slave throughout all phases of the exchange. Let us examine, for example, the case of input of information from a perforated tape reader. When the user needs a piece of information which is on perforated tape, he must do as follows:

- Send the reading order. This will cause the tape reader motor to start up.

- Await the ending of reading. Since a peripheral of this sort is very slow in relation to the speed of the processor, it will be necessary for the user to programme a holding loop until the information is available.

- Store the information read by the peripheral into the central processing unit.

This principle of functioning of an input defines the structure of the interface itself. The reading order instruction will be decoded in the form of an order transmitted on the control 'bus'. Now there can be several peripherals there and therefore several interfaces connected to the three 'buses' of the central processing unit. It will therefore be necessary to accompany this reading order with the number of the peripherals concerned. This number will actually be an address transmitted on the address 'bus'. Since the computer memory is also connected to the address 'bus', a special bit will accompany the peripheral address on the address bus so as to prevent the memory not understanding that this address is intended for itself.

All interfaces must therefore contain an address decoder. The function of the decoder is to compare any address transmitted on the address 'bus' with its proper address; if the two addresses are identical then the interface will understand that the signals circulating on both the control 'bus' and the data bus are intended for itself. The read command circulating on the control bus will then be recorded by the interface which will interpret it by releasing a logical signal towards the peripheral.

Holding the end of reading also conditions the hardware structure of the interface; for the user to know for each programme the end of reading of information carried out by the peripheral, it is necessary for this peripheral to release a special end-of-work signal. The holding loop programmed by the user will therefore consist of the repeated testing of this signal. As long as the signal says "read unfinished" the programmer must begin the test again. When the read is finished the user can programme the following phase. The interface must therefore include a device which recognises this signal and its release on one of the 'buses' (control bus or data bus depending on the calculator).

The data read by the peripheral is automatically stored as it is read in a special register called the buffer register found in the interface. The phase of storing information in the central processing unit therefore consists of transferring the information from the buffer register to one of the general registers of the processing unit. The interface should therefore have a mechanism for recognising this transfer command. The structure of the interface can therefore be represented as follows (Fig. 2.18):

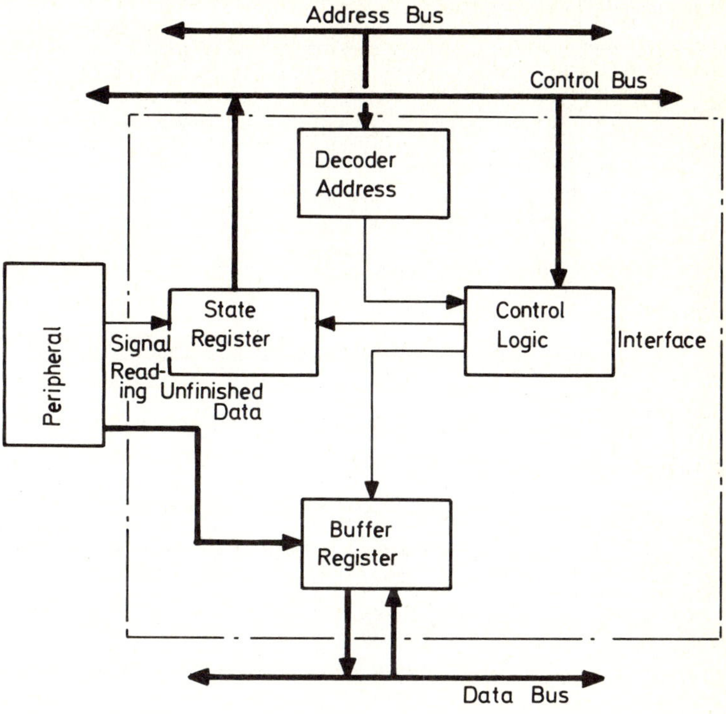

Figure 2.18

So the exchanges of commands and information between peripheral and central processing unit in programmed mode can be represented by the following chronogram:

(A) Read command: (B) Test of the state, contained in the interface of the signal "reading unfinished"; (C) Transfer command; (D) Transfer of information (Fig. 2.19)

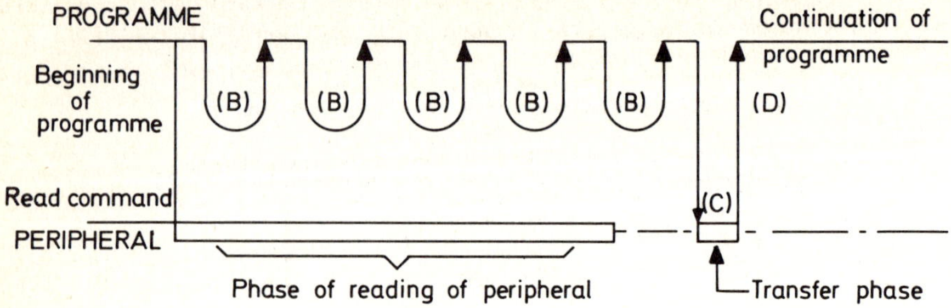

Figure 2.19

2.8 STRUCTURE OF PROGRAMMABLE INTERFACES FOR MICROPROCESSORS

2.8.1 Parallel Interfaces

A parallel interface is a device which makes possible the transfer of information to a parallel peripheral (one line per bit). Many manufacturers offer a parallel input/output interface with its own characteristics. The most widely used are Intel's PP 8255 Programmable Peripheral Interface) and Motorola's PIA 6820 (Peripheral Interface Adaptor).

Without going into precise details, we can assume that these interfaces contain:

- two data registers which temporarily restore the information coming from the peripheral (in the case of an input) or from the microprocessor (in the case of an output);

- a control register which determines the mode of operation of the interface (input and output);

- a state register which receives the information from the peripherals on their state (for example information read, peripheral without voltage, etc.);

- a logic control block containing the address decoder and which generates and synchronises the exchanges between the peripheral and the 'bus' of the central processing unit.

2.8.2 Series interface

Certain peripherals transmit or receive their data in series on one single wire. This is so in the case of teletypewriters, visualisation consoles, cassette recorders, etc. The interest of this type of link lies in the economy of conductors realised, which is greater by as much as the link is longer. A disadvantage is the rhythm of transmission which is obviously slower than that of a parallel link, since the bits of the information transmitted are sent one after the other on the same transmission line. Therefore this method is only applied to slow exchanges.

There are two different methods of series transmission:

- the synchronous method which consists of sending a whole block of characters (a character is equal to an actet) following one another. Two special characters called framing characters synchronise the beginning and end of release of the block.

- In the asynchronous method the characters are sent separately. A heading bit is added to each character and this allows the receiver to synchronise the reception (Fig. 2.21)

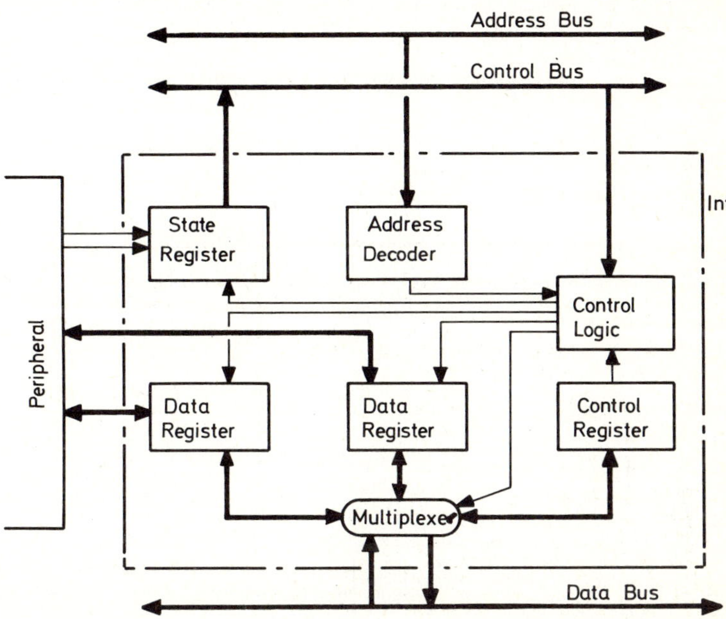

Figure 2.20

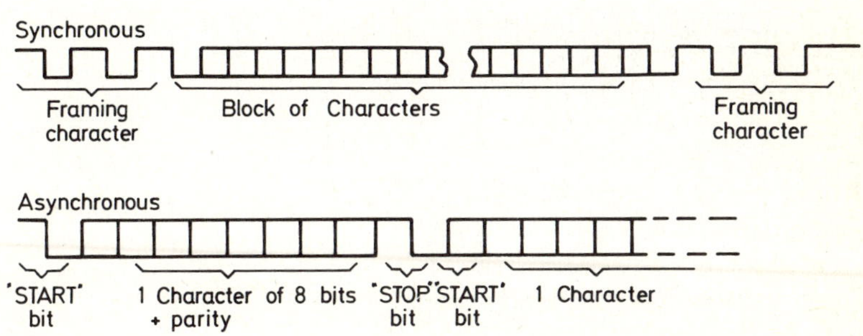

Figure 2.21

The integrated programmable circuits now being offered by manufacturers which all allow for series transmission are called UART (Universal Asynchronous Receiver Transmitter) in the case of asynchronous links and USART (Universal Synchronous Receiver Transmitter) in the case of synchronous links.

Without going into precise details we can assume that a standard UART structure contains:

- a transmission module which receives the information in parallel from the microprocessor on the data bus and converts it into series information with its framing characters.

- a reception module which receives the series information coming from the peripheral eliminates the framing characters in it and converts the data into parallel information.

- a command module which starts up the circuit and synchronises the exchange.

- a control module which contains the address decoder and supplies information concerning the transmission (checking of the parity at the end of reception, etc.), to a state register linked to the data bus (Fig. 2.22)

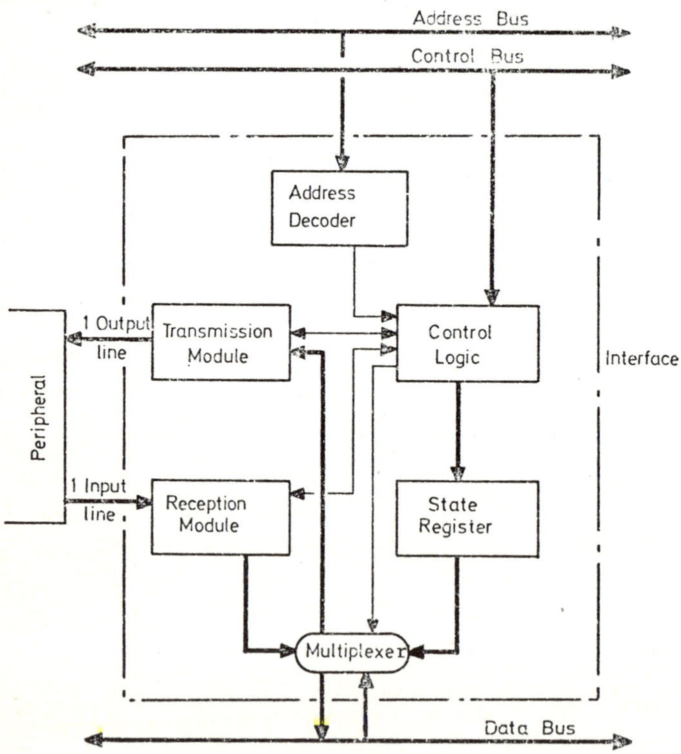

Figure 2.22

2.9 EXCHANGE IN INTERRUPT MODE

The disadvantage of the input/output method in programmed mode lies in the fact that for each programme the microprocessor must await the end of any input/output. Naturally the user can only continue his programme when he has received the data coming from the peripheral (in the case of an input), and in the same way he can only send information to an output peripheral if the peripheral has finished the output of the preceding information. However, instead of waiting for the end of the input/outputs for each programme one can think in terms of using this loss of time to carry out other operations: this is the case in real time where several programmes are stored simultaneously in the memory: the waiting time of and input/output in a programme can then be recovered to carry out part of another programme.

This is why microprocessors are furnished with devices which enable them to make exchanges on the initiative of the peripherals, which is called the interruption mode. In this mode it is the peripheral itself which informs the processor of the end of an input/output, and so it is no longer the processor who will test whether this input/output is finished. This principle of operation is even essential in cases where alarm type signals are connected to the computer, that is, signals whose moment of arrival is completely unforseeable. These alarm signals are, for example, logical signals indicating the appearance of an operational anomaly on a process controlled by the computer.

The interruption procedure is represented on the figure which follows (Fig. 2.23). When a peripheral releases an interruption,an alarm signal for example, the calculator is usually in the middle of carrying out a programme which is totally foreign to that interruption. The arrival of the interruption therefore causes the following operations:

- the processor finishes the execution of the instruction in hand.

- it tests a special bit of its state register called the interruption inhibiting bit; this bit, which can be modified for each programme, makes it possible for the processor to authorise or forbid the arrival of all the interruptions (that is, coming from all the peripherals).

- if the value of this bit authorises the arrival of interruptions, the processor then proceeds to what is called the safeguard of the context of the interrupted programme; this safeguard consists of storing in an internal memory called File the value of the ordinal counter at the time of interruption (this then contains the address of the next instruction to be carried out in the interrupted programme), together with the accumulator and the contents of the different general registers (the contents of which must be restored when the interrupted programme is resumed).

- the processor then reads the address of the beginning of the programme processing this interruption into the ordinal counter. The processor then carries out the programme.

- at the end of the programme the return to the interrupted programme will be done by refilling of the safeguarded registers, including the ordinal counter.

All these operations are carried out by hardware in some processors. In others, some of them are the responsibility of the programmer. This is so in the case, for example, of the safeguard and of the restoration of the general registers of the interrupted programme.

The automatic procedure for jumping to the processing programme of the interruption is carried out in different ways depending on the processors. In some processors the jump address is the same for all the interruptions. The programme which starts at this address must therefore begin by recognising the interruption. In order to effect this recognition the programmer must go and test the contents of the state register of all the interfaces to determine the origin of the interruption. In the event of simultaneous interruptions, it is the programmer who chooses which to process first. In other processors the recognition is automatic, that is to say that 2 different interruptions cause the jump to different addresses of the ordinal counter. In the case of simultaneous interruptions it is therefore by hardware that the order of processing of these interruptions will be chosen. The latter case is called a vector interrupt.

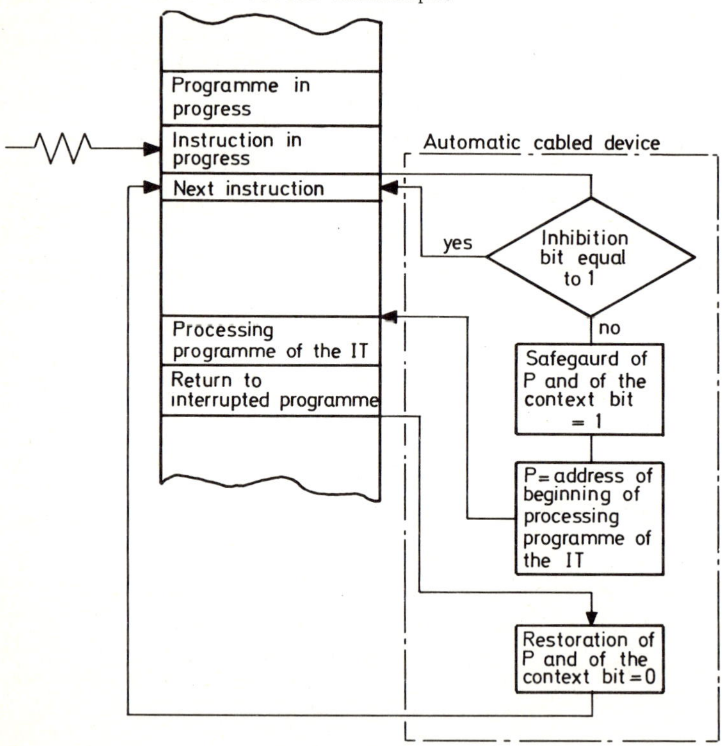

Figure 2.23

2.9.1 Exchange in DMA Mode

When the rate of input/output transfer must be very rapid some microprocessors accept a direct connection between 1 peripheral and the memory. The procedure of an input/output in DMA (Direct Memory Access) is as follows:

- the peripheral sends a transfer demand signal in DMA,

- the processor then completes its execution of the instruction in hand.

- the address, data and control bus are then disconnected from the microprocessor by setting the gates which command them at the "high impedance" state,

- the peripheral then takes over the buses and sends its own signals which specify - the addresses in which the transferred information (in the case of an input) will be stored, the direction of the exchange (input or output), the rhythm of the exchange, etc.

- then the data moves directly from the peripheral to the memory or conversely according to the direction of the exchange.

2.10 "INDUSTRIAL" PERIPHERALS

One of the main applications for microprocessors is in the control of industrial processes. Information taken from or transmitted to such processes is usually of three types:

- analog information: this is the case, for example, of voltage delivered by measurements (of level, pressure, flow, temperature).

- logic information: open or closed contacts, detection of high or low levels, etc.

- digital information: information supplied by digital output measurements such as angular position measurements or movement measurements.

Analog Input Interface

This type of interface is one of the most widely used when a processor is connected in an industrial environment. The following figure (Fig. 2.24) shows schematically the various parts of such an interface, which are:

- a multiplexer, whose role is to allow only one of the analog signals applied on the interface to pass through at a time. The signal recorded will be selected by the address transmitted for each programme to the interface decoder. Usually relay multiplexers are used for low frequency signals and for rapid signals MOS multiplexers are normally used.

- a digital-analog convertor. The analog voltage chosen must be converted into digital information before being transmitted on the data bus. For this a sampler-blocker is used, this being an instrument which follows the development of an analog signal and makes it possible to block the value of the signal for short periods. The conversion itself takes place during these periods. It is carried out by an analog-digital convertor, which is initiated by a signal sent by the sampler-blocker. When the conversion is finished the convertor gives out an "end of conversion" signal. This signale is a special instance of the "end of Input/Output" signal.

Analog Output Interface

An analog output module is much simpler than an input module, as is demonstrated by Fig. 2.2.5. On the diagram we can see:

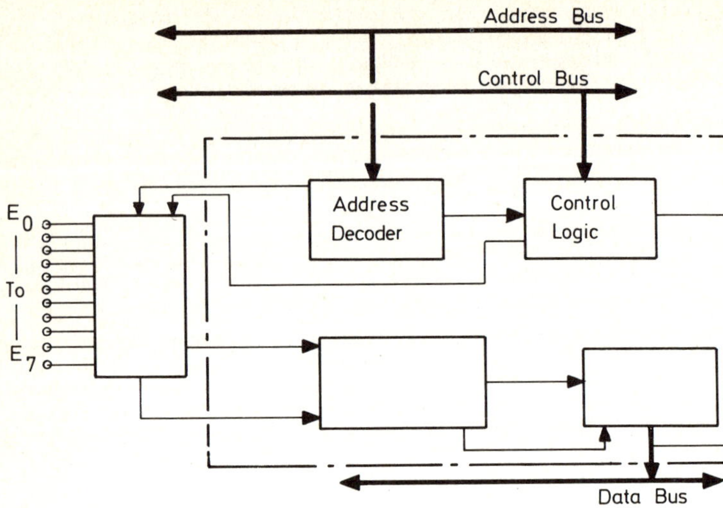

Figure 2.24

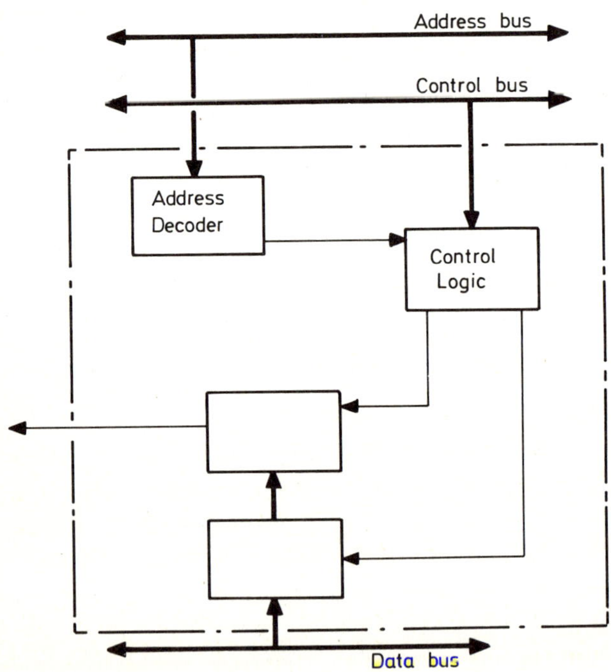

Figure 2.25

- a buffer memory placed on the data bus which contains the digital information between two successive outputs.

- the analog-digital converter which changes the coded information supplied by the processor into an analog electric signal.

2.11 GENERAL VIEW OF MICROPROCESSORS

Since the first Intel 4004 microprocessor appeared in 1971, a considerable number of new components have been introduced on the market. In design, all these microprocessors obey one basic principle - they communicate with other modules (memories, interfaces) through two-directional buses whose logic has three states. Any one bus can thus be multiplexed between several boxes and the "high impedance" state of all three state logic enables any number of boxes to be connected on the same bus. This principle therefore reduces the number of interconnections between circuits, although this is to the detriment of the complexity of each circuit which must itself be able to recognise whether information circulating on the bus is intended for itself.

Aside from this principle of common design the different microprocessors which are at present available are distinguishable by their organisation (i.e. their internal structure), the length of their words, their rapidity. All these characteristics closely depend on the technology used; thus the length of the words depends on the level of integration. The same applies to internal structure: until 1974 microprocessors were built around general registers (Motorola 6800, Intel 8080). Since then incorporated memory microprocessors have appeared (TMS 9900), together with circuits with specialised and more sophisticated functions than those of traditional microprocessors. The latter development is, moreover, one of the most significant today. So, instead of classifying microprocessors according to their technology, it is without doubt better to do so according to their field of application.

2.11.1 "Elementary" Microprocessors

In this group are placed microprocessors with words of 1 or 4 bits which are intended to control elementary systems: replacement of cabled logic, electrical household gadgets, traffic lights, small instrumental systems, weights, data logging, cash tills, etc.

Also, units which are strictly intended for pocket calculators must also be included. This is a special case, for the inputs/outputs in these cases are specially intended for keyboard coding without output on 7 segment display; exceptions apart, it is difficult to divest them from their original use by making them create other inputs/outputs. The comments below are related to the table that follows:

1. The ordinal counter and the addresses of the inputs/outputs are contained in other boxes.

2. Intended to replace cabled logic.

3. The MM 57109 is intended for scientific calculation (square root, exponential, trigonometric functions).

4. The general registers are of 32 bits.

5. There are many versions which differ in the size of the memory incorporated and in the numbers of inputs/outputs.

6. It is possible to extend the RAM memory by external boxes.

7. It is possible to extend the ROM memory by external boxes.

The TMS 1000 is one of the most widely used microsystems (microprocessor with incorporated memory). Its architecture is representative of that of most of the Intel 4004's successors. It contains three registers: an accumulator and two address registers which can act as data registers in some instructions. Its operation is scarcely any different from that of conventional 8 bit microprocessors except at the input/output level. The

Manufacturers and types of microprocessors	Technology	Length of word Data/instruction	Clock included	RAM included	ROM included	Number of inputs/outputs	Number of register accum./gen.use	Number of instructions	
Motorola MC 14500B	Cmos	1/4	Yes	No	No	1 bus I/S bit	1	16	(1)
General Instruments SBA	Nmos	1/8	Yes	120x1	1023x8	30 I/O		24	(2)
National Semiconductor MM5799	Pmos	4/8	Yes	96x4	1536x8	.5 I, 12 0, 3 I/O, 1I/O	1/1	41	(3)
Western Digital CR1872	Pmos	4/10,20	Yes	No	512x10	1x70, 4x40 1x30, 1x4I	1/4	20	(4)
Texas Instruments TMS 1000	Pmos	4/8	Yes	64x4	1024x8	41 + 19 0	1/2	43	(5)
NEC PD345C	Pmos	4/	No	96x4	512x10	10 0, 2x8 0, 1x4I, 1x4 I/S	1/1	72	(6)
AMI S2000	Nmos	4/8	Yes	64x4	1024x8	8I, 13 0, 8 I/O	1/64	51	(7)

TABLE 2.1

outputs are, in fact, separated into two groups, R and O: the R outputs are the decoding of the contents of register Y (decoding of 4 bits in 11 or 13 logical signals; the O outputs are the decoding of the state and of the accumulator contents (decoding of 5 bits in 8 logical signals). This configuration thus allows the user to programme these inputs/outputs in coded form without worrying about their decoding.

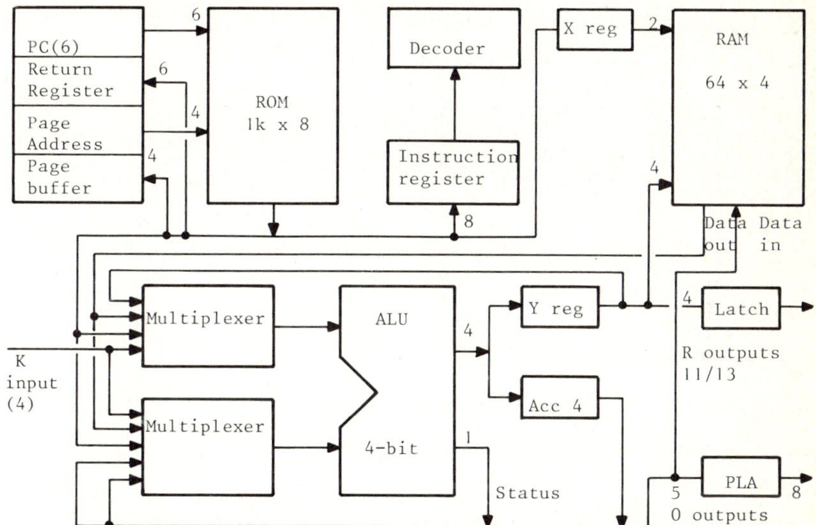

TMS 1000 architecture

2.12 Microprocessors for Process Control

Some monolithic microsystems with better performance than preceding ones are intended for the control of relatively complex systems. It is often a question of versions of high performance microprocessors to which ROM and RAM memories have been added in the same box.

The following are comments relating to Table 2.2.

1. It contains 32 registers, 9 of which are specialised; the remaining 23 play the part of RAM memory.

2. Designed for use with another box (ROM or RAM or Input/Output) and thus to constitute a complete system with 2 boxes.

3. Because of its many inputs/outputs, it is specially designed to conduct sequential processes.

4. Available in many variants : 8748 (with EPROM (21 input/output lines), 8022 (with A/D converter incorporated).

5. This is the F8 Mostek system integrated in only one box.

6. Some input/output lines can be made to specialise and they then become address buses and data buses and the addressing capacity at 64 K RAM and 64 K ROM is extended by adding extra boxes.

Manufacturers of types of microprocessors	Technology	Length of word data/ instruction	Clock included	RAM included	ROM included	Number of inputs/ outputs	Number of registers accum/gen. use	Number of instruc- tions	
General Instruments PIC 1650	Nmos	8/12	Yes	No	512x12	4x8 E/S	1/23	31	(1)
Rockwell PPS 4/2	Pmos	4/8.16	Yes	No	No	4Ex4S 1x4E/S 1bus8 bitsE/S	1/1	50	(2)
NEC PD 546C	Pmos	4/8.16	Yes	96x4	2000x8	2x4E,4x4S 1x3S, 2x4E/S	1/6 in Ram	80	(3)
Intel 8048	Nmos	8/8.16	Yes	64x8	1024x8	3E,2x8E/S 1bus 8 bits E/S	1/8 + 8	90	(4)
Mostek 3870	Nmos	8/8.16.24	Yes	64 registers for general use	2048x8	1E 4x8E/S	1/64	70	(5)
Zilog Z8	Nmos	8/8	Yes	96x8 of which 76 reg. for general use	2048x8	3x8E/S 4E,4S	/76		(6)
Texas Instruments TMS 9940	Nmos	16/16.32	Yes	128x8	2048x8	32E/S	4x16 gen.reg. in RAM	68	(7)
National Semi- conductor SC/MP 11	Nmos	8/8.16	Yes	No	No	1E,3S. 1E/S + 1 bus 8 bits E/S	1/3	46	(8)

TABLE 2.2

7. 16 bit microprocessor. Its input/output lines are adapted to the constitution of multi-processor systems.

8. Specially adapted to limited applications (sales points terminal, games, etc.).

The Intel 8048 microsystem is derived directly from the 8080 microprocessor. Although the arrangement of pins on the two modules is not identical, the 8048 can use the interface modules designed for the 8080. In the 8048 the programme is stored in EPROM memory with 1 K x 8 bits directly addressable by the ordinal counter. The address bus is not directly accessible on the pins of the 8048. Nevertheless, the addressable space can be increased using the three input/output ports of 8 bits which are present on the 8048. Moreover, one of the ports can be used to gain access to the data bus (which can transfer not only data but also the low weight octet of an address) and at the same time an address of 4 bits can be transferred through one of the other two ports and this will assist in the selection of a ROM or RAM external case.

In the absence of such cases these ports can be used in the normal way as input/output interface.

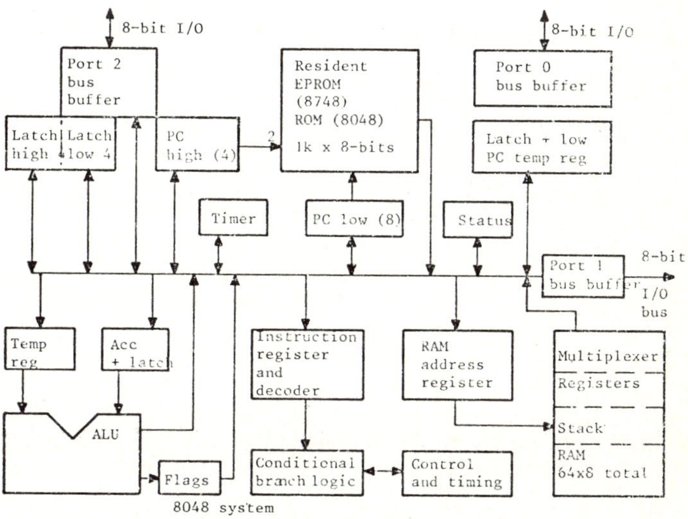

Figure 2.27

2.13 GENERAL PURPOSE MICROPROCESSORS

General purpose microprocessors are circuits with words of 8, 12, or 16 bits which have a significant processing power, a comprehensive set of instructions, varied modes of address and numerous possibilities for taking account of interruptions. Their field of application is relatively wide; it extends from the relatively limited activities of microsystems to small individual computers, through instrumentation, control of peripherals, process control, distributed systems, etc. Thus, the same box can be a basis for complex systems, thanks to an address capacity of 64 K octet, or for simple systems in two or three cases. The comments that follow relate to the table on the next page.

1. From PPS 8/2 a system can be made in two boxes by joining it up with the A21 or A22.

2. Its principle elements are spread over several boxes. A minimal system can be constructed with the aid of the CPU 3850 and the PSU 3851.

3. Integrated version of the IMP 16 which works on 8 or 16 bits; the addresses and the data are multiplexed on one single bus.

4. The best known microprocessor. In order to use it one must add to it an 8224 clock and an 8228 bus controller.

5. The disadvantages of the 8080 do not occur in the 8085 which has in addition 4 extra interruption lines.

6. The Z 80 is a sophisticated component whose set of instructions is compatible with those of the 8080. It has, in particular, a dynamic memory refreshment system, two index registers and two sets of 7 common registers.

7. The Motorola 6800 is the greatest rival of the 8080 although its architecture is quite different.

8. The 6802 is a more integrated version of the 6800. It has, in addition in the same box, a clock and 128 RAM octets. A minimal system in two cases can be made by linking it to the 6846.

9. The 6809 has a set of instructions which include multiplication and which make it possible to work on a bit, an octet or a word of 16 bits. Adapted to multi-processor systems.

10. The 2650 has a great variety of modes of address. A minimal system can be obtained by linking it with the 2656.

11. The 6502 is the most widely used microprocessor for the realisation of individual micro-computers. Its bus is compatible with that of the 6800.

12. The TMS 9980 is actually a slower version of the 16 bits 9900 microprocessor. It has the latter's set of instructions in their entirety.

13. Th 1802 is an 8 bits microprocessor of low consumption with high immunity to noise.

14. The 6100 is compatible with the PDP 8/E mini-computer and has the same set of instructions, although they are rather limited.

Manufacturer and types	Technology	Length of word data/instruction	clock included	addressing capacity	number of registers accum/index general purpose	Number of instructions	
Rockwell PPS 8/2	Pmos	8/8.16.24	No	16K	1/0/3	109	(1)
Fairchild F8 (3850 CPU + 3851 PSU)	Nmos	8/8.16.24	Yes	64K	1/0/64	70	(2)
National Semiconductor PACE	Pmos	8/16.16	No	64K	2/2/0	45	(3)
Intel 8080 A-1	Nmos	8/8.16.24	No	64K	1/0/6	74	(4)
Intel 8085	Nmos	8/8.16.24	Yes	64K	1/0/6	76	(5)
Zilog Z 80 A	Nmos	8/8.16.24.32	No	64K	2/2/12	158	(6)
Motorola 6800	Nmos	8/8.16.24	Yes	64K	2/1/0	72	(7)
Motorola 6802	Nmos	8/8.16.24	Yes	64K	2/1/0	72	(8)
Motorola 6809	Nmos	8/16.	Yes	64K	2/2/0	80	(9)
Signetics 2650	Nmos	8/8.16.24	No	32K	0/0/7	75	(10)
Mos-Technology 6502	Nmos	8/8.16.24	Yes	64K	1/2/0	56	(11)
Texas Instruments TMS 9980	Nmos	16/16.32	Yes	8K x 16 bits	16 registers for programme in RAM	54	(12)
RCA 1802	Cmos	8/8.16.24	Yes	64K	1/10/11	91	(13)
Intersil 6100 A	Cmos	12/12	Yes	4K x 12 bits	1/0/1	50	(14)

TABLE 2.3

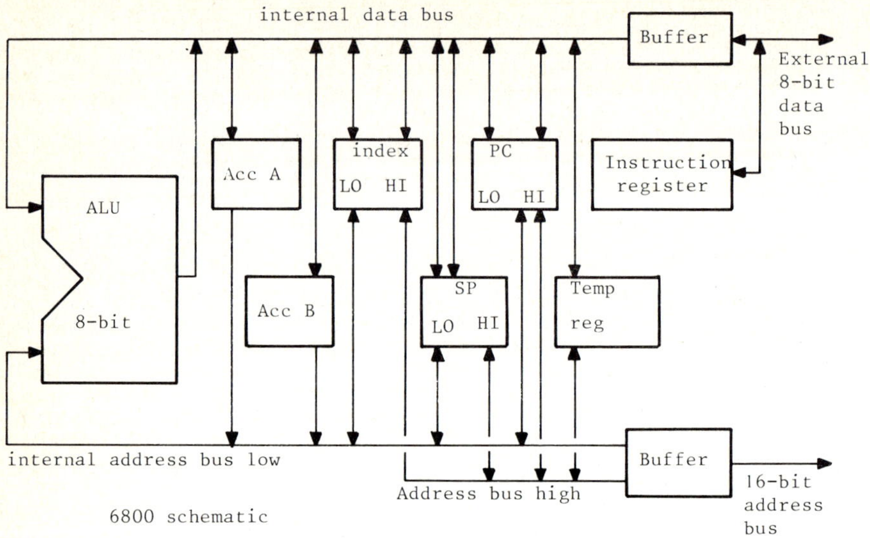

Figure 2.28

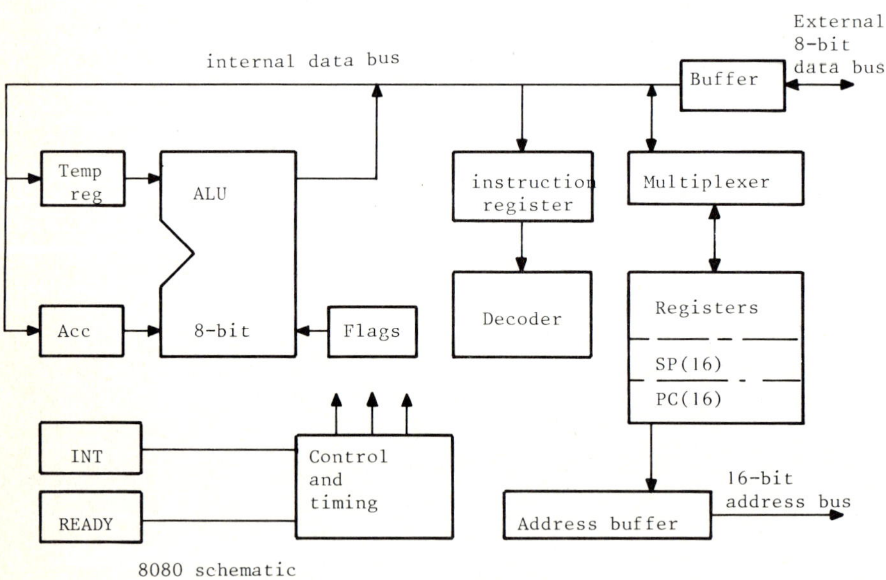

Figure 2.29

2.14 High Performance Microprocessors

We refer here to 8 or 16 bits models in bipolar technology, NMOS or VLSI (Very Large Scale Integration). Some of these are integrated versions of central processing units of existing mini-computers. They are characterised by a wide ranging set of instructions, a usually large number of registers, a high address capacity, high speed and great possibilities for coping with interruptions. They are used in all areas requiring high performance-instrumentation, process control, multi-microprocessor system etc.

In this category we find:

1. The CP 1600 produced by General Instruments. This microprocessor constitutes the core of the TDC 2000 process control system developed by Honeywell.

2. Texas Instruments' TMS 9900 has an uncommon architecture, its 16 work registers do not form a part of the central processing unit but are implanted in RAM memory. Thus each programme has its own working area and any change of context is expressed simply by the safeguard of three registers (ordinal counter state register and working area pointer). It is therefore particularly well suited to applications where interruptions are frequent.

3. Ferranti's F100L is a 16 bit microprocessor entirely designed and manufactured in Europe. It takes programmes compiled using Coral 66, a high level real time language.

4. The mN 661 manufactured by Data General is a copy of the central processing unit of the Nova 3 minicomputer. It has a real time clock, a dynamic memory refreshment device and also a set of instructions which include multiplication and division. Fairchild's 8440 is a 16 bit model which is also inspired by the Nova minicomputer. It is more rapid but more stripped than the Data General microprocessor.

5. The Intel 8086 can operate according to two modes. In its maximum version it addresses 1 mega octets. The programmes, data zones and batteries are relatively addressed to 4 basic registers in the central processing unit. The 8086-1 version operates very rapidly with an 8 MHz clock. It is a 16 bit microprocessor.

6. Motorola's MC 68000 is also a 16 bit model. It comes in a 64 pin case. It can address 16 mega-octets (addressing on 24 bits). It has 16 general registers of 32 bits. Its bus structure enables it to accept direct connection of the peripherals of the 6800 family, but a number of other specific 16 bit cases are now provided.

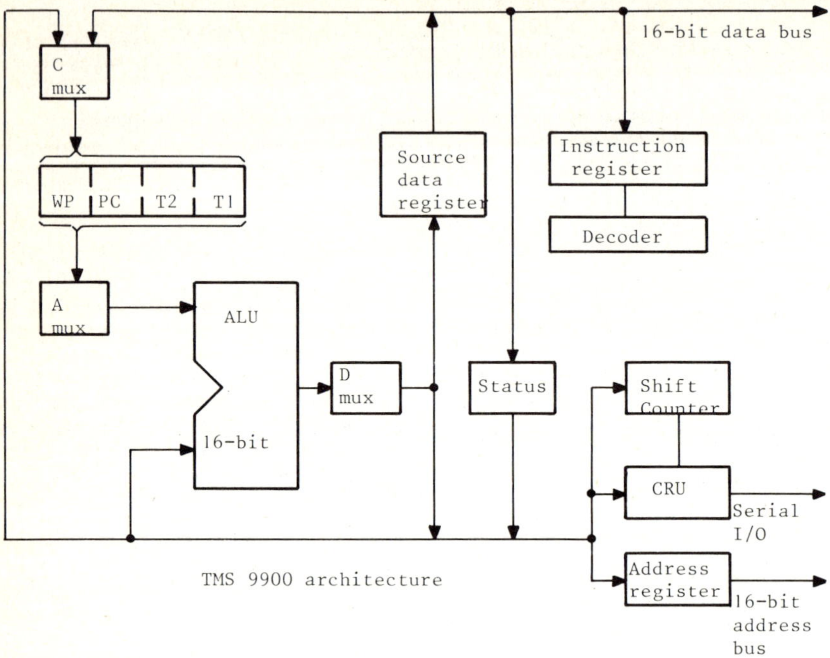

Figure 2.30

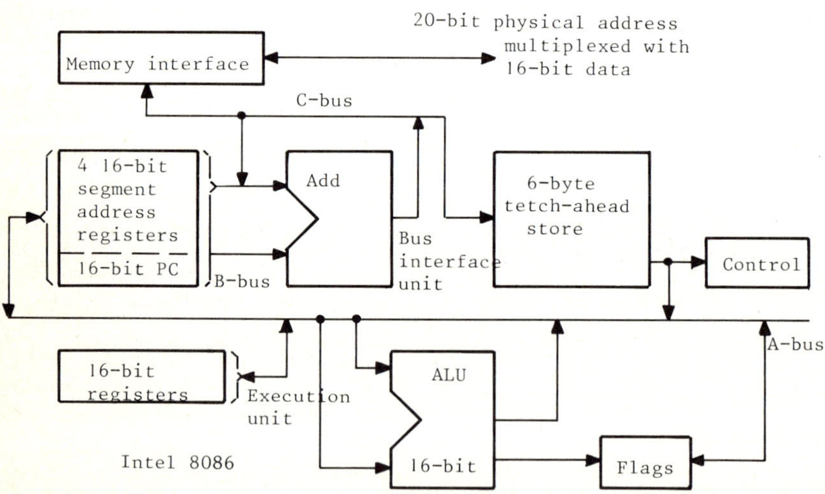

Figure 2.31

2.15 CHOICE OF A MICROPROCESSOR

There are, of course, many criteria to be considered when choosing a microprocessor. Some of these criteria are:

Length of word: This must be adapted to the data to be processed. If one wishes to carry out the logical command of a sequential process, one will usually choose a 4 bit microprocessor. For a complex system of data acquisition or process control it is tempting to use a 12 or 16 bit microprocessor. Let us note, however, that on certain 8 bit microprocessors some operations are carried out with twice the accuracy on 16 bits.

Associated boxes: Except in unusual cases, a microprocessor cannot work alone. A certain number of boxes have to be added to it which perform particular functions. Therefore, it is often useful to opt for a microprocessor with a "large family".

Rapidity of the central processing unit: The frequency of the clock controlling the Timing of a microprocessor does not necessarily signify the rapidity of the microprocessor. It is preferable to examine the time of execution of specific instructions. The basis is generally the time of addition of two registers. The number of internal registers also influences the time of execution of a programme. The more registers of this type there are, the less the programmer will use the central memory, hence more rapid the execution.

Rapidity of exchanges: The possibilities of interruption differ enormously from one microprocessor to another. Some do not have an interruption line whilst others have one or more. The most sophisticated models have "vector" interruptions which indicate to the processor the origin of the interruption and the treatment to be carried out. The speed of the exchanges may be an important criterion in the case of a microprocessor which communicates with a large number of peripherals.

Arithmetic possibility: In addition to elementary operations of arithmetic and logic, some microprocessors can carry out operations of BCD arithmetic and maybe even multiplication or division. They make it possible to avoid BCD-binary or binary-BCD transcoding which can be the source of errors. These instructions are very useful on microprocessors which must process limited data. In the case of complex processing, as in scientific calculation, it is useful to be able to execute a multiplication or a division in one simple instruction. The resulting programme is less cumbersome and more rapid. This possibility exists on some high performance microprocessors.

Address capacity: This corresponds to the maximum memory size addressable by the microprocessor. This criterion must be examined carefully in the case of monochip microprocessors and 4 bit microprocessors. The choice of this type of component must be rejected if the application to be used involves writing a programme whose size exceeds the address capacity.

Next we consider micro-computer software.

2.16 MINI AND MICRO COMPUTERS SOFTWARE

In typical computing applications, the cost of the programming of algorithms is often higher than the cost of the hardware. Figure 2.32 shows this very clearly. This is a result of the steady decrease during the seventies of the cost of computing equipment with the parallel increase in the comparative cost of labour. The percentage represented by the software cost, of the total cost, varies as a function of the application, of the hardware used and of the number of units produced. For example, the cost of the software for an "automatic" washing machine programme is soon amortised over the large number of machines sold whereas for some other special applications, the control software costs could well be very much larger than the hardware costs. This may constrain somewhat the designer of control systems into using more conventional solutions not requiring microprocessors. This is the reason why computer manufacturers usually sell along with the machine a set of utility programmes to help the programmer to write, test and correct his programme. In this section, we describe the basis of the different utility programmes that one comes across most frequently.

2.16.1 Machine Language and Assembler Language

Machine Language

Let us assume that we have a microcomputer with its microprocessor CPU (central processing unit), its memories and an input/output peripheral (for example a teletype). Let us assume that we need to execute an application programme that we have written by using the reference manual of the microprocessor that we have chosen. This programme is obviously coded in binary (or its equivalent hexadecimal representation). In this case we say that the programme is written in machine code. This is the only language that is directly executable by the machine.

Example Let us suppose that we need to multiply two positive integers which are initially in the memory on the 2 registers called A and X of the microprocessor. The result, which will be assumed to be less than 255 will be put in A. In order to write this programme, we use the following instructions which are coded on 2 octets (the first octet contains the operation code and type of addressing and the second octet contains the address of the operand) or on one octet (instruction without operand).

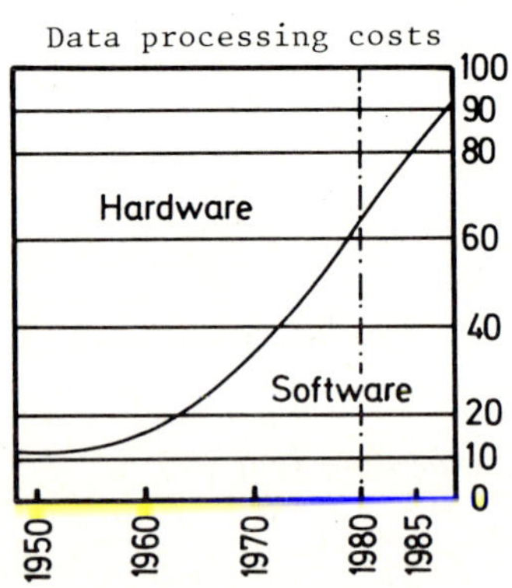

Figure 2.32

Hexadecimal code	Meaning
00 xx	xx ← (X): Putting the contents of register X at the address xx
01 xx	xx ← (A): Putting the contents of the register A at the address xx
02 xx	A ← (xx): Memorisation of the contents of xx in A
03 xx	A ← (A) + (xx): Additions to A of the contents of the memory address xx
04	A ← (A) - !: decrementation of A
05 xx	P ← (xx): unconditional jump to the address xx
06 xx	P ← (xx) if A = 0
07	A ← 0
OF	Halt instruction

The program consists of calculating S = NxM by adding N times the operand M to itself.

Address (octet)	Memory contents	comments
ID	initially arbitrary	s: will contain A
1E	00 33	m ← (X)
20	01 32	n ← (A)
22	07	A ← 0
23	03 33	A ← (A) + (m)
25	01 1D	S ← (A)
27	02 32	A ← (n)
29	04	A ← (A) - 1
2A	06 34	Jump if A = 0
2C	01 32	n ← (A)
2E	02 1D	A ← (S)
30	05 23	Jump
32	initially arbitrary	n: will contain N
33	initially arbitrary	m: will contain M
34	02 1D	A ← (S)
36	OF	Stop

(The programme starts in the memory from the address 1D)

In order to execute this programme, it is necessary to introduce this binary code into the machine. If the microcomputer has a console, then it is by the manipulation of the keys of this console that we could introduce the programme into the machine octet by octet. In order to check that the programme has been correctly introduced into the machine, the memory contents could be displayed on the console screen. This procedure is long and fastidious and it is subject to errors. In addition, it is not applicable in the case of microcomputers which don't usually have a console (a console being a peripheral).

Basic Monitor

In microcomputers, the console was advantageously replaced by a basic monitor. The basic monitor is the software provided by the manufacturer and it resides in the read only memory (ROM). A simple pressing of the RESET button causes the charging of a fixed value in the ordinal counter. If the ROM is connected to the microprocessor at this address, the pushing of the button causes the execution of the monitor.

The function of the monitor is to wait for an elementary order typed in by the user at the teletype and to respond with a piece of hexadecimal coded information (contents of a memory cell or of a register) or by the execution of a simple function (starting of a programme).

Examples of orders:

M: display and modification of the contents of a memory cell.

L: Listing of the contents of a group of memory cells.

A: display and modification of the contents of the register A.

X: same for X

G: Loading of a value into P.

For example, the typing of the order M_{22} will cause the writing out in hexadecimal form of the contents of the address octet 22 (hexadecimal). These contents could subsequently be modified by the user by typing consecutively the two hexadecimal digits: M22 00 07.

Such a monitor thus fulfills exactly the same roles as a console although in a much more readable fashion. However, instead of a console, its use requires the connection of an input/output peripheral like a teletype or a display screen.

Assembler Language

If we examine the multiplication programme in this section we conclude that it produces an incorrect result if initially the contents of A (i.e. N) are zero. Modifying the programme in order to take account of this possibility implies the displacement of the test "Jump if A=0" in particular and thus of the shifting in the memory of a whole section of the programme. Such a case could often lead to the modification of many addresses in the programme's instructions. All these modifications need to be done by the user if he or she only has the basic monitor without an assembler. The Assembler is a user software whose main function is to replace the hexadecimal coded instructions as well as the addresses by a symbolic equivalent.

The assembler is thus a translator which converts this symbolic representation into machine language. Figure 2.33 shows the relationships.

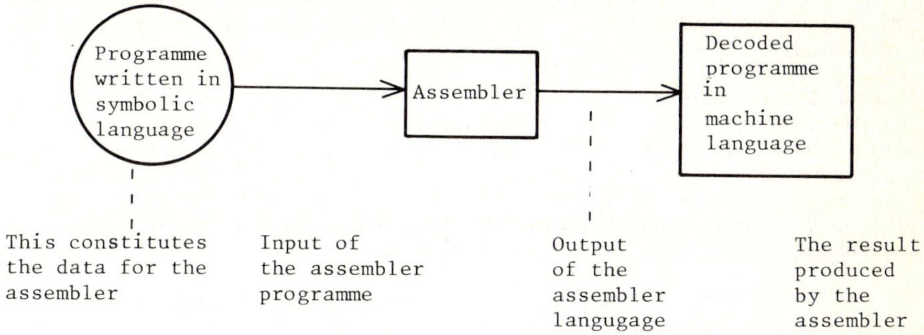

Figure 2.33

We will denote by a programme written in the Assembler language a programme which obeys the rules defined by the Assembler. These rules consists essentially of decomposing the writing of each instruction, or the definition of all data into four fields:

(i) The "address" field which expresses the address of the instruction or of the data in a numerical or symbolic form. In the latter case, this field takes on a label. It is not necessary that each instruction be labelled, only those instructions which are referred to in other instructions must have labels.

(ii) The "operational" field which contains a symbolic representation of the instructions and possibly the addressing mode that is used.

(iii) The field "operands address" which could contain either the symbolic address of the operand manipulated by the instruction or an immediate data (numerical value) coded in hexadecimal or decimal or possibly even ASCII as well as the sign of the operand or a special character implying the type of coding in the case of a piece of data.

(iv) The "comment" field which the user can fill in as he or she likes in order to facilitate the comprehension of the programme. The assembler is oblivious of this field.

The obvious advantage of this type of programming is its ease of writing as well as the symbolic representation of the addresses. It is the Assembler which thus calculates the actual addresses. Displacing an instruction thus no longer requires the programming of a new calculation of addresses as in the case of Machine Language.

2.16.2 The Pseudo-Instructions of the Assembler Language

In addition to the Assembler Instructions which are a symbolic representation of instructions executable by the microprocessor, all assemblers also accept another category of instructions called pseudo-instructions. These Pseudo-Instructions represent information reserved for the Assembler itself and which are therefore not translated into machine language. Their purpose is to direct the translation of symbolic language to machine language which is done by the Assembler. Amongst these directives, we can distinguish the following:

(i) The pseudo-instructions for the beginning and end of the programme which are aimed at indicating to the Assembler the beginning and end of the module to be assembled (A module could group together several programmes or sub-programmes).

(ii) The pseudo-instructions for reserving a zone of data. These avoid the necessity of reserving each octet separately.

(iii) The pseudo-instructions for translatable or absolute assembly. A directive for Absolute assembly specifies the definitive address for the beginning of the programme. The Assembler thus calculates all the addresses that enter into the programme starting from this absolute address. A directive for translatable assembly causes the Assembler to calculate all the addresses in the programme with respect to the beginning of the programme. We note that the address of the first instruction of the programme is fixed at zero. The final calculation of addresses will be done at the moment of loading the programme in the machine using a special utility called the loader (cf section 2.17).

(iv) The pseudo-instructions of external symbols which are aimed at putting together the calls by a principal programme and sub-programme assembled separately. These external symbols will be used by the loader at the moment of loading into the memory of the programme and sub-programme in order to replace the calls of the principal programme to its sub-programmes by their actual addresses.

2.17 THE DIFFERENT PHASES OF THE OPERATION OF A PROGRAMME

A programme which has been written in machine language can be directly installed in the microcomputer and executed with the help of a basic monitor. When the programme is written in Assembler language or in a higher level language, many manipulations are necessary in order to go from the writing to the execution of the programme. These manipulations are essentially the same whether they are performed on the microprocessor or on a big computer. In the latter case, these manipulations are relatively clear to the user since these are managed by a special software called the supervisor. In the case of microprocessors, these manipulations often need to be performed by the user; they put into operation a number of special pieces of software, i.e. text editor, the absolute loader, the relative loader, the line editor and the debugging facility. We will now examine the functions and uses of these different utility softwares.

The absolute Loader

The absolute loader is a piece of utility software which is often integrated into the basic monitor. Its function is to read an arbitrary programme written in machine language from a standard peripheral and to load this programme into its appropriate address. It is this loader which allows one to successively load all the other pieces of utility software. The starting of its execution is obtained by a special order of the basic monitor.

The text editor

This is a utility programme loaded into the memory from a standard peripheral (for example a punch tape reader) by the absolute loader. The function of the text editor is to read a source programme communicated by the user on the teletype. We denote by the source programme the algorithm of a programme written in any language which could be the Assembler language or equally a high level language (Fortran, Pascal, etc.). Indeed, the text editor will accept without verification all that is supplied to it by the user. It does not therefore verify the syntax of the sentences that have been produced. It only allows the user to introduce, modify, or print at his convenience the sentences of the source programme. This source programme is organised into the memory as and when it is produced in code ASCII. Once the programme has been prepared, the user could print out the final version or store for example on punched tape. In the latter case, this tape will constitute the data for the Assembler. The user has at his disposal a series of orders which allow him or her to specify to the text editor: Read a statement of the source programme, eliminate a statement, replace a statement, insert one or more statements into an existing programme, list a corrected source programme, etc.

Assembly or compilation

The Assembly is the translation of a source programme written in Assembler which has been produced by the text editor, into object language. This translation is achieved by the Assembler which is loaded into the memory with the help of the absolute loader. The compilation achieves the same function for a source programme written in a high level language; the translation programme is therefore called the compiler. The object language that we have just mentioned is not necessarily identical to the machine language. We have indeed shown previously that a program written in Assembler could have calls to external sub-programmes. This implies that the address of the location of these sub-programmes is not known at the moment of Assembly. The Assembler is therefore not able to translate these calls (which are all jump instructions) into their equivalent machine code. It is for this reason that it is important to distinguish between machine language and object language. Indeed, in the case that we have just described, the object language produced by the Assembler is not directly executable by the machine. After the translation, the user can ask the Assembler, by using a specific order, to output the object programme onto, for example, punched tape.

Relative loader

If the source programme written in Assembler is preceded by a directive of absolute assembly the addresses in the object programme are the final addresses. If this programme does not call any external sub-programme, the code produced by the Assembler is then directly executable. It could thus be loaded into the memory by the absolute loader (phase 7).

If the source programme written in Assembler is preceded by a directive of translatable assembly the address in the object programme should be incremented by the address of the location of the programme. The relative loader is a utility (loaded into the memory by the absolute loader) which achieves this incrementation. Thanks to this loader, it is possible to memorise anywhere in the memory the object programme produced by the Assembler. This property is important when the user wishes to use the same programme for different applications, i.e. in different memory configurations (phase 8). The execution of the object programme is then carried out by loading in the ordinal counter the address of the first instruction of the programme (the address 1E in the case of the multiplication programme).

Link Editor

The link editor is a relative loader which, in addition, achives the liaison between a programme and its external sub-programme, i.e. assembled separately in the relative mode. At the moment of Assembly, the Assembler recognises the external symbols with the help of a pseudo-instruction placed at the head of the programme. Every time it comes across an external symbol in a programme, the Assembler replaces it by a special code. When the link editor sees this symbol during the relative loading of a programme or sub-programme, it writes the name of the external programme thus designated on the control terminal. At the end of loading, the user should thus provide to the editor the links of the named programmes and sub-programmes. The link editor then proceeds to load them following the programmes already in the memory and replaces the external symbols in the programme called by the address of the location of the programmes that have been called.

Figure 2.34 shows the various phases and the subsequent manipulations within the memory configuration.

Debugging

When one assembles or compiles a programme, it is fairly certain that it will have syntactical or even semantic errors. Syntax errors are detected at the assembly (or compilation) stage whilst semantic errors can only be detected once the programme is executed. If the programme is written in a high level language, the source and object programmes are too similar for one to be able to follow the step by step execution of the programme starting from the source text. On the other hand, if the programme is written in Assembler, there is a close link between the source programme instructions and those of the object programme. In case of error, it is thus possible to follow step by step the execution of the object programme. The Debug programme is a utility software which could be loaded into the memory with the help of the relative loader and which allows one to follow the execution of an object programme.

This utility has a series of orders which allow it to ask to debug, e.g.:

(i) The execution of one or more instructions of the object programme.

(ii) The display and modification of the contents of any memory cell or register.

(iii) The insertion of traps in the object programmes which allows one to add in a number of instructions into the object programme without modifying the existing addresses.

Phase 1:
Initial representation
of the memory

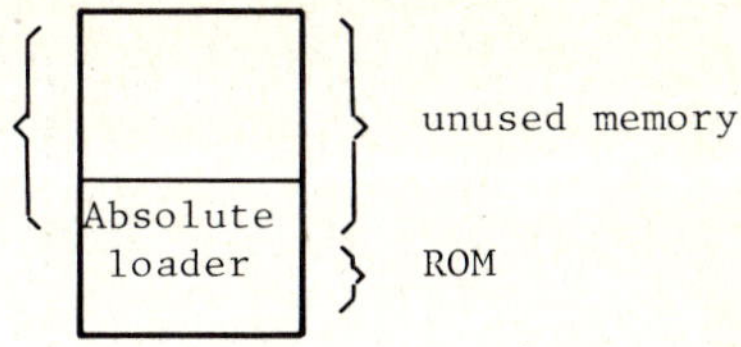

Phase 2

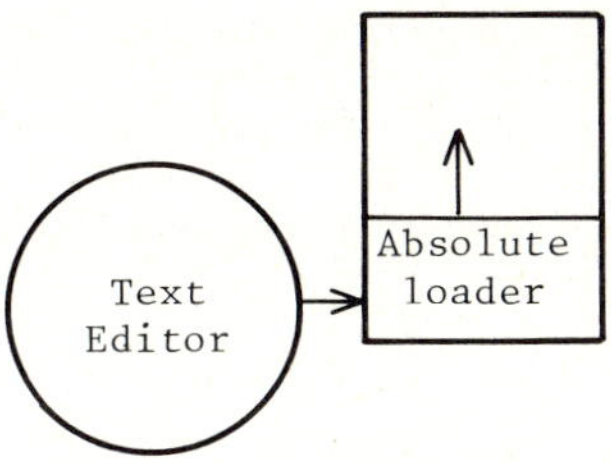

Phase 3

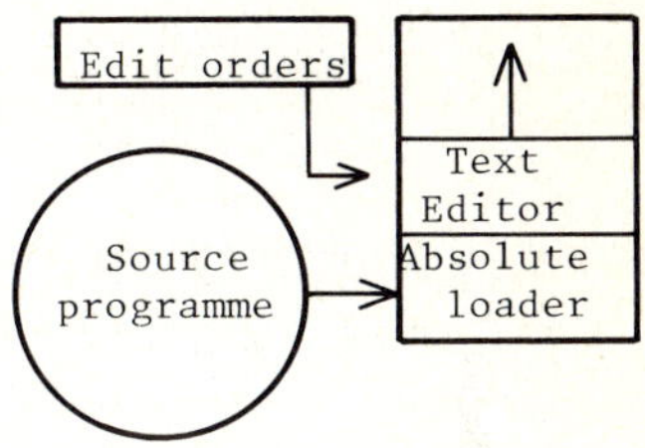

Phase 4

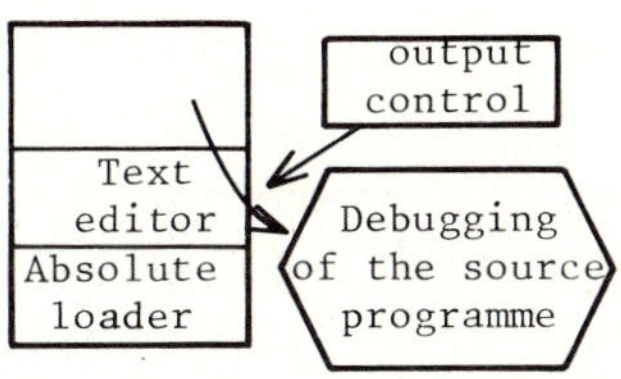

Phase 5

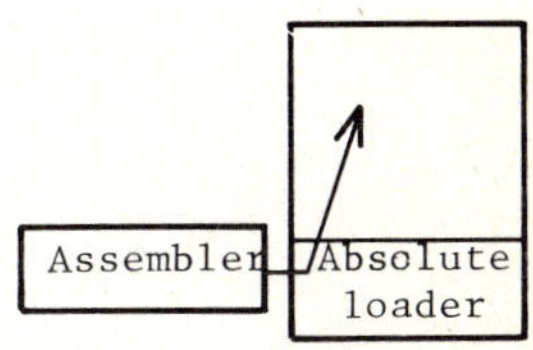

Phase 5a

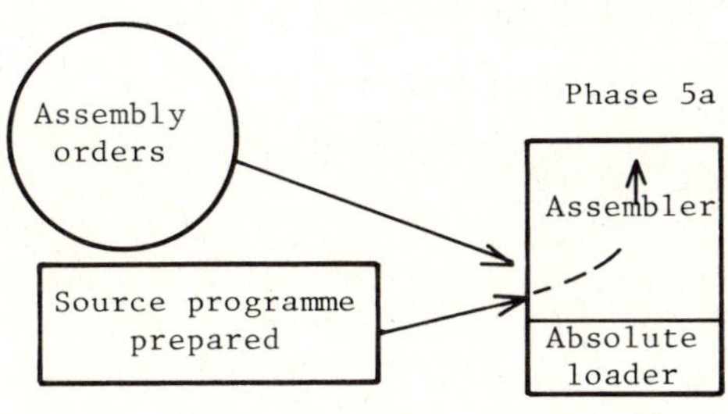

We note that during assembly, the Assembler could detect syntax errors in the source programme (forgotten labels, undeclared data, etc.). In this case the user must repeat phases 2,3,4 in order to correct the source programme.

Phase 6

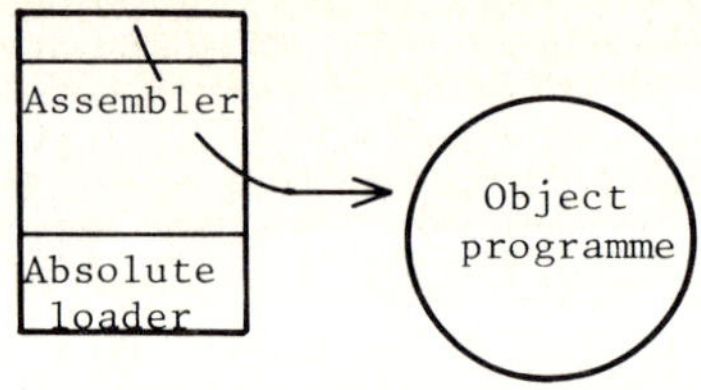

Phase 7

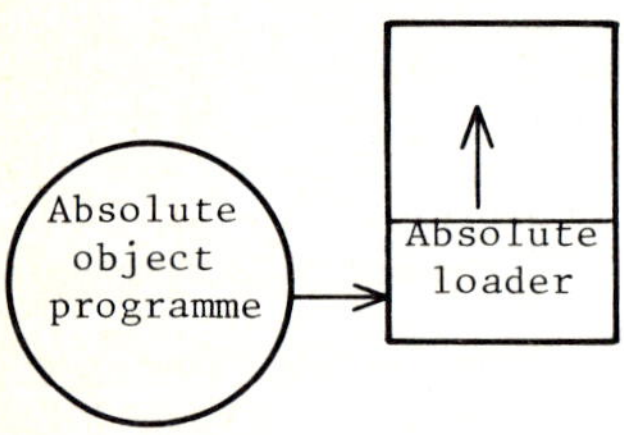

Phase 8

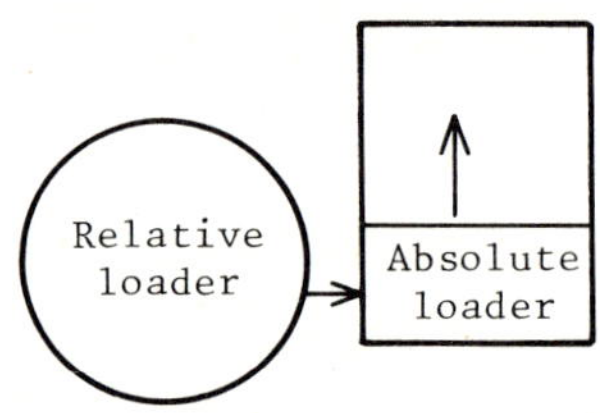

Phase 9

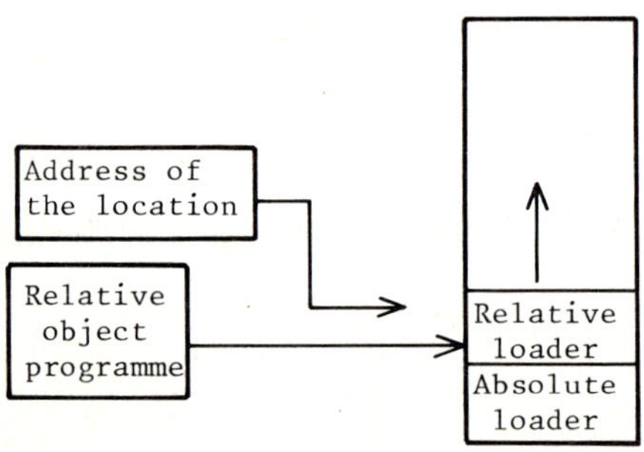

(note that the phases 10 and 11 are the same as the phases 8 and 9)

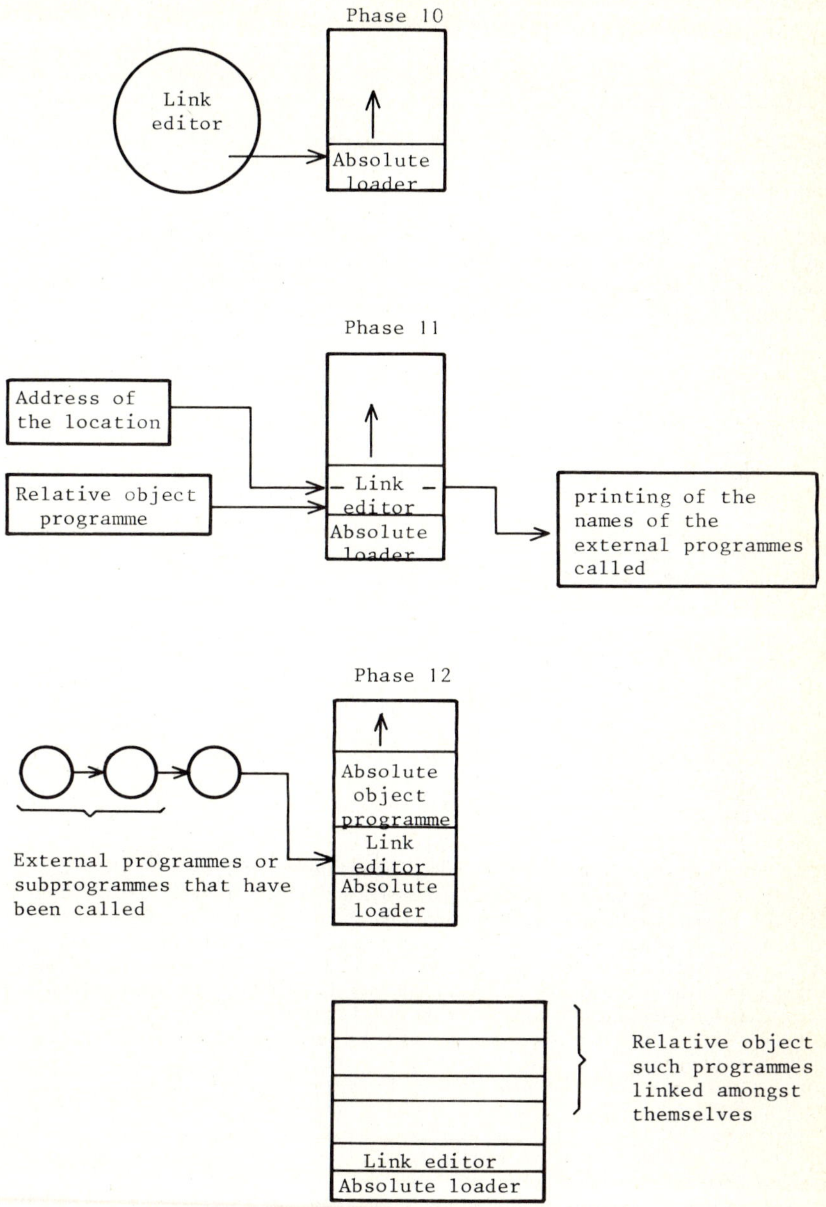

Figure 2.34

(iv) The displacement of a block of memory octets from one zone to another.

Figure 2.35 shows the position of debug in the memory configuration.

Utility sub-programmes

In section 2.15.1 we presented a programme for the multiplication of two integers. Such a programme will obviously be useful for virtually all computer users. It is therefore reasonable that such a programme exists in the library provided by most manufacturers. Such a sub-programme is called a utility. There are many such utility sub-programmes. We mention in particular the sub-programmes for addition, multiplication, subtraction, multiplication and division in floating point. It would thus be useful to have such sub-programmes in order to programme a complex task.

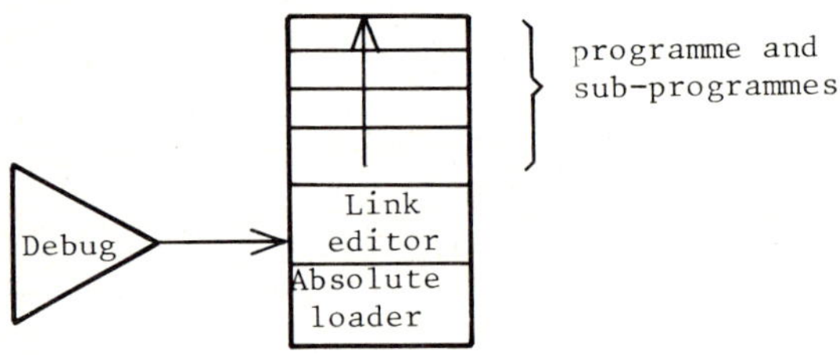

Figure 2.35

2.18 OPERATING SYSTEMS

A basic monitor (cf 2.15.1) is a programme which requires little space and which facilitates all the operations of loading of programmes and their listing. It is however restricted to programmes written in machine code. If the user wishes to programme in Assembler or in a high level language, he will be constrained to the use of the different phases described in section 2.15.2 in order to execute it. However, since all these phases are the same in any case, it is reasonable to develop an extra utility whose single function will be to ensure the automatic following of phases. Such a possibility exists only if the computer has a big secondary memory (floppy disc, etc.) in which could be placed the utility programmes as well as the source and object programmes of the user. This extra utility is called the supervisor. This constitutes one part of the so-called operating system.

The functions of the operating system are used to manage computer resources, i.e. CPU, the memory and the peripherals. In order to appreciate the importance of this management, it is desirable to examine certain special features of three major fields where computers are used.

	Scientific	Management	Real time
Environment (no. of peripherals)	Few	Many	Many
size of the algorithms	Medium to high	very high	low
Structure of the algorithms	Serial	Series and parallel	parallel in real time

We deduce that in management as well as in real time, there could be a large number of programmes which reside in the memory and all of which need to be executed. Since these programmes are likely to use the same resources (e.g. the CPU), it is necessary to arrange a division of their use. It is for this reason that a user system contains a special software called the scheduler whose job is indeed to divide the resources. We will next examine the basic principles of operation of the different types of operating systems that exist at present.

Monoprogramming

Consider a computer system which is to be used to do a succession of tasks whose average characteristics are as follows:

Time of input of data : 1 minute
Time of printing : 1.5 minutes
Execution time : 0.5 minutes

We assume that this computer has an additional large memory and a supervisor and that the only way for the computer to work is in a strictly sequential mode i.e.:

(i) It could only execute one programme at a time.

(ii) When a programme asks for an input/output, its execution is stopped until the end of transfer of the data.

The simplest form of scheduling is the "open loop": Each user has for example 20 minutes to work on his programme. For the user, this is an ideal mode of operation; he/she could immediately correct syntactical errors in his/her programme as soon as it is assembled or compiled and then follow the step by step execution of the programme.

Unfortunately, in this case, the rate of use of the processor, i.e. the ratio of the execution time (the time for the work of the CPU) to the total time, is very low: of the order 0.5/20 = 2.5%. However, the turnaround time for a user is zero.

This large waste of the CPU time which is due to the manual intervention of the user, could be significantly reduced by a very simple form of automatic scheduling. The scheduling is based on the following principle: the user submits his work to an operator who piles up all the work onto an input peripheral (card reader for example) in the order of arrival. The jobs thus enter directly into the machine and are successively executed without an intermediate loss of time. In this form of scheduling, an average piece of work will use the computer for 3 minutes (sum of the execution time and the time for input/output) and this implies that the rate of usage will go up to $\frac{0.5}{3} = 16.6\%$. However, even this simple form of scheduling creates new problems: how to protect the user system itself from the errors in the programmes of the user? How to ensure that a user's programme hands back control to the operating system when the programme is finished? In order to solve the first problem, many manufacturers have introduced new hardware mechanisms in the CPU: Thus the base/limit registers ensure that all addresses calculated in a user programme are contained between two bounds fixed by the computer. In order to ensure that the contents of these registers are not accidentally modified by a user programme, there exists, in many machines, two modes of execution of instructions: In the "Master" mode all the instructions are executable. In the "slave" mode, all attempts at executing a forbidden instruction will cause the outputting of an interruption which will be handled by a special programme of the operating system. This will cause an error message to be transmitted to the operator. As far as the second problem is concerned, i.e. the return of control to the system at the end of a user's programme, it is merely necessary to replace the Stop instruction which nominally finishes this programme by a jump instruction to the scheduler.

Such an operating system is often called "Batch". In Fig. 2.36 we show a brief flow chart for the functioning of such a system.

Multiplexing or Multiprogramming

The principal disadvantage of the above operating system is that it is subjected to the slow speed of the input-output peripherals. The CPU loses more than 83% of its time during the execution of a job. The problem could therefore be formulated in the following terms: Is it necessary to stop a computer during 100 milliseconds in order to print a character on the teletype whereas during this period it could perform many tens of thousands of elementary operations? The answer is obviously no. The technique of input/output interrupts allows parallel work on the CPU and on the peripherals. Indeed, it is the peripheral itself which informs the CPU, using an interrupt, when it has finished the transfer of data. In this case, the CPU can get on either following the current programme if the data is not immediately required or executing another programme. Hence the importance of keeping simultaneously in the memory a number of programmes which are ready to be executed. This technique is called multiprogramming.

The scheduler of an operating system for multiprogramming should be able to manage the simultaneous execution of many programmes using the same processor. This scheduling technique which consists of dividing a single resource (CPU) between many jobs is called multiplexing. There exists many types of multiplexing which can be distinguished essentially by the rules of movement from one programme to another: At what point to make this move? What is the criterion for the choice of the new programme to be executed? To take some examples of multiprogramming systems we have:

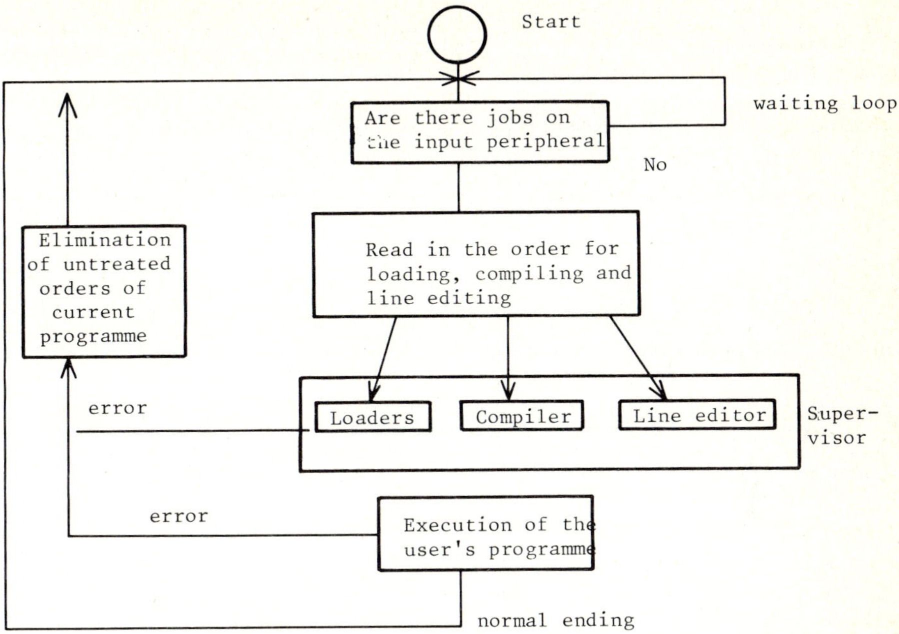

Figure 2.36

(i) Multiprogramming in divided time is a technique of scheduling which consists of changing the programme at regular periods. All the Jobs have the same degree of "priority" and the time for using the CPU is equally divided between them. This technique is usually used in installations which have many consoles (teletypes or display systems) connected to the computer. It has the advantage of giving the illusion to each user on his/her console that the computer is working only for him/her.

(ii) Multiprogramming with interruption of programme: When a programme is being executed, it uses the CPU until it is finished or until it is obliged to await an end to an input/output operation. In both these cases, the scheduler starts the execution of the higher priority task which is not waiting for an end of an input/output operation. This type of scheduling is, in fact, seldom used since here a low priority task could occupy the CPU for as long as it wants.

As an example, let us examine Fig. 2.37 where T1 has higher priority than T2 which itself has higher priority than T3.

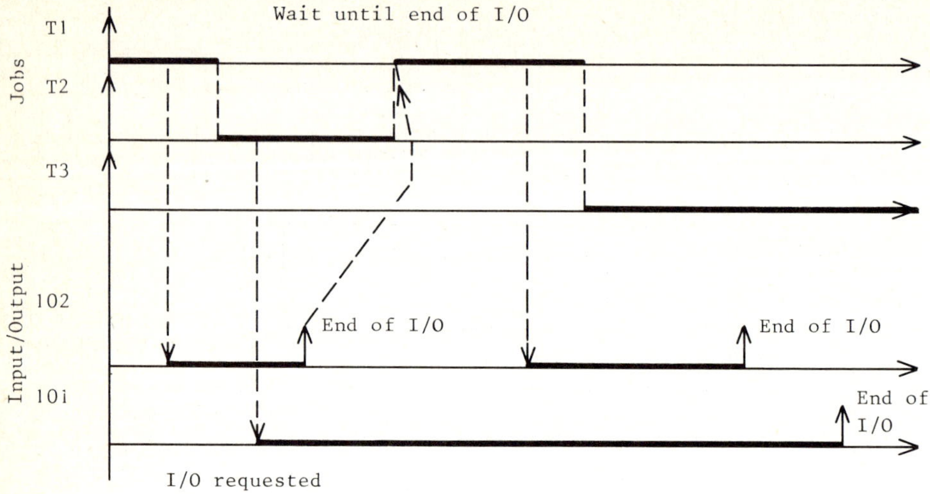

Figure 2.37

(iii) Multiprogramming with programme interruption: In this technique, the scheduler chooses the next task to be executed either when a programme is finished or when a programme asks for an input/output or finally when an input/output operation is terminated. The scheduler thus seeks the highest priority programme which is not currently awaiting an end to an input/output for starting its execution. This gets around the difficulty of letting the CPU be occupied over lower priority Jobs. In addition, it extends the functions of the scheduler to other events (interrupts) than those sent by peripherals as for example the request for user dialogue, the sounding of an external alarm, etc. It is this type of an operating system that is used in real time applications and particularly in control applications. Figure 2.37 shows the use of such an operating system.

2.19 REAL TIME OPERATING SYSTEM

Definition, priority and representation of Jobs

Definition of a Job or task:

When a computer is connected within an industrial environment, the application software could in general be decomposed into modules each of which is to be executed at points in time depending on the industrial environment. For example a data capture job will be periodical and will therefore be executed at the time when a clock signal is emitted; an alarm signal needs to be executed when there is an appropriate external signal (interrupt). Such modules will be called tasks or jobs. A task is thus a procedure which is assumed to reside in the memory and which brings together all the operations that need to be performed at the same instant, i.e. following the appearance of the same event.

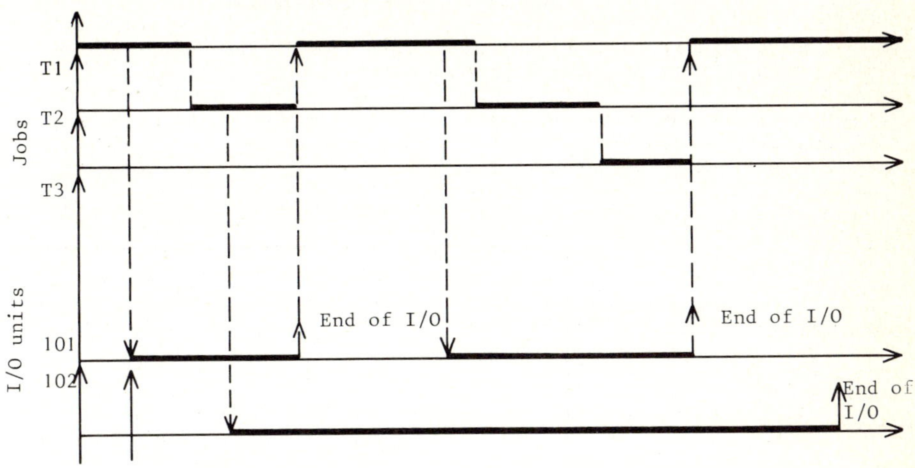

Figure 2.38

Priority of a task:

Since the starting of a task depends upon an event external to the computer, it is possible that there may arise multiple quasi-simultaneous tasks which need to be executed simultaneously. Such tasks are called concurrent. In order to allow the scheduler to choose the task which will next be performed, we give to each task a certain priority. At any instant, the task having the highest priority will be performed. The choice of task priority assignment is not arbitrary. This depends on the particular application and it should be fixed as a function of the order of urgency of the tasks with respect to the environment for the cases where many tasks require simultaneous attention.

States of a task:

In order for a task to be executed, it is necessary that the event which is supposed to trigger the start of its treatment has already taken place and in addition, it has all the necessary resources for its execution. During its life cycle, a task will pass through many states, e.g.

(i) A state which doesn't exist as long as it is not declared to the scheduler which manages the triggering of the execution of tasks following the appearance of an external event.

(ii) An activable state is one where the event which triggers its execution has taken place and it has available all its resources (peripherals, memory zone, utility sub-programmes, etc.) except for the CPU.

(iii) A blocked state is one where one or the other (or both) of the conditions of the activable state have not been satisfied.

(iv) The active state or execution state. This is the state with an activable task of currently the highest priority.

We could represent the different states of a task by Fig. 2.39 as well as the transitions between them. We will find in our study of primitives the means of causing these different transitions.

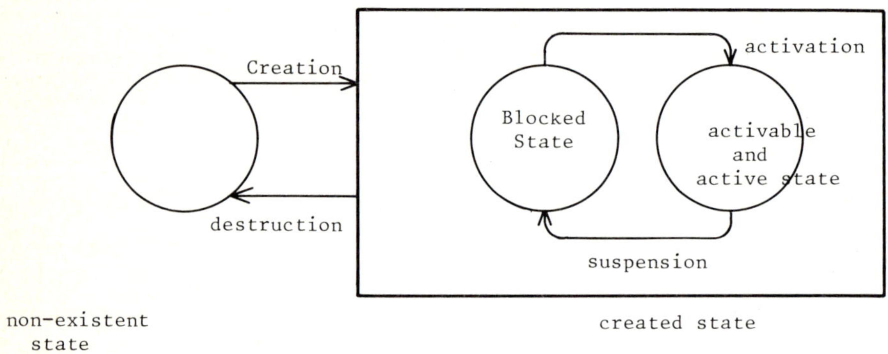

Figure 2.39

Context of a task:

An active task can be suspended at any moment in favour of a task of higher priority which becomes activable. The suspended task remains activable, it could therefore take up the continuation of its execution as soon as the higher priority task has been executed. It is therefore necessary to safeguard the contents of the general registers of the computer during this suspension. This is what we call the context. The context of a task will be safeguarded in a table managed by the scheduler. It is therefore the scheduler which must itself load this table of contents of the general registers of the computer at the instant when a task is suspended and subsequently to reinstate the contents of this table into the general registers when the execution of this task is taken up once again.

The physical implementation of the parameters of a task:

It is important to distinguish between the task itself which is a procedure that is stored anywhere in the memory and its parameters which are managed by the scheduler. These parameters are the following: The state of the task, its priority and its context.

All the parameters of the same task are grouped together in the same table and the tables of the different tasks are linked together by pointers or flags. This is called a list form of data structure (see Fig. 2.39).

List of tasks:

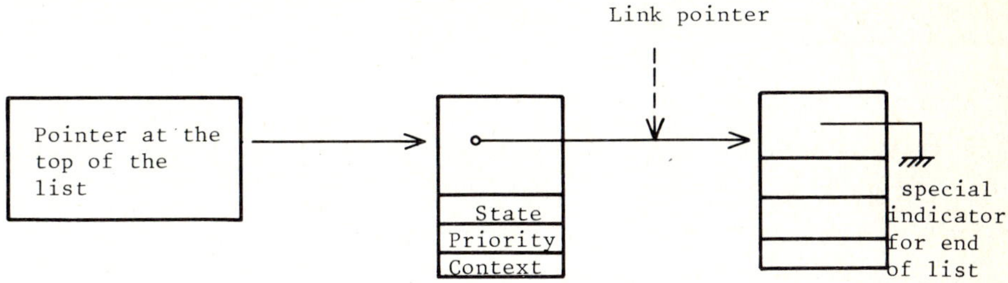

Figure 2.40

If we organise the elements of this list in the order of task priority, it is possible to eliminate the "priority" information from these tables provided of course that all the priorities are different.

Primitives for the changing of the state of a task

According to Fig. 2.39 there exist 4 transitions of a task, i.e. creation, destruction, activation and suspension. Since the state of a task is the information memorised in the scheduler, these state transitions should be made by specialised programmes of the operating system. Such sub-programmes are called primitives.

Creation primitive (CREER):

This primitive allows the user to declare a new task to the scheduler. This primitive has the consequences of inserting a new table in the list described previously and to load its contents using priority parameters, and the initial context supplied as an argument of the primitive by the user. The initial state of a task will always be the activable state. In many real time operating systems all these task creation primitives are grouped together into a very special task which is executed systematically on initialising the operating system.

Destruction Primitive (END):

This primitive has the effect of withdrawal of the list from the appropriate table. Since the scheduler chooses at every instant the task to be executed by going through this list, a destroyed task could not therefore be executed anymore even if its procedure still resides in the memory.

Activation Primitives:

This primitive has the effect of making activable the state of a task by modification of the "state" information in the table of the task. Since, at every instant, the scheduler chooses the task to be executed amongst the activable tasks i.e. by consulting the list of tasks, this simple modification of the state information will ensure the execution of the task concerned when it will have the highest priority (i.e. when tasks with a higher priority are suspended).

Suspension primitives (WAIT):

Many types of events could cause the suspension of a task being executed:

(i) The unavailability of resources e.g. when a task needs to use a peripheral which is already being used by another task.

(ii) An external interrupt which will cause the activation of a higher priority task.

(iii) A clock signal which restarts a higher priority periodical task.

When a task is suspended, the context of the task must be safeguarded in the list of tasks.

All the primitives that we have described are sub-programmes which could be called not only by an external event but also by the task itself. Thus a task could create another task; it could destroy itself (the destruction of a task is not usually tolerated); a task could trigger another task by freeing a resource (this is an implicit activation); a task could suspend itself (while awaiting the freeing of a resource).

Remark

There exists a last transition amongst the states which has not been depicted on Fig. 2.39. This is the passage of an activable to an "active" state. It is only the scheduler which can perform this state change. The procedure that is followed is very simple: in going through the list of tasks, the scheduler seeks the first table containing the activable state. The context of the task thus found is put into the general registers of the computer (amongst which is the ordinal counter). This task is then executed.

We will next see how we can use these primitives.

Use of the Primitives

Tasks executed on the appearance of an external event: All the interruptions (except for those of the real time clock) are masked by the initialisation of the operating system. If a task requests to be executed on the appearance of an external event, it must make the request using the suspension primitive (WAIT) to the scheduler with the argument being the number of the appropriate interrupt. The scheduler then suspends the task, unmasks the designated interrupt and adds into the table of this task the interrupt number which will cause its reactivation. The appearance of all interrupts is treated by a specialised scheduler programme which achieves the following:

(i) Safeguarding of the context of the interrupted task in the task list.

(ii) Scrutiny of the task list, activation of the task awaiting an interrupt and elimination of the number of this interruption from the table.

(iii) Search in the task list for the new task with the highest priority (which could equally be the task that has been interrupted or a newly activated one).

Example:

```
  Start :  Wait(i)
procedure for starting
          i
     Go to start
```

We note that this example is for the case of events which occur many times. If the same event occurs during the time that this task is being treated, the event is ignored.

Periodic Tasks

Since a clock signal is an external event, a periodic task could just as easily be constructed as the above case. However, if the periodicity of the task is some multiple of the clock cycle, the task must itself count this number of cycles before starting the execution of its own procedure.

```
  INIT:  N : = 0
START:  WAIT (j)                j: clock interrupt

  N : = N + 1

IF N < PERIOD GO TO START

Procedure for the treatment

     GO TO INIT
```

In order to avoid the task having to measure the number of clock cycles, many operating systems offer the user an additional primitive (DELAY) whose argument is the period of the task. This primitive initialises a counter to the value of this period. This counter is subsequently decremented after each clock cycle. When it becomes zero the periodic task becomes activable.

```
     Start:  DELAY(10)            Periodic task
                                  of periodicity
Procedure of treatment            10

       GO TO Start
```

From the above description, the functioning of the DELAY primitive is somewhat unsatisfactory. Indeed, it isn't the two instants of start of consecutive execution which will be separated by a period but the end of one execution and the start of the next

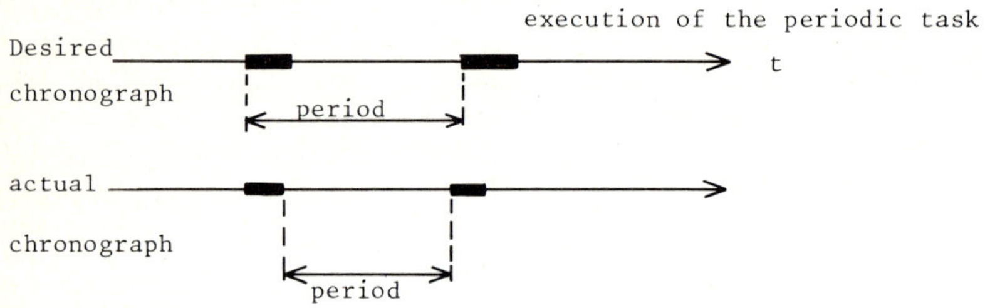

In order to improve the functioning of this primitive, it is necessary for the counter to be reinitialised to the value of the period as soon as it becomes zero. This modification of the primitive has two consequences:

(i) The last instruction of the higher priority task should now be a new primitive (PAUSE) whose function is to cause the suspension of the task that is awaiting the end of the period.

(ii) It is possible that a period may pass whilst the periodic task has not finished the previous treatment. (This will especially be the case if this periodic task had been temporarily suspended to the benefit of a higher priority task like an alarm task.) It could become necessary in this case to memorise these trigger signals for the periodic tasks which could not be treated. We see thus that the construction of a DELAY primitive is closely linked to the application. Thus according to the case being treated we will use one or other of the methods of functioning of the primitive.

```
START:  DELAY(10)
Treatment procedure
   PAUSE
```

Access to a common resource

On using the preceding primitives, we deduce that the scheduler allows the separation of the function of the tasks from their temporal management and this obviously is a great help in writing them. Each task thus informs the scheduler of its execution conditions with the help of primitives whilst the scheduler takes charge of the multiplexing of the CPU between the different concurrent tasks. But this multiplexing could cause a particular type of conflict: it is indeed possible that two or more tasks require simultaneous access to the same resource during their execution. These communal resources could be of hardware (peripherals for example) or software (variables or sub-programmes). Thus if a task is interrupted at the moment when it manipulates a communal resource, for example a variable, the new task being executed may try to use this resource whose value is thus inconsistent. The CPU itself is a communal resource and its management is organised by the scheduler. But when two or more tasks wish to access the same variable or table or sub-programmes, i.e. sources which should not be split, it will become necessary to synchronise the tasks between themselves i.e. to suspend all tasks which request a resource which is currently occupied by somebody else until the resources are free. The standard mechanism for synchronisation is the semaphore.

Semaphore:

A semaphore is a positive negative or zero integer variables followed by a queue which will contain the name of the blocked tasks awaiting the same resources. This variable s is manipulated by two primitives P(s) and V(s) which should be called by all tasks which need to access and free a communal resource.

Primitive P(s):

The primitive P(s) is used for requesting access to a common resource. It causes the decrementation of s. If the resource is not free, i.e. if s is negative, the task which executes P(s) is suspended while waiting the freeing of the resource.

```
P(s)        s : s - 1;
            IF s < 0 THEN BEGIN
            Comment: we assume that this primitive is executed
            by the task r.

            State (r): = Blocked
            Put the name of the task, i.e. r in the queue f(s) of
            the semaphore s; END

V(s)        s : = s + 1
            IF s ≤ 0 THEN BEGIN
            output a task of the queues f(s);
            Comment: Let q be the name of this task;

            State (q): = activable
            END
```

Primitive V(s):

The V(s) primitive is executed on the freeing of a common resource. It increments the value of s. If, at the moment of execution of V(s) a number of tasks are blocked while awaiting the freeing of a resource, i.e. s is < 0, the first amongst them is reactivated. We deduce thus that the value of the semaphore has the following meaning:

(i) if s < 0, its absolute value equals the number of tasks blocked in its queue awaiting the freeing of the resource.

(ii) if s < 0, its value is equal to the number of tasks which could access the resource without being blocked.

It is important to note that there will be one semaphore for each common resource and that a single task could use the primitives P and V of different semaphores.

<u>Examples of the use of semaphores</u>:

a) Suppose that there resides within the memory 3 tasks having the following functions:

(i) The first task's function is the periodic acquisition of measurements from the industrial environment of the computer. These measures are arranged in B.

(ii) The second task is a task which is triggered by an external event and whose function is to read memorised data into B and to deduce from this the value of the analogue control signals sent to the process.

(iii) The third task is a periodic one (with a period different from that of task 1) whose function is to read the data stored in B and to print it out.

We note that B is a resource divided between the three tasks. It is imperative that the tasks 2 and 3 do not access B when task 1 is in the process of modifying the value of the data it contains. Also, it is necessary to forbid task 1 to access B when tasks 2 and 3 are in the process of taking out its contents. In order to protect the access of B for the 3 tasks, we will surround the part of their procedure which accesses B by the primitives P and V of the same semaphore s as shown in Fig. 2.41. (note that initially s : = s-1).

Task 1	Task 2	Task 3
DELAY acquisition of the measurements P(s) Storing in V(s) PAUSE	START: WAIT P(s) Read B V(s) Calculation of the control GO TO START	DELAY P(s) Read B V(s) Write Pause

initially s = 1

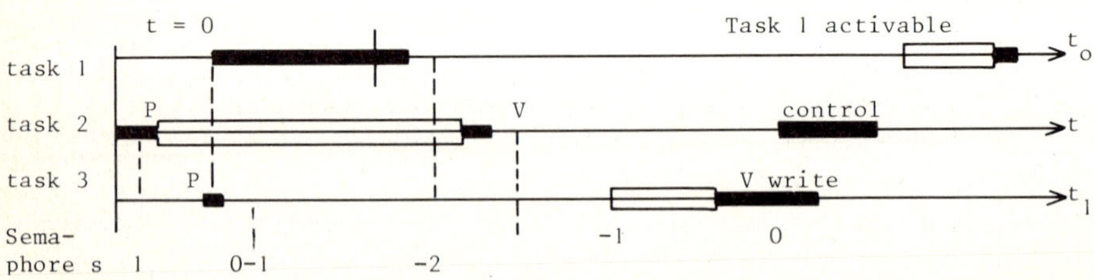

- task suspended (during the execution of another task of while waiting for a resource)

□ task using the communal resource B (thus in execution)

▬ task in execution, executing the parts of the procedure not calling on the communal resource

We assume that the order of priority of the tasks is as follows:

Priority of 3 > priority of 2 > priority of 1. and that at the instant t = 0, the three tasks are activable (start of period for the tasks 1 and 3 event arrived for task 2).

Figure 2.41

b) The use of semaphores could, under certain circumstances, cause certain dangers called Dead Lock. Suppose that a task 1 desires access to a communal resource 1 and that after having obtained this resource after the execution of a primitive P(s1) it requests access to a second communal resource 2 by P(s2). Assume also that a task 2 conversely requires access to the same resources but in a reverse order, i.e. execute P(s2) and then try to execute P(s1). If at this moment the task has already obtained the resource 1 but not resource 2, the task 2 which utilises the resource 2 and requests the resource 1 will not be able to obtain it (since it is already being used). The task 1 which requests resource 2 finds itself blocked for the same reason. It is impossible for the tasks to resolve this mutual blocking. It is necessary therefore to be very careful when defining access to communal resources. Figure 2.42 shows the situation that we have considered.

Watch Dogs

These are special counters which allow the monitoring of the execution time of a task. When a task is not terminated at the end of a period of time fixed by its counter, it is suspended using an Exit procedure. In order for a task to call for monitoring by a watchdog, any real time operating system provides a special primitive with 2 arguments: The time after which the watchdog should go into action and the address of the exit procedure. The scheduler thus associates with the task being monitored a counter decremented at each clock cycle; when its value falls to zero, the exit procedure becomes active. From an implementational point of view, a watchdog primitive is very close to a DELAY primitive, the only difference being that the task activated at the end of the delay does not constitute the normal procedure but rather a special procedure.

Conclusions

We have described most of the primitives used in many industrial applications. However, for a given application, it is not always necessary to use all these primitives. In addition, when the treatment of an application leads to a definitive software, the number of tasks becomes fixed. The implementation of a scheduler is then considerably eased. This is especially the case of software implemented on microcomputers. We could thus find for each installation a special implementation.

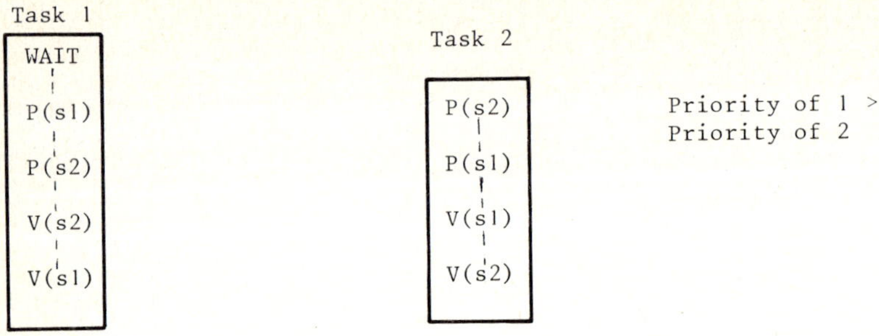

Initially s1 = s2 - 1

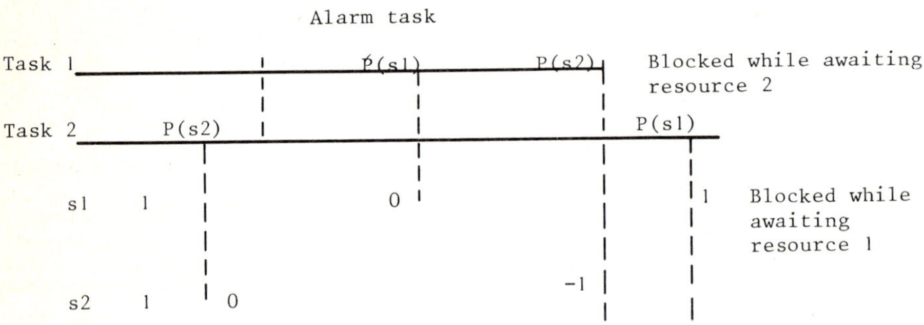

Figure 2.42

2.20 GENERAL CONCLUSIONS

In this chapter we have studies distributed computing and the hardware and software aspects of microprocessors which can be used in distributed computing networks. We have seen the state of the art of the microcomputer technology. In the next chapter we will go back to theoretical considerations by examining once again the decentralised control problem but this time we will consider the stochastic point of view.

2.21 REFERENCES

Specific References

1. Courcoubetis C. and Varaiya P. "A preliminary model for distributed algorithms" in Titli A. and Singh M.G. (editors) "Large Scale Systems Theory and Applications", Pergamon Press, Oxford, 1980.

2. Witsenhausen H. "on information structures, feedback and causality", SIAM J. Control Vol. 9, 2, pp. 149-160, May, 1971.

3. Merlin P. and Segal A. "A failsafe distributed routing protocol", IEEE Trans. on Communication, Vol. COM-27, 9, pp. 1280-1288, September, 1979.

General References

1. Singh M.G., Elloy J.P., Mezencev R. and Munro N. "Applied Industrial Control", Pergamon Press, Oxford, 1980.

2. Crocus, "Systeme d'exploitation des ordinateurs". Dunod, Paris, 1975.

3. Dijkstra, E.W. "The structure of the multiprogramming system", Communications of the ACM, Vol. 11, No. 5, pp. 341-346, 1968.

4. Brinch Hansen, P. "Operating Systems Principles", Eaglewood Cliffs: Prentice Hall, 1973.

5. Chaudouard, C. Systeme temps reel pour le commande de procemus. These doctorat 3e cycle, ENSM-Nantes, 1975.

6. Brie Cl. and Guivarch M. "Critere de choix des microprocesseurs existants, Le Nonvel automatisur, 1978.

2.22 ACKNOWLEDGEMENTS

In writing this chapter we have drawn heavily on the work of Varaiya (specific references *[1]*) and of Singh et al (general references *[1]*).

CHAPTER 3

NON-CLASSICAL INFORMATION PATTERNS AND DECENTRALISED CONTROL

3.1 INTRODUCTION

In this chapter we come back to the fundamental problem of decentralised control. This problem is particularly relevant within the context of stochastic control. The basic difficulty in decentralised control arises from the fact that such problems have a "non-classical" information pattern. In this chapter we will examine what non-classical information patterns are and we will illustrate the difficulties they cause by examining various problems of decentralised decision making in the theory of teams.

3.2 CLASSICAL AND NON-CLASSICAL INFORMATION PATTERNS

Let us consider a linear stochastic system described by

$$\dot{x}(t) = Ax(t) + \sum_{i=1}^{N} B_i u_i(t) + \xi(t) \tag{3.2.1}$$

$$y(t) = Cx(t) + \sum_{i=1}^{N} D_i u_i(t) + \theta(t) \tag{3.2.2}$$

Here (3.2.1) is the state equation for a linear dynamical system being controlled from N control stations, (3.2.2) is the output equation. $\theta(t)$, $\xi(t)$ are assumed to be independent Gaussian white noise processes. $y(t)$ in equation (3.2.2) corresponds to all the measured outputs of the system. The partition of these measurements available to a given controller is defined by the <u>information pattern</u> of the problem. For the decentralised control problem, the information pattern is defined by the set of matrices

$$\mathcal{H} = \{H_1, H_2, \ldots, H_N\} \tag{3.2.3}$$

where

$$z_i(t) = H_i(t)\, y(t) \tag{3.2.4}$$

$$i=1,2,\ldots N$$

For the case of the classical information pattern, let us define
B = Block diagonal matrix $[B_i]$ $i=1,\ldots N$ D = Block diag $[D_i]$ $i=1,\ldots N$ and H = Block diag $[H_i]$ $i=1,\ldots N$.

Then, the classical information pattern arises when

$$H = I \tag{3.2.5}$$

where I is the identity matrix. This implies that all the controllers have identical information.

Now, let us consider the quadratic performance index

$$J = \lim_{T} \{ \frac{1}{T} \int_0^T \left[x^T(t) Q x(t) \sum_{i=1}^{N} u_i^T R_i u_i(t) \right] dt \} \tag{3.2.6}$$

where Q is a positive semi-definite matrix and R = Block Diag $[R_i]$ is a positive definite matrix.

A non-classical information pattern arises if the controls are restricted to be of the form

$$u_i(t) = \gamma_i(z_i^t)$$

where

$$z_i^t = \{z_i(\tau): \quad 0 \leqslant \tau \leqslant \} \tag{3.2.7}$$

i.e. the controls at the ith control station are chosen as some function γ_i of the past and present information available to it. The difficulties associated with non-classical information patterns were first pointed out by Witsenhausen [1]. We know that in the case of the classical information pattern (H=I), the well known separation theorem applies [2] i.e. the optimal control law is linear and it is therefore at least as good as any non-linear control law. Also, a control law based on a finite dimensional Kalman filter state estimates is at least as good as one based on an infinite dimensional record of all past measurements. Witsenhausen [1] considered a seemingly simple example to show that neither of these properties hold in the case of a non-classical information pattern. In view of the importance of these concepts, we will next examine the example of Witsenhausen [1].

3.2.1 The Example of Witsenhausen [1]

Consider the following 2 stage control problem where x_o and v are independent random variables with finite second elements and where all the variables are real scalers:

State equations

$$\left. \begin{aligned} x_1 &= x_o + u_1 \\ x_2 &= x_1 - u_2 \end{aligned} \right] \tag{3.2.8}$$

g) $\quad \text{Max}\,(0, 1 - E\{f^2(x)\}) \leqslant I(D_f) \leqslant 1$ and for $E\{(x - f(x))^2\} \leqslant \sigma^2$ one has $J_2^*(f) \leqslant k^2\sigma^2 + \min\,(1, 4\sigma^2)$

For proof, cf. Witsenhausen *[1]*.

We note that all functions f which give the same distribution to f(x) also give the same optimal cost $1 - I(D_f)$ for the second stage. According to theorem 378 of Hardy et al *[3]*, among all these equimeasurable functions, the monotone non-decreasing rearrangement maximises $E\{xf(x)\}$ and hence minimises $E\{(f(x) - x)^2\}$. This establishes the existence of a minimising sequence $(f_n, g^*_{f_n})$ with monotone f_n.

Now, it is well known *[2]* that the search for an optimum can be limited to the class of affine (linear plus constant) functions if the information pattern is a classical one (i.e. all actions to be taken at a given time are based on the same data and any data available at time t will be available at any later time t' > t).

Next, we will see that here for our problem with a <u>non-classical information pattern</u>, <u>the best affine controller is not optimal over Γ.</u>

3.2.2 <u>Optimisation over the class of affine functions</u>

For the problem $\Pi(k^2, \sigma^2)$ let

$$J_a^* = \inf\,\{J(f,g)/f,g \text{ affine}\}$$

we note by transforming (f_1, g_1) in Lemma 3.2.1(b) into (f,g) we can map the class of affine pairs into itself. Thus we only need to consider $E\{f(x)\} = 0$ or

$$f(x) = \lambda x.$$

for such f,

$$g_f^*(y) = \mu y$$

where

$$\mu = \frac{\sigma^2\lambda^2}{1+\sigma^2\lambda^2}$$

and $\quad J_2^*(f) = J_{2a}^*(\lambda) = k^2\sigma^2\,(1 - \lambda)^2 + \dfrac{\lambda^2\sigma^2}{1+\sigma^2\lambda^2}$

Since this expression is non-negative, analytic and radially unbounded optimal values of λ exist and are the stationary points of J^*_{2a}. Thus we have the lemma

<u>Lemma 3.2.3</u>

Optimal affine solutions exist and are of the form f(x) = λx, g(y) = μy where

$$\mu = \frac{\sigma^2\lambda^2}{1+\sigma^2\lambda^2}$$

and $t = \sigma\lambda$ is a real root of the equation $(t-\sigma)(1+t^2)^2 + \frac{1}{k^2} t = 0$.

However, the main contribution of Witsenhausen *[1]* was to show that the separation theorem *[2]* does not hold in the case of systems with a non-classical information pattern. This result is given by the next theorem.

Theorem 3.2.1 *[1]*

There exist values of the parameter k and σ for problem $\Pi(k^2,\sigma^2)$ such that J^*, the optimal cost, is less than J^*_a, the optimal cost achievable in the class of affine designs.

3.2.3 Remarks

The above results show that once we have a non-classical information pattern (which is the case of decentralised control), then the optimal solution is non-linear and may depend on an infinite dimensional record of past measurements. Such solutions are extremely difficult to find. In fact, so far no solution has been found to Witsenhausen's example that we have described above. However, a great deal of insight has been obtained. In the next chapter we will examine fixed structure controllers for decentralised control. Here we continue by considering dynamic team problems as an example of decentralised decision making problems.

3.3 DYNAMIC TEAMS

In the decentralised decision making situation there may be many decision makers each controlling different decision variables at different times. A team is an organisation in which there is a single goal or pay off function common to all the members. Ho *[4]* points out that the main ingredients of a theory of team decisions are:

(i) The presence of different but correlated information for each decision maker about some underlying uncertainty and

(ii) The need for coordinated actions on the part of all decision makers in order to realise the payoff.

If one or more of these ingredients are missing, the problem simplifies, decouples or becomes trivial.

It should be emphasised that the decision makers are allowed to communicate beforehand and reach any agreement, i.e. they can agree on taking any coordinated actions as a function of the information they separately have. This is in contrast to the traditional non-cooperative game where no pre-game arrangement can be enforced.

Next, let us consider a formal model for teams decisions *[4]*.

3.3.1 A model for team decisions *[4]*

In the model suggested by Ho *[4]*, there are five basic elements:

(i) A vector of random variables $\xi=[\xi_1,\ldots,\xi_m]$ with given distribution $p(\xi)$. This random vector represents <u>all</u> the uncertainties associated with the problem e.g. measurement noise, random disturbances, uncertain initial conditions, etc. ξ is often referred to as "states of nature" or "nature's decision".

(ii) A set of observations $z = [z_1,\ldots,z_N]$ which are given functions of ξ i.e.

$$z_i = \eta_i\,(\xi_1,\ldots,\xi_m) \qquad i=1,\ldots n \tag{3.3.1}$$

z_i is in general a vector and it is known as the "<u>information</u>" or observation available to the ith decision maker (DM); the set $\{\eta_i/i=1,\ldots n\}$ is called the <u>information structure</u> of the problem.

(iii) A set of decision variables $(u_1,\ldots u_n) \equiv u$, one per decision maker. There is no loss of generality in assuming that u_i is a scaler, in that if u_i is a vector, it can always be decomposed into more decision makers who are assumed to have access to the same observation z_i.

The variables ξ, z, u are all assumed to take values in appropriate spaces Ξ, Z, U.

(iv) The <u>strategy</u> (decision rule or control law) of the ith DM is a map

$$\gamma_i : Z_i \to U_i$$

This represents a contingency plan of what decision to take based on observations. Thus we could write

$$u_i = \gamma_i(z_i) \tag{3.3.2}$$

The cost function of the problem is a map $L : \Xi \times U \to R$

i.e.

$$\text{cost} = L(u_1,\ldots u_n,\ \xi_1,\ldots,\xi_m) \tag{3.3.3}$$

Note that for a given set of strategy tuples γ_i i=1,...n, L is a well defined function of the ξ's i.e.

$$L(u,\xi) = L(\ldots,u_i = \gamma_i(\eta_i(\xi)),\ldots,\ldots,\xi_j,\ldots)$$

Thus the expectation of L w.r.t. $p(\xi)$ is well defined.

We can therefore state the team decision problem as

$$\left\{\begin{array}{l} \text{Find } \gamma_i \in \Gamma_i\ \forall_i \text{ in order to} \\ \text{Minimise } J = E_\xi\,[L(u = \gamma(\eta(\xi)),\xi)] \\ \qquad\text{or} \\ \min\limits_{\gamma \in \Gamma} J(\gamma) \end{array}\right\} \tag{3.3.4}$$

(3.3.4) is known as the strategic form of the team problem. We note that conceptually (3.3.4) is a deterministic optimisation problem although it is a difficult one in general.

Let us next look at the team problem from the point of view of the ith decision maker. Let $\bar{\gamma}_i$ denote the strategy of all the other decision makers. Let us assume further that $\bar{\gamma}_i$ is fixed and that DM_i knows this.

Then the problem facing DM_i is

$$\min_{\gamma_i \in \Gamma_i} J(\gamma_i, \bar{\gamma}_i) \tag{3.3.5}$$

$$= \min_{\gamma_i \in \Gamma_i} E_\xi [L(u_i = \gamma_i(\eta_i(\xi)), \bar{\gamma}_i, \xi)]$$

Since η_i's are fixed, z_i is a well defined random variable, we can replace E_ξ by $E^i_{z_i} E_{\xi/z_i}$ where E^i_{ξ/z_i} denotes the expectation conditional on z_i.

Then

$$\min_{\gamma_i \in \Gamma_i} J(\gamma_i, \bar{\gamma}_i)$$

$$= \min_{\gamma_i \in \Gamma_i} E_{z_i} E_{\xi/z_i} (L(\gamma_i, \bar{\gamma}_i, \xi))$$

$$= E_{z_i} \min_{u_i \in u_i} F_{\xi/z_j} (L(u_i, \bar{\gamma}_i, \xi)).$$

because determining the optimal u_i for each z_j is the same as choosing γ_i. Thus we arrive at the so-called extensive form of (3.4.5):

$$\min_{u_i \in U_i} E_{\xi/z_i} [L(u_i \bar{\gamma}_i, \xi)] \quad \forall i \tag{3.3.6}$$

(3.3.6) is called the person-by-person optimality requirement and we note that it represents a parametric (as opposed to functional) optimisation problem for each z_i and fixed $\bar{\gamma}_i$. We emphasise that (3.3.6) is necessary for (3.3.4) to occur but not vice-versa. We note that the difficulties of the strategic form have not been completely eliminated since (3.3.6) is still parametrised by $\bar{\gamma}_i$. Figure 3.1 shows the basic iterative loop which is necessary in order to solve (3.3.6).

This is to be contrasted with a one-person decision problem in the extensive form where no dependence on any part of γ is present in (3.3.6). For this reason, (3.4.6) is often called the semi-strategic form of the problem.

With this model in mind, we come to the first result in team decision problems [5].

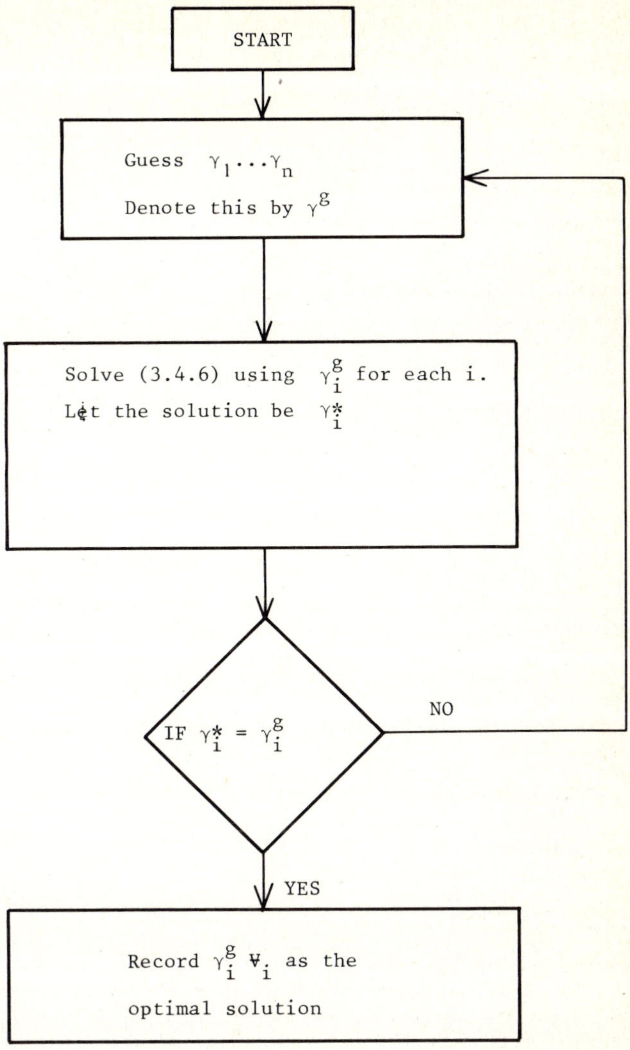
START
Guess $\gamma_1 \ldots \gamma_n$
Denote this by γ^g
Solve (3.4.6) using γ_i^g for each i.
Let the solution be γ_i^*
IF $\gamma_i^* = \gamma_i^g$
NO
YES
Record $\gamma_i^g \ \forall_i$ as the
optimal solution

Figure 3.1

Proposition 3.3.1

In linear-quadratic-Gaussian team with $Q > 0$,

$$L = \tfrac{1}{2}\, u^T Q u + u^T S \xi, \quad Q > 0, \quad \xi \sim N(0, \Sigma)$$

and

$$y = H\, \xi$$

the unique optimal solution is linear in the information and is obtained by solving the person-by-person optimality condition (3.3.6).

In order to illustrate proposition (3.3.1), let us consider the two person team problem where the decision variables u_1 and u_2 must be chosen so as to minimise cost function [4].

$$L = \tfrac{1}{2}\,(x + au_1 + u_2)^2 + hu_1^2 + gu_2^2 \tag{3.3.7}$$

where $a, g \geqslant 0$ and $h > 0$

Two observations are possible for the system i.e.

$$\left.\begin{aligned} y_1 &= x + bv_1 \quad ; \quad b > 0 \\ \text{and} \quad y_2 &= x + cu_1 + dv_2 \quad ; \quad d > 0,\ c > 0 \end{aligned}\right\} \tag{3.3.8}$$

The three random variables x, v_1 and v_2 are assumed to have independent Gaussian distributions ($x \sim N(0,\sigma^2)$, $v_1 \sim N(0,1)$ and $v_2 \sim N(0,1)$). Ho [4] chooses appropriate values of the unspecified constants a, b, c, d, k and g in order to illustrate different aspects of the team problem.

For the first problem (P1) which we use to illustrate proposition (3.3.1), let us choose:

$$\left.\begin{aligned} &a = b = d = 1 \qquad c = 0 \\ &h = g = \tfrac{1}{2} \\ &z_1 = y; \qquad z_2 = y_2 \end{aligned}\right\} \tag{P1}$$

One interpretation of the problem is that there is a random initial state x; u_1 acts based on noisy observations $z_1 = x + v_1$ in order to bring the state to $x + u_1$; u_2 now acts based on $z_2 = x + v_2$ resulting in the final state $(x + u_1 + u_2)$; The aim is to minimise the value achieved of the final state and the energy required for achieving it.

Following the flow diagram in Fig. 3.1, we guess

$$u_1 = k_1 z_1, \quad u_2 = k_2 z_2 \tag{3.3.9}$$

Specialising (3.3.6) for these controls yields

$$\left.\begin{aligned} u_1 &= \tfrac{1}{2}\, E_{/z_1}\,(k_2 z_2 + x) \\ u_2 &= -\tfrac{1}{2}\, E_{/z_1}\,(k_1 z_1 + x) \end{aligned}\right\} \tag{3.3.10}$$

Using the standard relationship for conditional expectation of Gaussian random variables and requiring that (3.3.9) and (3.3.10) hold yields the consistency condition

$$\begin{pmatrix} 1 & \frac{\sigma^2}{2(\sigma^2+1)} \\ \frac{\sigma^2}{2(\sigma_1^2+1)} & 1 \end{pmatrix} \begin{pmatrix} k_1 \\ k_2 \end{pmatrix} = - \begin{pmatrix} 1 \\ 1 \end{pmatrix} \frac{\sigma}{2(\sigma^2+1)} \tag{3.3.11}$$

This is a set of linear equations with the intuitively reasonable solution $k_1^* = k_2^* = -\frac{\sigma^2}{3\sigma^2+2}$. Is this solution optimal, i.e. does it satisfy (3.3.4) in addition to the person-by-person optimality condition (3.3.6)? In order to answer this question, let us denote the person-by-person optimal solution that we have just obtained by $u_i^* = \gamma_i^*(z_i) = k_i^* z_i$ $(i=1,2)$; and any other strategy by $u_i = \gamma_i(z_i)$. Then by the strict convexity of L in u_1 and u_2 we have

$$L(u_1,u_2,\xi) > L(u_1^*,u_2^*,\xi) + \left.\frac{\partial L}{\partial u_1}\right|_{u_1^*,u_2^*} (u_1 - u_1^*) + \left.\frac{\partial L}{\partial u_2}\right|_{u_1^*,u_2^*} (u_2 - u_2^*) \tag{3.3.12}$$

Now taking the expected value for both sides of (3.3.12) and substituting for u_i, u_i^* by γ_i, γ_i^* (we recall that u_1, u_2 are well defined random variables only if γ_1, γ_2 are specified) we obtain:

$$\begin{aligned} J(\gamma_1,\gamma_2) &\equiv E[L(u_1 = \gamma_1(z_1),\, u_2 = \gamma_2(z_2),\xi)] \\ &> J(\gamma_1^*,\gamma_2^*) + E\{(\gamma_1 - \gamma_1^*) \left.\frac{\partial L}{\partial u_1}\right|_{u_1^*,u_2^*}\} \\ &\quad + E\{(\gamma_2 - \gamma_2^*) \left.\frac{\partial L}{\partial u_2}\right|_{u_1^*,u_2^*}\} \\ &= J(\gamma_1^*,\gamma_2^*) + \left[E_{z_1} (\gamma_1 - \gamma_1^*) E_{/z_1} \{ \left.\frac{\partial L}{\partial u_1}\right|_{u_1^*,u_2^*} \}\right] \\ &\quad \left[E_{z_2} (\gamma_2 - \gamma_2^*) E_{/z_2} \{ \left.\frac{\partial L}{\partial u_2}\right|_{u_1^*,u_2^*} \}\right] \end{aligned} \tag{3.3.13}$$

However if we use the condition (3.3.6) which states that

$$\frac{\partial}{\partial u_i} E_{/z_i}(L) = E_{/z_i} \left.\frac{\partial L}{\partial u_i}\right|_{u_1^*,u_2^*} = 0$$

the interchanging of expectation and partial differentiation being permitted since u_i is z_i-measurable. Then (3.3.13) reduces to

$$J(\gamma_1,\gamma_2) = J(\gamma_1^*,\gamma_2^*)$$

which illustrates Radner's theorem *[5]* (proposition 3.3.1).

3.4 PARTIALLY NESTED INFORMATION STRUCTURE

Next, let us consider the key idea of partially nested information structures which was developed by Ho and Chu *[6]* . We saw in section 3.2 that where we have a non-classical information pattern (as is the case with decentralised control which is a special case of the team decision problems being considered here), the basic optimal control law is non-linear. Ho and Chu *[6]* have shown that we can get linear controls if the information structure is partially nested. In order to see what a partially nested information structure is, let us consider once again Ho's thematic example of section (3.3.1) where the parameters and information structure are specified as follows:

Problem P2

$$\left.\begin{aligned} z_1 &= y_1 \equiv x + v_1 \\ z_2 &= \begin{pmatrix} y_1 \\ y_2 \end{pmatrix} \equiv \begin{pmatrix} x + v_1 \\ x + u_1 + v_2 \end{pmatrix} \end{aligned}\right\}$$

$$a = b = d = 1,\ c = 1,\ h = g = \tfrac{1}{2}$$

The difference between P2 and P1 in section (3.3.1) lies in the information available to u_2 i.e. u_2 knows what u_1 knows and c = 1. We notice that z_2 is not in the form of the model specified in section (3.3.1). However, if we consider

$$z_2 \equiv \begin{pmatrix} y_1 \\ y_2 \end{pmatrix} \equiv \begin{pmatrix} y_1 \\ y_2 - \gamma_1(z_1) \end{pmatrix} = \begin{pmatrix} y_1 \\ x + v_2 \end{pmatrix} \tag{3.4.1}$$

then knowing z_2 we can generate z_2' and similarly for z_2 from z_2' and γ_1. Thus any $u_2 = \gamma_2(z_2)$ can be realised by $u_2 = \gamma_2(z_2')$ and vice-versa. Thus the information structure z_2 is equivalent to z_2 and the problem P2 comes under the class of LQG problems covered in proposition 3.3.1 and P2 can be solved by exactly the same method with the optimal strategies linear in the information

$$u_1 = k_1 z_1 \ ; \qquad u_1 = k_2^T z_2 \tag{3.4.2}$$

where

$$k_1 = \frac{-\sigma^2}{3\sigma^2+1} \quad , \quad k_2 = \begin{pmatrix} \frac{-\sigma^2}{3(1+2\sigma^2)} \\ \frac{-\sigma^2}{2(1+2\sigma^2)} \end{pmatrix} \tag{3.4.3}$$

The interpretation of P2 is the same as that for P1 except that y_2 (with $c = 1$) is to be considered a noisy measurement of the state of the system after u_1 has acted, i.e. we are considering a dynamic team evolving in time with initial state x. In this sense the information structure z_2 has perfect recall since latter decision makers always know what earlier ones know. Thus if u_1 and u_2 are regarded as the decisions made by the same person at different times then the above is an illustration of the result that:

Proposition 3.4.1

In one person LQG stochastic control problems with perfect memory, the optimal control law is linear.

More generally, let us consider the observations

$$y_i = H_i \xi + D_i u \tag{3.4.4}$$

where D_i needs to satisfy the causality conditions:

$$\text{If } u_i \text{ acts before } u_j \text{ then } [D_i]_{j^{th} \text{ component}} = 0 \tag{3.4.5}$$

Let us consider the information structure

$$z_i = \begin{pmatrix} z_k ; & k \in E_i \\ y_i & \end{pmatrix} \tag{3.4.6}$$

where $\quad E_i = \{k \mid D_{ik} \neq 0\}$

The information structure (3.4.6) implies that if u_k affects the observation y_i (i.e. $D_{ik} \neq 0$) then u_i knows what u_k knows. This structure is a generalisation of perfect recall and defines a partial order of inclusion on what the various decision makers know. This is called a partially nested information structure (PN).

It is clear from the development of P2 that partial-nesting is sufficient to enable us to generate an equivalent information structure z' of the form specified in (3.4.1). Hence we have

Proposition 3.4.2

In a LQG team problem with PN information structure, the optimal strategies are linear and can be solved by solving a system of linear equations (generalisation of equation (3.3.11)).

To illustrate a PN information structure let us consider the dynamic team example of Ho and Chu [6]

$$z_1 = H_1\xi$$

$$z_2 = H_2\xi$$

$$z_3 = \begin{bmatrix} H_1 \\ H_2 \\ H_3 \end{bmatrix}\xi + \begin{bmatrix} 0 \\ 0 \\ D'_{31} \end{bmatrix} + \begin{bmatrix} 0 \\ 0 \\ D'_{32} \end{bmatrix}$$

$$z_4 = \begin{bmatrix} H_1 \\ H'_4 \end{bmatrix}\xi + \begin{bmatrix} 0 \\ D'_{11} \end{bmatrix} u_1$$

$$z_5 = \begin{bmatrix} H_1 \\ H_2 \\ H'_3 \\ H'_5 \end{bmatrix}\xi + \begin{bmatrix} 0 \\ 0 \\ D'_{31} \\ D'_{51} \end{bmatrix} u_1 + \begin{bmatrix} 0 \\ 0 \\ D'_{32} \\ 0 \end{bmatrix} u_2$$

$$+ \begin{bmatrix} 0 \\ 0 \\ 0 \\ D'_{53} \end{bmatrix} u_3$$

$$z_6 = H_6\ \xi$$

$$z_7 = \begin{bmatrix} H_6 \\ H'_7 \end{bmatrix}\xi + \begin{bmatrix} 0 \\ D'_{76} \end{bmatrix} u_6$$

$$z_8 = H_8\xi$$

Figure 3.2 shows the information structure diagram for this example.

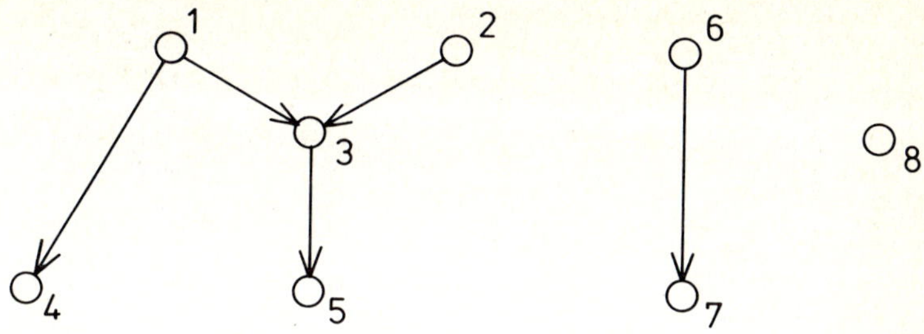

Figure 3.2

It is clear from Fig. 3.2 and from the above information structure that what precedents have been known will always be known by their followers. For example, member three's precedents are one and two; however, z_1, z_2 are respectively the first and second components of z_3.

Ho and Chu's result *[6]* (proposition 3.4.2) was obtained by converting the dynamic team problem to a static one.

3.5 WITSENHAUSEN'S PROBLEM REVISITED

As another special case, we have problem P3 where

<u>Problem P3</u>

$$a = c = d = h = 1; \qquad b = g = 0$$

$$z_2 = y_2' = x + u_1 + v_2$$

We recognise this to be a version of Witsenhausen's problem *[1]* which we so extensively studied in section 3.2. The crucial difference between this and the previous problem is that z_2 does not include y_1. The interpretation of P3 is that Decision Maker 1 knows x perfectly (b = 0) but it costs DM_1 energy to act ($h \neq 0$). DM_2 has no cost (g = 0) on u_2 but his information about x is corrupted by noise and is dependent on u_1.

Since here we do not have perfect memory, we can not reduce this problem to P2 or P1. The <u>dynamics</u> or <u>order of action</u> are of importance here. The type of information structure in P3 is a special case of the general form

$$z_i = \eta_i(\xi, u) \tag{3.5.1}$$

subject to obvious causality requirements.

Equation (3.5.1) is usually called a <u>dynamic information structure</u> and team problems with such an information structure are called dynamic team problems. The crucially important aspect of the dynamic team problem is that the information used by latter decision makers depends on the decisions of earlier ones. Thus, unless we specify the decision rule or strategy of the earlier decision makers, for example γ_1 in (P3), the information variable z_2 is not a well defined random variable. In fact $z_2 = x + \gamma_1(x) + v_2$ need not be a Gaussian random variable even if x and v_2 are unless γ_1 is linear in x. Thus the calculation of $E(x/z_2)$ could be complicated.

Next, if we look at the objective function $J = E_\xi[(x + \gamma_1(x) + \gamma_2(x + \gamma_1(x) + v_2))^2 + h\gamma_1^2(x)] \equiv J(\gamma_1, \gamma_2, (\gamma_1))$ we note that γ_1 enters J not only directly (through ku_1^2) but also indirectly through γ_1 (via z_2). Thus, unless γ_2 is linear (or some appropriate function), J need not be convex in γ_1 even though L is convex in u_1. The important point to emphasise as Ho *[4]* notes is that we have an optimisation problem in the γ space so we require appropriate properties (convexity, compactness, etc.) in this space. Very little work has been done in tackling such optimisation problems.

One way of tackling this problem, as we saw earlier, is to use an iterative guessing of γ as in Fig. 3.1.

Let us use as our guess $u_1 = k_1 z_1$ and $u_2 = k_2 z_2$. Then consistency and equation (3.3.6) requires that k_1 and k_2 satisfy a set of non-linear algebraic equations, i.e.

$$k_1 = -\frac{(1+k_2)^2}{2h+(1+k_2)^2}$$
$$k_2 = -\frac{(1+k_1)^2}{(1+k_1)^2\sigma^2+1} \qquad (3.5.2)$$

Although roots of these constitute a person-by-person optimal solution for P3, they are not team-optimal as we saw when we considered Witsenhausen's problem in section 3.2.

Ho *[4]* points out that a lower bound to the optimal cost can always be obtained for non-partially nested LQG dynamic team problems by "nestification" of the information structure (for example let $z_2 = [z_1, y_2]$ in P3). Since adding information to any decision maker does not make the optimal payoff worse in the sense that any strategy which could be realised before can still be realised with the enlarged information structure, proposition 3.4.2 applied to the nestified version of P3 will yield a lower bound to P3. Thus if we denote by J_L^* the linear person-by-person optimal solution of P3 in (3.5.2) and J_{PN}^* as the solution of the nestified version of (P3), we have the general result:

<u>Proposition 3.5.1</u>

In non-PN dynamic LQG team problems the optimal J* is bounded above and below by

$$J_{PN}^* \leq J^* \leq J_L^* \qquad (3.5.3)$$

In the above discussion about P3 we also saw the signalling effect in action. u_1 could modify the information z_2 received by u_2 and this could be used to obtain a superior non-linear solution to P3.

To see this signalling effect more clearly, let us consider a relatively degenerate case, i.e. P3 with d = 0 and $z_2 = y_2 = u_1$.

Here we can choose $u_1 = \varepsilon x$ and $u_2 = -(1+\varepsilon)u_1$ and J* can be made arbitrarily close to zero by taking a small enough ε. Here u_1 is used purely to inform u_2. Because d = 0 and $y_2 = u_1$ are equivalent to infinite precision in the transmission of information, we can think of the number u_1 having two parts with the most significant digits being used for control whilst the least significant digits carry coded information. These least significant digits should not affect the payoff in any significant way. This phenomenon is called "transparency" of information *[7]*. Basically, Bismut *[7]* has shown that by accurate infintesimal signals which perturb the value of the criterion as little as possible and conform to a system of conventions known to the other decision makers in the team, it is possible for any decision maker to transfer very good information on his own observations to the other decision makers. In this context, Ho *[4]* points out that in traditional control problems the concept of "dual control" summarises the idea that the purposes of control are twofold, one being to reduce error and the other to reduce our own knowledge about any uncertainty which exists. In decentralised control, the control actions can also have a third purpose, i.e. to reduce the uncertainties of the other decision makers through signalling.

3.6 DECOMPOSITION OF DYNAMIC TEAM DECISION PROBLEMS *[8]*

As we have seen, dynamic team decision problems (and consequently decentralised control problems) have immense informational complexity. For the practical resolution of these problems, there is some incentive in using decomposition. Yoshikaura *[8]* has developed the concepts of independent and sequential partitions in order to do this decomposition.

Let us begin our analysis by recapitulating the basic dynamic team design problem. The N member team of DM_i has a common payoff function which is the expected value of:

$$w = w(\xi, u) \tag{3.6.1}$$

where $\xi \in R^n$ is a random vector with probability distribution $F(\xi)$ which represents all the uncertainties in the system, $w(.,.)$ is a (Borel) measurable function bounded from below and $u_i \in R^{m_i}$ is the decision vector of DM_i. The information $z_i \in R^{r_i}$ is given by

$$z_i = \eta_i(\xi, u) \tag{3.6.2}$$

where η_i is measurable. $\eta = (\eta_1, \eta_2, \ldots \eta_N)$ is called the information structure of the team.

The admissible decision rule is of the form

$$u_i = \gamma_i(z_i) \tag{3.6.3}$$

where γ_i is an arbitrary measurable function. All the DMs are assumed to know the functions w, η and the probability distribution $F(\xi)$.

The team decision problem involves finding a decision rule $\gamma \underline{\Delta} \{\gamma_1, \gamma_2, \dots \gamma_N\}$ which minimises the expected valued of w, $E_\gamma[\omega(\xi,u)]$ where $E_\gamma[.]$ is the expectation operation w.r.t. ξ for any given strategy ξ.

A team is called <u>static</u> if η_i, i=1,2,...N in (3.6.2) is independent of the decision u, i.e.

$$z_i = \eta_i(\xi) \tag{3.6.4}$$

Otherwise, it is said to be <u>dynamic</u>. A team is <u>finite</u> if ξ, u_i, z_i take one of finite possible values. For a finite team, there is always an optimal decision rule γ^* whilst this may not be true for a team which is not finite.

The <u>precedence</u> relation [6] among DMs is defined as:

If there exists ξ, $u^a \underline{\Delta}$ col $(u_1, u_2, \dots, u_{i-1}$, $u_i^a, u_{i+1}, \dots, u_N]$ and $u^k \underline{\Delta}$ col $[u_1, u_2, \dots, u_{i-1}$, $u_i^b, n_{i+1}, \dots u_N]$ such that

$$\eta_j(\xi, u^a) \neq \eta_j(\xi, u^b) \tag{3.6.5}$$

then we say that u_i affects z_j and write i Rj.

<u>Definition 3.6.1</u>

DM_i is a precedent of DM_j ($DM_i \rightarrow DM_j$) if (a) i Rj or (b) there exist r,s,...t such that i Rr, rRs,...tRj.

In order for the precedence relation to satisfy causality and be deterministic, we assume that if $DM_i \rightarrow DM_j$ then DM_j does not precede DM_i.

<u>Definition 3.6.2</u>

Information z_i of DM_i is said to be nested in information z_j of DM_j (denoted $DM_i \dashrightarrow DM_j$) if there exists a measurable function f_{ji}^γ and that for any ξ,

$$z_i = f_{ji}^\gamma(z_j) \tag{3.6.6}$$

To represent the precedence and nestedness relationships graphically, let us use node i to represent DM_i, an arrow from node i to j to represent the relation $DM_i \rightarrow DM_j$ and a broken arrow to represent the relation $DM_i \dashrightarrow DM_j$. An arrow (or broken arrow) is omitted in case node j can be reached from node i by tracing a set of arrows (or broken arrows).

With this notation and definitions, we are now in a position to examine the crucial concept of independent partitions [8] which will enable us to decompose certain team decision problems.

3.6.1 Independent partitions and sequential partitions

Let $G \triangleq \{DM_i,\ i=1,2,\ldots N\}$ and $G_i,\ i=1,2,\ldots K$ be subsets of G.

Definition 3.6.3

$(G_1,G_2,\ldots G_K)$ is a partition of G if

$$\bigcup_{i=1}^{K} G_i = G, \quad G_i \cap G_j = \emptyset \ i,j=1,2,\ldots K,\ i \neq j \tag{3.6.7}$$

A partition of G merely divides G into groups of DMs.

Definition 3.6.4

Let $G_i,\ G_j \subset G$, $G_i \cap G_j = \emptyset$. G_j is said to be a <u>non precedent group</u> of G_i (denoted $G_j \nrightarrow G_i$) if DM_j is not a precedent of DM_i' for any $DM_i' \in G_i$ and $DM_j' \in G_j$.

Figure 3.3 gives an example of $G_2 \nrightarrow G_1$. Clearly if $G_j \nrightarrow G_i$, $u_{G_j} \triangleq col[u_j;\ DM_j' \in G_j]$ does not affect $z_{G_i} \triangleq col\ [z_i^j;\ DM_i' \in G_i]$

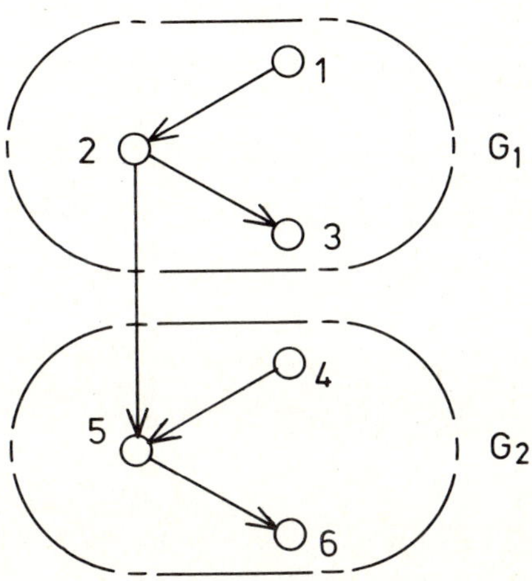

Figure 3.3

Definition 3.6.5

A partition $(G_1,G_2,...G_k)$ is an independent partition (denoted i- partition) of G if (1) $G_i \rightarrow G_j$, $i \neq j$; $i,j=1,2,...k$ (3.6.8)

and (2) $w = \sum_{i=1}^{K} w_i(\xi,u_{G_i})$

This definition implies that an i-partition is a partition for which there is no precedence relation between any pair of groups and the overall payoff is a sum of the payoffs for each group.

We note that the random vector ξ is common to all the groups and there is no assumption on the form of $F(\xi)$.

An example of an i-partition is shown in Fig. 3.4 which shows the precedence diagram of a team $G = \{DM_i,\ i=1,2,3,4\}$ with

$$w = w_1(\xi,u_1,u_2) + w_2(\xi,u_3,u_4)$$

Here $G_1 = \{DM_1,DM_2\}$ and $G_2 = \{DM_3,DM_4\}$

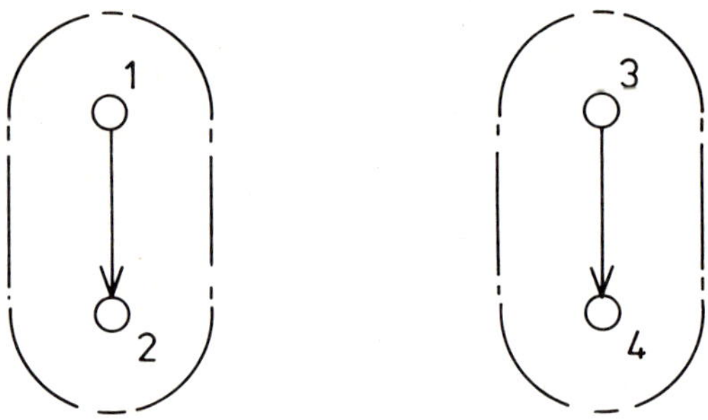

Figure 3.4

We are now in a position to state the basic theorem which allows decomposition.

Let $\gamma_{G_i} = \{\gamma_i^1, DM_i, \varepsilon\ G_i\}$

Theorem 3.6.1

Let a team G have an i-partition $(G_1,G_2,...,G_k)$. If subproblem i {minimise $E_{\gamma_{G_i}}[w_i(\xi\ u_{G_i})]$ w.r.t. γ_{G_i}}, $i=1,2,...K$, has an optimal solution

$\gamma^*_{G_i}$ then $\gamma^* = \{\gamma^*_{G_i}$ i=1,2,...K} is an optimal solution of the original problem.

This theorem is proved in *[8]*.

Now, it is not always possible to decompose a team problem using i-partitions.

Next, we introduce the notion of sequential partitions (s-partitions) which enlarges the possibilities of decomposition.

<u>Definition 3.6.6</u>

Let G_i, $G_j \subset G$, $G_i \cap G_j = \emptyset$. The information of G_i is <u>nested</u> in the information of G_j, $G_i \dashrightarrow G_j$, if $DM_i \dashrightarrow DM_j$, for any $DM_i' \varepsilon G_i$ and $DM_j' \varepsilon G_j$.

As in the case of the nesting relationship, $DM_i \dashrightarrow DM_j$, $G_i \dashrightarrow G_j$ does not necessarily mean $G_j \nrightarrow G_i$.

<u>Definition 3.6.7</u>

A partition $(G_1,G_2,\ldots G_k)$ is an s-partition of G if

1) $G_j \nrightarrow G_i$, $\quad j = i+1, i+2,\ldots K$ (3.6.9)

2) $G_i \dashrightarrow G_{i+1}$ (3.6.10)

for i=1,2,...,k-1

An s-partition represents an ordering of the groups such that the information of any group is nested in that of the group immediately after and which does not contradict the precedence relation of the groups.

In order to use the notion of s-partitions in order to decompose dynamic team decision problems, let $\gamma_i(z_i, z_G, u_G)$ be arbitrary measurable functions and $E[.|.]$ be the conditional expectation operator. Note also that if (G_1,G_2) is an s-partition then there exists a function $f_j^{\gamma_{G_j}}$ $G_1, j \,\varepsilon\, G_2$ such that for and γ_{G1}

$$z_{G1} = f_j^{\gamma_{G1}}\, G_1(z_j) \tag{3.6.11}$$

Then we have

<u>Lemma 3.6.1</u>

Let a team G have an s-partition (G_1,G_2). Assume that

a) Subproblem 2 { minimise $E_{\hat{\gamma}_{G2}}[w|z_{G1},u_{G1}]$

w.r.t. $\hat{\gamma}_{G2}$, where $u_j = \hat{\gamma}_j(z_j,z_{G1},u_{G1})$, $j \,\varepsilon\, G_2$ has an optimal solution $\hat{\gamma}^*_{G2}$ and the minimum $w^*_{[2]} = E_{\hat{\gamma}^*_{G2}}[w|z_{G1},u_{G1}]$ and

b) Subproblem 1 {minimise $E_{\gamma_{G1}}[w^*_{[2]}]$ w.r.t. γ_{G1} has an optimal γ^*_{G1}.

Then

$$\gamma^*_j(z_j) = \hat{\gamma}^*_j(z_j, f_j^{\gamma^*_{G_1}} G_1(z_j)\ \gamma^*_{G1}(z_j))),\ j \in G_2 \tag{3.6.12}$$

The proof of this Lemma is given in [8].

The next theorem is an extension of this Lemma.

Theorem 3.6.2

Let a team G have an s-partition $(G_1, G_2, \ldots G_K)$. Let us assume that

a) Subproblem K {minimise $E_{\hat{\gamma}_{GK}}[w|I(K-1)]$ w.r.t. γ_{GK}, where $u_j = \hat{\gamma}_j(z_j, I(k-1))$, $j \in G_K$} has an optimal solution γ^*_{GK} and the minimum $w^*_K \triangleq E_{\hat{\gamma}^*_{GK}}[w|I(K-1)]$,

b) Subproblem i {minimise $E_{\hat{\gamma}_{G_i}}[w^*_{[i+1]} \mid I(i-1)]$ w.r.t. $\hat{\gamma}_{G_i}$, where $u_j = \hat{\gamma}_j(z_j, I(i-i))$, $j \in G_i$}, i=k-1, k-2,...,2 has an optimal solution $\hat{\gamma}^*_{G_i}$ and the minimum $w^*_1 \triangleq E_{\hat{\gamma}_{G_i}}[w^*_{[i+1]} \mid I(i-1)]$, and

c) Subproblem 1 {minimise $E_{\gamma_{G1}}[w^*_{[2]}]$ w.r.t. γ_{G1} has an optimal solution γ^*_{G1}.

Then $\{\gamma^*_{G_i}\ i=1,2,\ldots K\}$ is an optimal solution of the original problem where $\gamma^*_{G_i}$ i=1,2,...K, is a function of z_i only which is obtained from $\gamma^*_{G_i}$ by using f^{γ}_{ij}, $\gamma^*_{G1}, \gamma^*_{G2}, \ldots \gamma^*_{G(i-1)}$.

This theorem shows that it is possible to apply dynamic programming to a team with an s-partition. If there is no precedence relation among the DMs in G_i, the subproblem is a static team problem; otherwise it is a dynamic one.

Next, we have the corollary:

Corollary 3.6.1

Let a team have an s-partition $(G_1, G_2, \ldots G_K)$. Let us assume that

$$w = \sum_{i=1}^{K} w_i(\ , u_{G1}, u_{G2}, \ldots u_{Gi}) \triangleq \sum_{i=1}^{K} w_i \tag{3.6.13}$$

Then w in subproblem K of theorem 3.6.2 can be replaced by w_K and $w^*_{[i+1]}$ in subproblem i by $[w^*_{[i+1]} + w_i]$, i=k-1, k-2,...1.

This corollary is of importance when considering decomposition in decentralised control.

Yoshikawa *[8]* points out that conceptually an s-partition is an extension of a partially nested information structure where the s-partition $(G_1, G_2, \ldots G_K)$ can be reduced to a sort of a PN information structure by regarding each G_i as a new DM_i $(i=1,2,\ldots K)$, the relation "not $G_i \nleftrightarrow G_j$" as $DM_i \rightarrow DM_j$. However, partial nestedness is neither necessary nor sufficient for the existence of an s-partition.

In the case of a finite team, the subproblems in theorems 3.6.1 and 3.6.2 always have optimal solutions. This may not be true in the case of non-finite teams. However, there one could use the notions of ε-optimality or u-measurability as is done in *[8]*.

Next, let us see under what conditions an L-Q-G static team has an i-partition or an s-partition.

Consider a static team problem where $F(\xi)$ is given by a Gaussian distribution with mean $\hat{\xi}$ and a covariance Σ which is non-negative definite. The payoff is given by

$$w = u^T R_a u + 2u^T R_b \xi + 2u^T c$$

Here R_a is a symmetric positive definite matrix, R_b is a given matrix and c is a given vector. DM_i obtains the information

$$z_i = H_i \xi$$

For this problem, the condition (a) of definitions 3.6.5 to 3.6.7 is always satisfied by any partition. It is easy to see that $(G_1, G_2, \ldots G_K)$ is an i-partition if and only if

$$R_{ai'j'} = 0 \qquad i' \in G_i, j' \in G_j, \ i \neq j$$

where $R_{ai'j'}$ is the $(L',j')^{th}$ block (of dimension $(r_i' x r_j')$) of $R_a = -[R_{ai'j'}]$ $(i',j' = 1,2,\ldots N)$. This means that a team has an i-partition if and only if R_a is block diagonal. As for the s-partition we have the result: $(G_1, G_2, \ldots G_K)$ is an s-partition if and only if

$$\text{Ker}\,(H_i') \supset \text{Ker}\,(H_j') \qquad i' \in G_i, \ j' \in G_{i+1},$$
$$i = 1,2,\ldots K-1$$

where Ker (H) denotes the null space of H.

3.6.2 Application of decomposition to decentralised control *[8]*

We will consider only discrete time stochastic decentralised decision problems. Such problems will be treated in much greater detail in the next chapter. Here, however, we will see the emergence of a basic result on decentralised control with one-step delay sharing information structure using the decomposition in teams that we have discussed in this section. Let us consider a discrete-time stochastic system with two control stations which is described by

$$x(t+1) = f(t,x(t),\ u_1(t),u_2(t),v_s(t)) \qquad i=1,2,\ldots T$$

$$y_j(t) = g_j(t,x(t),v_m(t)) \qquad j=1,2)$$

where as usual $x(t)$ is the state at time t, $u_j(t)$ the control of station j, $y_j(t)$ the measurement of station j. $x(1)$, $v_s(t)$ and $v_m(t)$ t=1,2,...T are mutually independent random vectors.

It is assumed that the two control stations can share information about their observations after one time step. Thus the station j can utilise the data

$$\tilde{I}_j(t) = \{y_j(t),\ \tilde{I}(t-1)\} \qquad j=1,2$$

for the determination of control $u_j(t)$ where

$$\tilde{I}(t) = \{y(\tau),y_2(\tau),u_1(\tau),u_2(\tau),\ \tau=1,2,\ldots t)\}$$

The payoff function is given by

$$w = \sum_{i=1}^{T} w_t(x(t+1),x(t),u(t).$$

The problem is to find the control law which minimises the expected value of w.

By the elimination of $x(2), x(3),\ldots,x(t+1)$, this problem is reduced to a team problem with 2T DMs where ξ consists of $x(1)$, $v_s(t)$ and $v_m(t)$, t=1,2,...T and $u_j(t)$ is decided by DM $[2(t-1)+j]$ based upon the information $I_j(t)$. The precedence diagram of such a system is given in Fig. 3.5. It is clear from the diagram that $(G_1,G_2,\ldots G_T)$, $G_j = \{DM(2t-1),\ DM(2t)\}$ is an s-partition.

It is also easy to show that w satisfies 3.6.13. Hence corollary 3.6.1 is applicable resulting in the subproblems:

<u>Subproblem T</u>: { minimise $E_{\gamma_{GT}}[w_T|I(T-1)]$ w.r.t. $\gamma_{GT}\}$. Let the solution of this problem be γ^*_{GT} and the minimum cost $w^*_{[T]}$.

<u>Subproblem t</u>: minimise $E_{\gamma_{Gt}}[w^*_{[t+1]} + w_t|\tilde{I}(t-1)]$ w.r.t. γ_{Gt} ξ, t=T-1,T-2,...1. Let the solution be γ^*_{Gt} and the minimum cost $w^*_{[t]}$.

$\{\gamma^*_{Gt}$ t=1,2,...T} is the optimal solution to the original problem.

We will examine this result from a different point of view, in much greater detail, in the next chapter.

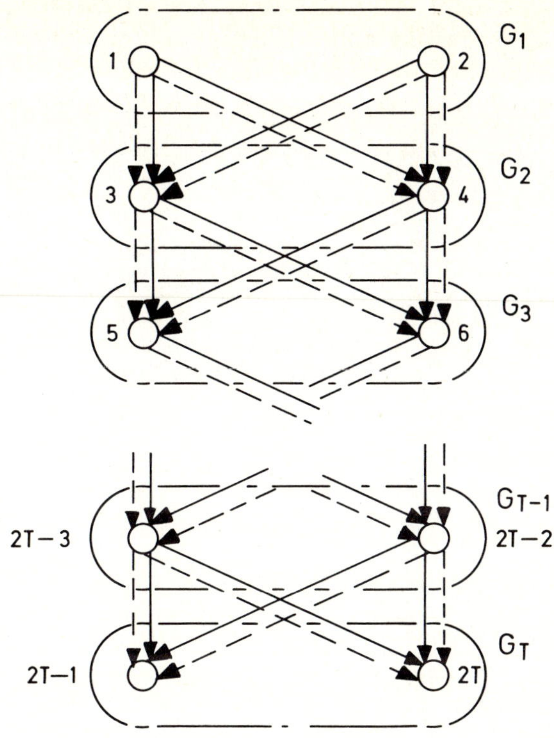

Figure 3.5

Note that each subproblem is static. Therefore, in the LQG case, these subproblems and hence the original problem, can be solved explicitly.

In the case of more than one-step delay sharing information structures there are in general no s-partitions.

Finally, in this chapter we outline some recent work of Ho *[9]* which considers the non-classical information structure problem from a computational point of view and sheds some new light on the difficulties associated with such structures.

3.7 COMPUTATIONAL CONSIDERATIONS IN THE NON-CLASSICAL INFORMATION STRUCTURE PROBLEM *[9]*

We have seen that the main difficulty in dynamic team decision problems arises from the non-convex nature of the associated optimisation problems where in general there are many local optimal points and good sufficient conditions for optimality do not exist. This can lead to significant computational difficulties. In the previous section we outlined the use of decomposition to overcome some of the computational difficulties. Here, let us examine the original Witsenhausen problem from a computational point of view in order to highlight the basic problems.

We follow Ho and Chang [9] by considering the control law as a function of the state of nature ξ as opposed to the information variable. The information structure can then be used as a constraint to determine the admissible region of the control laws. This makes the cost function convex although admissible region becomes extremely complex.

For simplicity, we will consider only two person problems. Let us formulate a quantized version of Witsenhausen's problem of section 3.2.

Problem P4

Minimise $J = E\{k^2u_1^2 + (x + u_1 - u_2)^2\}$ with the information structure

$$\begin{cases} z_1 = x \\ z_2 = x + u_1 + v \end{cases}$$

where x and v are discrete random variables with symmetric probability density functions, i.e.

$$p_i = p_{-i} = \text{Prob}(x = i\delta = x_i); \quad x \in D$$

$$q_j = q_{-j} = \text{Prob}(v = j\delta = v_j); \quad v \in E$$

$$D = \{-N\delta,\ldots, \delta,0,\delta,\ldots N\delta\}$$

$$E = \{-M\delta,\ldots,-\delta,0,\delta,\ldots M\delta\}$$

$$\begin{cases} u_1 = \gamma_1(z_1) \\ u_2 = \gamma_2(z_2) \end{cases}$$

We can restrict γ_1, γ_2 to be symmetric without any loss of generality, i.e.

$$\gamma_1(z_1 = i\delta) = a_i$$

with $a_i = - a_{-i}$

$$\gamma_2(z_2 = k\delta) = c_k$$

with $c_k = c_{-k}$

where a_i and $c_k \in D$.

This is a discrete time problem where choosing the two strategies is equivalent to specifying the 2N parameters a_i's and c_k's.

Now the relationship of u_2 to the underlying variables x and v depends on what has been chosen as the strategy γ_1. Therefore γ_1 enters the payoff both directly through u_1 and indirectly through u_2.

When doing the discretisation, we exclude the possibility of choosing N, M and δ so as to make a one to one mapping between x and z_2 since this will trivialise the problem.

Next let us reformulate this problem so as to make the cost function convex.

Problem P5

We will consider u_1 and u_2 to be a function directly of x and v, i.e.

$$\begin{cases} u_1(x_i) = a_i \\ u_2(x_i, v_j) = b_{ij} \end{cases}$$

$$\forall\ i,j$$

Then, the payoff function is convex in a_i and b_{ij} (being quadratic in this case). However, a_i and b_{ij} cannot take arbitrary values in their domains since the values they can take are restricted by their information structures.

Since u_2 is a function of z_2,

$$z_2(x_i, v_j) = z_2(x_h, v_k)$$

implies that

$$b_{ij} = b_{hk}$$

Therefore the constraint is of the form:

If

$$x_i + a_i + v_j = x_h + a_h + v_k$$

then (3.7.1)

$$b_{ij} = b_{hk}$$

i.e. the constraint depends on the choice of a_i.

Ho and Chang *[9]* introduce a tabular method to maintain the constraint set. Let us consider a particular choice of γ_1:

$$\gamma_1 = \{a_1 = 2\delta,\ a_2 = 3\delta,\ a_3 = 2\delta, \ldots\}$$

The constraints for this choice are shown in table 3.1 which is constructed as follows:

The row at the top of the table defines a particular choice of γ_1. The z_2 column comprises 2(2N+M)+1 consecutive integers from (2N+M) descending to -(2N+M) i.e. row it represents the value of $z_2 = i\delta$. b_{ij} is put into column i and row k (i.e. $z_2 = k\delta$) if $k\delta = i\delta + a_i + j\delta$. For example we put $b_{2,0}$ into column 2 and row 5 becomes $2\delta + 3\delta + 0\delta = 5\delta$.

TABLE 3.1

z_2	$a_1=2\delta$	$a_2=3\delta$	$a_3=2\delta$	$\ldots a_N=$ ..	C_k
(2N+M)	.	.	.	. . . .	$C_{(2N+M)}$
.	.	.	.	. . . .	.
.	$b_{1,M}$	.	.	. . . .	.
.	.	.	.	. . . .	. .
5	.	$b_{2,0}$	$b_{3,0}$	. . . .	$C_5=9\delta$
4	$b_{1,1}$	$b_{2,-1}$	$b_{3,1}$	. . .	$C_4=6\delta$
3	$b_{1,0}$	$b_{2,-2}$	$b_{3,-2}$	. . .	$C_3=5\delta$
2	$b_{1,-1}$	$b_{2,-3}$	$b_{3,-3}$	. . .	$C_2=4\delta$
1	$b_{1,-2}$	.	.	. . .	$C_1=5\delta$
0	.	.	.	. . .	.
.	.	.	.	. . .	.
.	$b_{1,-M}$	.	.	. . .	.
.	.	.	.	. . .	.
-2(N+M)	.	.	.	. . .	.

Basically, specifying the heading of the table determines γ_1. The information structure constrains those b_{ij}'s which must assume equal numerical values and hence fixes the shape of the table; finally specifying the numerical entry in the column C_k determines γ_2. Any change of a_i (i.e. γ_1) will change the position of entries in column i.

Since the payoff J is convex in a_i and b_{ij}, we can write the gradient of J w.r.t. a_i and b_{ij} by using the symmetrical condition of γ_2 as:

$$\Delta J = \sum_{i=1}^{N} A_i \, \Delta a_i + \sum_{i=1}^{M} B_{oi} \, \Delta b_{oi} + \sum_{i=1}^{N} \sum_{j=-M} B_{ij} \, \Delta b_{ij} \tag{3.7.2}$$

where

$$A_i = 4p_i \, \{(k^2+1)a_i + x_i - \sum_{j=-M}^{M} q_j b_{ij}$$

$$B_{oi} = 4p_i q_j$$

$$B_{ij} = -4p_i q_j (x_1 + a_i - b_{ij})$$

Unfortunately, it is not possible to use this gradient directly in order to perform an optimisation because of the information structure constraints on the a_i and b_{ij}.

To see the problems these information structure constraints cause, let us consider first the case with fixed γ_1. Here the entries of the table are fixed. Let S(k) be the set of b_{ij}s in the same row. Then we need to change all the $b_{ij} \varepsilon S(k)$ together because of the information constraints. Let $\Delta b_{S(k)}$ represent the common change. Then

$$\Delta J = \sum_{k=1}^{2N+M} \left\{ \sum_{b_{ij} \varepsilon S(k)} B_{ij} \right\} \Delta b_{S(k)} \tag{3.7.3}$$

We note that $\Delta b_{S(k)}$ in (3.7.3) is equal to ΔC_k. Therefore, for fixed γ_1, we can use (3.7.3) to compute the corresponding material $\gamma_2^*(\gamma_1)$.

Next, let us consider the case with fixed γ_2. Now, even though we have fixed γ_2 (and hence the C_k column), any change Δa_i of a_i will induce certain changes of b_{ij} because the change of a_i will reorganise the entries in the table. With a fixed C_k column, an appropriate set of Δb_{ij} needs to be generated in order to satisfy the information structure constraints.

Let $I(\Delta\gamma_1)$ denote the induced set of b_{ij} due to $\Delta\gamma_1$. Then

$$\Delta J = \sum_{i=1}^{N} A_i \, \Delta a_i + \sum_{b_{ij} \varepsilon (\Delta\gamma_1)} B_{ij} \, \Delta b_{ij} \tag{3.7.4}$$

In general, these two effects (3.7.3) and (3.7.4) are mixed up. For any Δa_i we are only free to choose certain Δb_{ij}'s: The rest must take on specific values induced by the information structure. This constraint <u>changes</u> with different Δa_i's.

From this whole discussion we conclude that the gradient of a particular Δa_i can not be easily determined merely by examining the coefficients A_i in (3.7.4). It is necessary to simultaneously work out the induced changes in Δb_{ij} using the B_{ij} coefficients in (3.7.4) and the table. Thus, instead of doing a straightforward hill climb, it is necessary to go over the whole neighbourhood of the a_i's to make sure that nothing is missed out. This of course makes the problem very complicated.

Intuitively, therefore, it would be attractive to solve problem P5 numerically instead of P4 even though the search procedures in solving P5 need to go over the whole neighbourhood of b_{ij}. There is a difficulty here. As Ho and Chang *[9]* point out, if we map a neighbourhood of P4 into P5 then the size of the equivalent neighbourhood in P5 <u>depends</u> on γ_2. Thus we can not specify a priori a neighbourhood in P5 that must be searched in order to guarantee a relative minimum in P4.

3.7.1 Solutions for certain information structures

We recall that two kinds of information structures are particularly simple in team decision problems, i.e. <u>static</u> and <u>partially nested</u>. Static structures are interesting because the decisions are only a function of the state of nature ξ and do not interact with the decisions of other decision makers.

If the information structure is static then clearly the table always keeps the same shape no matter what values we choose for the a_i's. Hence the constraint set does not change. Such problems are easy to solve since by combining all the b_{ij}'s on the same row (i.e. by equating all the b_{ij}'s on row h to C_h), we can get a quadratic cost function on a_i and C_h's without any constraints on the value they make. This is of course a standard problet which is easy to solve.

As we saw earlier in this chapter, PN information structures ultimately reduce to static structures. Thus these structures could be solved in the same way.

Ho and Chang *[9]* have shown that if the information structure is linear then only the partially nested information structure is the one which can be reduced to a static one and thus be used in the solution procedure outlined above.

3.8 CONCLUDING REMARKS

In this chapter we have discussed non-classical information patterns and we have illustrated the difficulties they cause by examining the theory of dynamic teams. Dynamic teams are interesting not only for illustrating non-classical information patterns, but also because the basic decentralised control problem which is of interest in this book could be seen as a subclass of the dynamic team decision problem. In the next chapter, we deal with the decentralised problem explicitly by considering discrete decentralised stochastic control problems.

3.9 ACKNOWLEDGEMENTS

I have relied heavily on the pioneering paper of Ho, Witsenhausen and Yoshikawa in writing this chapter.

3.10 REFERENCES

Specific References

1. Witsenhausen H., "A counter example in stochastic optimal control", SIAM J. Control 6, 1, pp. 131-147, 1968.

2. Striebel C., "Sufficient statistics in the optimum control of stochastic systems", J. Math. And Appl., 12, pp. 576-592, 1965.

3. Hardy G., Littlewood J., and Polya G., "Inequalities", 2nd Edition, CPU, 1952.

4. Ho Y.C., " Team Decision Theory and Information Structure", Proc. IEEE, June, 1980.

5. Radner R., "Team decision problems", Annals of Math Statistics 33, pp. 857-881, 1962.

6. Ho. Y., and Chu K., "Team decision theory and information structures in optimal control problems - Part I", IEEE Trans AC 17, pp. 15-22, 1972.

7. Bismut J., "An example of interaction between information and control: the transparency of a game", IEEE Trans AC 18, 5, pp. 518-522, 1973.

8. Yoshikawa T., "Decomposition of dynamic team decision problems", IEEE Trans AC 23, 4, pp. 627-632, 1978.

9. Ho Y.C. and Chang T.S., "Another look at the non-classical information structure problem", IEEE Trans AC, June, 1980.

General References

1. Ho Y.C. and Mitter S., "Directions in Large Scale Systems, many person optimisation and decentralised control", New York, Plenum Press, 1976.

2. Chu K.C., "Team decision theory and information structures in optimal control problems - Part II", IEEE Trans AC 17, pp. 15-22, 1972.

3. Ho Y.C. and Chu K.C., "Information structures in dynamic multi-person control problems", Automatica 10, pp. 341-351, 1974.

4. Kurtaran B. and Sivan R., "Linear-quadratic Gaussian control with one step delay sharing pattern", IEEE Trans AC 19, pp. 571-574, 1974.

5. Kurtaran B., "Dynamic two person two objective control problems with delayed sharing information patterns", IEE Trans AC 22, pp. 659-661, 1977.

6. Schoute F., "Symmetric team problems and multi access wire communication", Automatica 14, pp. 255-269, 1978.

7. Yoshikawa T and Kobayashi H., "Separation of estimation and control for decentralised stochastic control systems", Automatica 14, pp. 623-628, 1978.

8. Witsenhausen H., "Separation of estimation and control for discrete time systems", Proceedings of the IEEE 59, pp. 1557-1566, 1971.

CHAPTER 4

DECENTRALISED STOCHASTIC CONTROL

4.1 INTRODUCTION

In the previous chapter we saw that in a stochastic environment, the interplay of information and control makes the decentralised control problem an exceedingly difficult one to solve. There are however, certain special classes of problems where we could obtain some kind of a solution. For example, we have already mentioned in the previous chapter that certain delay sharing patterns allow for a separation between estimation and control and here clearly a practical solution is possible. Again, instead of seeking to find the best control law within the class of non-linear control laws, we could seek the best <u>linear</u> control law (which is of course suboptimal). Thus we could consider certain fixed structure controllers. A final possibility is to somehow assess the informational needs of the various subsytems and try to provide for some limited information exchange. We will deal with all these three possibilities in this chapter. We begin our analysis by examining delay sharing patterns.

4.2 ONE-STEP-DELAY SHARING PATTERNS *[1]*

We saw in the previous chapter when we analysed Witsenhausen's problem *[2]* that for the case where the information is shared between the controllers after one step, it is possible to obtain an optimal solution which is linear. We will now examine this important special case in some detail.

For ease of exposition, let us consider the case of two control stations with their corresponding observation posts. The dynamic behaviour of the system could be described by the discrete linear equations:

$$\left.\begin{aligned} x_{k+1} &= A_k x_k + B_k^1 u_k^1 + B_k^2 u_k^2 + v_k \\ x_{ko} &= x_o \end{aligned}\right\} \qquad (4.2.1)$$

$$\left.\begin{aligned} y_k^1 &= C_k^1 x_k + w_k^1 \\ y_k^2 &= C_k^2 x_k + w_k^2 \end{aligned}\right\} \qquad (4.2.2)$$

Here x_k is the m-dimensional state vector at the time step k, and y_k^1, y_k^2 are the observations at the control stations 1 and 2 and at the time step k, u_k^1, u_k^2 are the corresponding controls. The initial state x_o is a

Gaussian random vector with

$$E(x_o) = \bar{x}_o,\ E[(x_o-\bar{x}_o)(x_o-\bar{x}_o)^T] = \Sigma_o \tag{4.2.3}$$

Let us define the augmented quantities:

$$\left.\begin{array}{l} u_k = \begin{pmatrix} u_k^1 \\ u_k^2 \end{pmatrix}, \quad w_k = \begin{pmatrix} w_k^1 \\ w_k^2 \end{pmatrix}, \quad \bar{v}_k = \begin{pmatrix} v_k \\ w_k \end{pmatrix} \\ y_k = \begin{pmatrix} y_k^1 \\ y_k^2 \end{pmatrix}, \quad B_k = [B_k^1, B_k^2] \\ C_k = \begin{pmatrix} C_k^1 \\ C_k^2 \end{pmatrix} \end{array}\right\} \tag{4.2.4}$$

Then we can rewrite equations (4.2.1) and (4.2.2) in the global form:

$$x_{k+1} = A_k x_k + B_k u_k + v_k \tag{4.2.5}$$

$$y_k = C_x x_k + w_k \tag{4.2.6}$$

Here $\bar{v}_k$ is a sequence of Gaussian mutually uncorrelated zero mean random vectors with

$$E[\bar{v}_k \bar{v}_k^T] \triangleq \begin{pmatrix} V_K & V_K^{12} & V_K^{13} \\ V_K^{12'} & W_K^T & W_K^{12} \\ V_K^{13'} & W_K^{12^T} & W_K^2 \end{pmatrix}$$

$$\triangleq \begin{pmatrix} V_K & \tilde{V}_{12} \\ \tilde{V}_K^{12T} & W_K \end{pmatrix} = \tilde{V}_K \tag{4.2.7}$$

where $$\tilde{V}_k \geq 0,\ V_k \geq 0,\ W_k \triangleq \begin{pmatrix} W_k^1 & W_k^{12} \\ W_k^{12T} & W_k^2 \end{pmatrix} \geq 0 \tag{4.2.8}$$

We also assume that x_o is uncorrelated with $\tilde{v}_k$.

The one step delay sharing information pattern is given by

$$\{\mathcal{D}_k^1, \mathcal{D}_k^2\} \quad k = k_0, \ldots N-1$$

where

$$\mathcal{D}_k^i = \{y_k^i, \delta_k\} \quad , \qquad i=1,2 \tag{4.2.9}$$

where δ_k is the common information of the two controllers at the instant k.

$$\delta_k = \emptyset \qquad \text{at } k = k_o$$

and

$$\left.\begin{array}{l} \delta_k = \{y_{ko}, \ldots, y_{k-1}; u_{ko}, \ldots, u_{k-1}\} \\ \qquad\qquad \text{for } k \geqslant k_{o+1} \end{array}\right\} \tag{4.2.10}$$

It is also assumed that both the controllers 1 and 2 know the statistics of the primitive random variables i.e. of x_o, $\{\tilde{v}_k\}$ $k = k_o, \ldots -1$. Thus a one step delay sharing information pattern means that each controller has all his own observations up to the current time step and all the previous observations and the controls applied by all the controllers.

In this case of a non-classical information pattern, the controls of each station depend only on the information available at the station i.e.

$$\left.\begin{array}{l} u_k^1 = u_k^1 \; (\mathcal{D}_k^1) \\ u_k^2 = u_k^2 \; (\mathcal{D}_k^2) \end{array}\right\} \tag{4.2.11}$$

The objective function is

$$J = E\left[\sum_{k=k_o}^{N-1} (x_k^T Q_k x_k + u_k^T R_k u_k) + x_N^T S x_N)\right] \tag{4.2.12}$$

where $Q_k \geqslant 0$, $R_k \geqslant 0$, $S_k \geqslant 0$ (4.2.13)

The optimal control problem involves finding the sequences

$$U^{1*} = \{u_k^{1*}\} \quad k = k_o, \ldots N-1,$$

$$U^{2*} = \{u_k^{2*}\} \quad k = k_o, \ldots N-1$$

such that

$$\left.\begin{array}{l} J(U^{1*}, U^{2*}) \leqslant J(U^1, U^{2*}) \\ J(U^{1*}, U^{2*}) \leqslant J(U^{1*}, U^2) \end{array}\right\} \tag{4.2.14}$$

where $U^1 = \{u_k^1\}_{k = k_o, \ldots N-1}$ and $U^2 = \{u_k^2\}_{k = k_o, \ldots N-1}$ are arbitrary control sequences.

We saw in the previous chapter that for this problem the optimal controller is unique and it is linear in its arguments. Thus the optimal controller has the form

$$u_k^i = L_1^{i*} (y_k^i) + L_2^{i*} (\delta_k) \qquad i=1,2 \tag{4.2.15}$$

where $L_1^{i*}(.)$, $L_2^{i*}(.)$ denote functions which are linear in all their arguments.

We will now follow Kurtaran and Sivan *[1]* in order to derive the precise form of this controller. Before we do this, we need to recall some elementary results in the case of centralised controls. These are stated in the form of the following Lemmas:

Lemma 4.2.1

For the stochastic system

$$x_{k+1} = A_k x_k + B_k u_k + v_k$$

where v_k is a sequence of discrete zero mean mutually uncorrelated stochastic variables and for a quadratic criterion

$$J = E \left[\sum_{k=k_o}^{N-1} (x_k^T Q_k x_k + u_k^T R_k u_k) + x_N^T S x_N) \right] \tag{4.2.16}$$

$$Q_k, R_k \geq 0, \quad S \geq 0$$

the optimal control law based on the measurement of the complete state

$$u_k = - H_k^* x_k \tag{4.2.17}$$

is identical to that for the deterministic system

$$x_{k+1} = A_k x_k + B_k u_k \tag{4.2.18}$$

for the same cost function.

(For proof, cf *[3]* pp. 502).

Lemma 4.2.2

For the system (4.2.5), (4.2.6), the conditional expectation of x_k given δ_k, denoted $\hat{x}_k$ is given by

$$\hat{x}_{k+1} = A_k \hat{x}_k + B_k u_k + K_k^* [y_k - C_k \hat{x}_k] \tag{4.2.19}$$

where K_k^* is the optimal gain. Also, the estimation error is orthogonal to the estimate $\hat{x}_k$ i.e.

$$E\,[e_k\hat{x}_k] = 0, \; k = k_o,\ldots,N-1 \tag{4.2.20}$$

where $e_k = x_k - \hat{x}_k$

(For proof, see any standard textbook e.g. *[4]* pp. 225-230 or *[11]*)

Lemma 4.2.3

The sequence

$$\phi_k = y_k - C_k\hat{x}_k, \; k = k_o,\ldots,N-1 \tag{4.2.21}$$

called the innovations process of y_k is a process of zero-mean uncorrelated stochastic vectors with variance matrices

$$E[\phi_k\phi_k^T] = C_kT_k^*C_k^T + W_k \tag{4.2.22}$$

where $T_k^* = E[(x_k - \hat{x}_k)(x_k - \hat{x}_k)^T]$ (4.2.23)

Also at each time step, ϕ_k is uncorrelated with past measurements i.e.

$$E[\phi_k y_j^i] = 0 \qquad \text{for } j < k \tag{4.2.24}$$

(For proof, cf *[5]*)

Now, we are in a position to give the first main result of Kurtaran and Sivan *[1]*.

Theorem 4.2.1

For the one step delay sharing pattern, the optimal controller is given by

$$u_k = F_k^*y_k + G_k^*\hat{x}_k \tag{4.2.25}$$

where F_k^*, G_k^* are gain matrices and $\hat{x}_k$ is the conditional expectation of x_k given δ_k^k.

In view of the information pattern, F_k^*, G_k^* are partitioned as follows:

$$\left.\begin{aligned} F_k^* &= \begin{pmatrix} F_k^{11*} & 0 \\ 0 & F_k^{22*} \end{pmatrix} \\ \text{and} \quad G_k^* &= \begin{pmatrix} G_k^{1*} \\ G_k^{2*} \end{pmatrix} \end{aligned}\right\} \tag{4.2.26}$$

(For proof of theorem 4.2.1 cf *[1]*)

We note that equation (4.2.25) exhibits a separation property between estimation and control i.e. as soon as a measurement is used to determine

$\hat{x}_k = E[x_k/\delta_k]$, it can be forgotten for control purposes. We also note that the separation property implied by equation (4.2.25) is different from the "classical separation" principle i.e. it is not true that

$$u_k^i = H_k^i \; E[x_k/y_k^i, \delta_k]$$

Next, we give the recurrence relationships which could enable us to compute these controls. Essentially, we are seeking recurrences for

$$[K_k^*, F_k^*, G_k^*] \qquad k = k_o, \ldots, N-1$$

where K_k^* is the gain matrix in equation (4.2.19).

The recurrence relationships for the gain matrices $\{K_k^*\}$ $k = k_o, \ldots, N-1$ are of course well known (cf [3] p. 528) and are given by

$$K_k^* = [A_k T_k^* C_k^T + \tilde{V}_k^{12}] \quad [G_k T_k^* C_k^T + W_k]^{-1}$$

$$T_{k+1}^* = [A_k - K_k C_k^*] T_k^* A_k^T - V_k - K_k^* \tilde{V}_k^{12T}$$

$$k = k_o, \; k_{o+1}, \ldots, N-1$$

with $\qquad T_{ko} = \Sigma_o$

We can thus compute the sequences $\{K_k^*\}$ $k = k_o, \ldots, N-1$ independently of the controls $\{u_k\}$ $k = k_o, \ldots, N-1$.

The computation of $\{F_k^*, G_k^*\}$ $k = k_o, \ldots, N-1$ can be done using theorem 4.2.2.

<u>Theorem 4.2.2.</u>

The optimal sequences

$\{F_k^*\}$ $k = k_o, \ldots, N-1$ and $\{G_k^*\}$ $k = k_o, \ldots, N-2$ are given by

$$F_k^* = -\,(R_k + B_k^T P_{k+1}^* B_k)^{-1} \, [B_k^T P_{k+1}^* K_k^* + \psi_k^*(C_k T_k C_k^T) + w_k)^{-1}]$$

and

$$G_k^* = -\,(F_k^* C_k + H_k^*)$$

where the sequences $\{P_k^*\}$ $k = k_{o+1}, \ldots, N$, $\{H_k^*\}$ $k = k_o, \ldots N-1$ are the solutions of

$$H_k^* = (R_k + B_k^T P_{k+1}^* B_k)^{-1} B_k^T P_{k+1}^* A_k$$

$$P_k^* = (A_k - B_k H_k^*)^T P_{k+1}^* (A_k - B_k H_k^*) + Q_k + H_k^{*T} B_k H_k$$

for $\quad k = k_o, \; k_{o+1}, \ldots, N-1$

where

$$P_N = S$$

Also ψ_k^* is a Lagrange multiplier matrix with

$$\psi_k^t = \begin{bmatrix} 0 & \psi_k^{12*} \\ \psi_k^{21*} & 0 \end{bmatrix}$$

where ψ_k^* has the same dimension as F_k^* and the partitioning is the same as that of F_k^*.

In addition, the optimal cost function value with this control law is given by:

$$J^* = \bar{x}_o^T P_{k_o} \bar{x}_o + \sum_{k=k_o}^{N-1} \mathrm{Tr}\,(Q_k T_k^*) + \mathrm{tr}\,(ST_N^*) + \sum_{k=k_o}^{N-1} \mathrm{tr}\,\{[F_k^* R_k F_k^*] + (B_k F_k^* + K_k^*)^T P_{k+1}^* (B_k F_k^* + K_k^*)].\ (C_k T_k C_k^T + w_k)\}$$

For proof cf [1].

This theorem enables us to compute in a simple way the controls for one step delay sharing information patterns.

Next, let us consider the separation of estimation and control.

4.3 SEPARATION OF ESTIMATION AND CONTROL [6]

Let us consider a discrete time system having J control stations. The system can be described by

$$x(t+1) = f(t,x(t),u_1(t),u_2(t),\ldots u_n(t),v_s(t)) \qquad t=1,2,\ldots T \tag{4.3.1}$$

$$y_j(t) = g_j(t,x(t),v_o(t)), \qquad j=1,2,\ldots J \tag{4.3.2}$$

Here $x(t) \varepsilon R^n$ is the state vector at time t, $u_j(t) \varepsilon R^{r_j}$ is the control vector applied at the station j and $y_j(t) \varepsilon R^{m_j}$ is the observation vector of the station j. $x(1)$, $v_s(t) \varepsilon R^{h_s}$ amd $v_o(t) \varepsilon R^{h_o}$, t=1,2,...T are random vectors with a given a priori probability distribution. Let

$$u(t) \triangleq [u_1^T(t),u_2^T(t),\ldots u_J^T(t)]^T \varepsilon R^r$$

$$r = r_1 + r_2 + \ldots + r_J$$

$$y(t) \triangleq [y_1^T(t), y_2^T(t), \ldots y_J^T(t)]^T \in R^{\bar{m}},$$

$$\bar{m} = m_1 + m_2 + \ldots + m_J.$$

Let us denote by $[t,k]$ the collection of vectors from time t to k. For example,

$$x[t,k] = \{x(t), x(t+1), \ldots x(k)\}$$

The cost function is written as

$$w = w(x[1,T+1], u[1,T]) \tag{4.3.3}$$

We assume that all the stations know the Borel measurable functions f, g_j, $j=1,2,\ldots J$, w and the a priori distributions of the random vectors. At time t, the station j observes y_j and then determines the control $u_j(t)$. In general, however, the information that station j can use to determine $u_j(t)$ consists not only of $y_j(t)$ but also the past data stored in the memory and data obtained through communication with other control stations.

Let us denote the information set of station j by $I_j(t)$. $I_j(t)$ is assumed to satisfy

$$\{y_j(t)\} \subset I_j(t) \subset \{y_j(t), I(t-1)\} \tag{4.3.4}$$

where $I(t)$ is the whole set of data up to time t i.e.

$$I(t) \triangleq \{y[1,t], u[1,t] \in R^{(r+\bar{m})t} \tag{4.3.5}$$

and $\qquad I(o) = \phi.$

As usual, the admissible control law of station j at time t is assumed to have the form

$$u_j(t) = \gamma_{ij}(I_j(t)) \tag{4.3.6}$$

where γ_{ij}^1s are arbitrary measurable functions.

Let $\gamma_t = \{\gamma_{ij}, j=1,2,\ldots J\}$,

$\gamma = \{\gamma_t, t=1,2,\ldots T\}$

$\Gamma = \{$The set of all admissible control laws $\gamma\}$

and $$W(\gamma) = E\{w/\gamma\} \tag{4.3.7}$$

where $E\{\omega/\gamma\}$ denotes the expectation of ω for a given control law γ. Then the optimal decentralised control problem is that of finding the control law γ^* such that

$$W(\gamma^*) \leqslant W(\gamma) \tag{4.3.8}$$

for any $\gamma \in \Gamma$.

Having defined the problem, let us consider the sufficient condition derived in [6] for separation of estimation and control. Before we do this, let us recall some definitions from chapter 3:

<u>Definition 4.3.1</u>

Information I(k) is said to be nested in $I_j(t)$ if there exists a measurable function h^{γ}_{tjk} such that

$$I(k) = h^{\gamma}_{tjk}(I_j(t) \tag{4.3.9}$$

for any given γ.

This nestedness relation means that I(k) can always be recovered from $I_j(t)$ for any given γ.

<u>Definition 4.3.2</u>

The information structure is said to have a sequential partition (s-partition) $\{[1,t_1], [t_1 + 1,]t_2, \ldots [t_M + \ell, T]\}$ $1 \leq t_1 < t_2 \ldots < t_M < T$ if $I(t_m)$ is nested in $I_j(t)$ for $t=1,2,\ldots T$, $j=1,2,\ldots J$ where t_m is the maximum of t_i such that $t_i < t$ and $t_i \in \{t_1, t_2, \ldots t_M\}$.

As we saw in the previous chapter, it is easy to check if an s-partition exists by examining the precedence diagram. For example, Fig. 4.1 shows the precedence diagram of the system

$$T = 5,\ J = 2$$

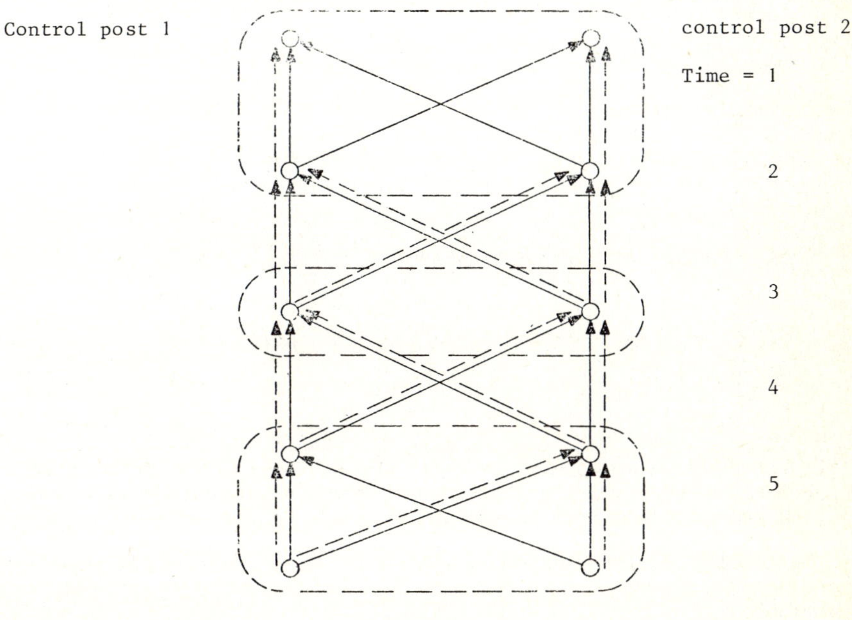

Figure 4.1

Here the information pattern is given by

$$I_j(1) = \{y_j(1)\},\ j=1,2$$

$$I_1(2) = \{y_1(1),y_2(1),u_1(1),u_2(1),y_1(2)\},$$

$$I_2(2) = \{y_2(2)\}$$

$$I_j(3) = \{I_1(2),I_2(2),u_1(2),y_j(3)\},\ j=1,2$$

$$I_j(4) = \{I_1(3),I_2(3),u_j(3),y_j(4)\},\ j=1,2$$

$$I_j(5) = \{I_j(4),u_j(4),y_j(5)\},\ j=1,2$$

From Fig. 4.1, we see that the system has the s-partition

$$\{[1,2],\ 3,\ [4,5]\},\ \{[2,3],\ [3,5]\} \text{ and}$$

$$[1,3],\ [4,5]\}$$

We note also that systems with m-step periodic sharing information patterns have an s-partition where the information structure is given by

$$I_j(t) = \{y_j(t),y_j(\tau),u_j(\tau),\tau = t-1,t-2,\ldots pm+1,\ I(pm)\}$$

where p is the largest integer such that

$$t > pm$$

Before we state the separation condition, let us outline three conditions which are needed for the separation result:

(1) The information structure has an s-partition

$$\{[1,t_1],\ [t_1+1,t_2],\ldots\ [t_M+1,T]\}$$

(2) The group of random vectors $\{x(1),\ v[1,t_1]\}$, $v[t_1+1,t_2],\ldots\ v[t_M+1,T]$ are mutually independent where $v(t) = [v_s^T(t)v_s^T(t)]^{TM}$

(3) The cost function is of the form

$$w = \sum_{m=o}^{M} w_m(x[t_m+1,t_{m+1}+1],u[t_m+1,\ t_{m+1}]) \tag{4.3.10}$$

where $t_o = 0$ and $t_{M+1} = T$.

Next, let us assume that Q is the set of all probability distributions on R^n. This set could be considered to be a complete separable metric space by introducing an appropriate distance. Let the conditional distribution of $x(t_m+1)$ given γ and $I(t_m)$ be denoted by

$$p[x(t_m+1]\,|\,I(t_m):\gamma)$$

if we consider

$$q_m \triangleq p([x(t_m + 1)] \mid I(t_m):\gamma)$$

to be an element of Q, we have a measurable function ψ^γ mapping from $\mathbb{R}^{(r+m)t_m}$ to Q i.e.

$$q_m = \psi^\gamma(T(t_m)) \tag{4.3.11}$$

Let us suppose that condition (1) is satisfied and let

$$\Delta I_j(t) = I_j(t) - I(t_m) \tag{4.3.12}$$

i.e. let $\Delta I_j(t)$ be the data set included in $I_j(t)$ but not in $I(t_m)$. Then we can assume without loss of generality that the information structure of the problem is given by

$$I_j^1(t) = \{\Delta I_j(t), I(t_m)\}, \; t \in [t_m + 1, t_{m+1}],$$
$$j=1,2,\ldots J, \; m=1,2,\ldots,M \tag{4.3.13}$$

Now let us define a control law

$$\pi = \{\pi_{ij}\} \text{ as}$$

$$u_j(t) = \pi_{ij}\,(\Delta I_j(t), q_m), \; t \in [t_m + 1, t_{m+1}] \tag{4.3.14}$$

where π_{ij} is an arbitrary measurable function. Using (4.3.14) and (4.3.11) we have

$$\begin{aligned} u_j(t) &= \pi_{ij}\,(\Delta I_j(t), \psi^\pi(I(t_m)) \\ &\triangleq \gamma^\pi\,(I_j^1(t)) \end{aligned} \tag{4.3.15}$$

Therefore π could be considered to be an admissible control law $\gamma^\pi = \{\gamma_{ij}^\pi\}$ and the set of all π, $\Pi\,(t_1, t_2, \ldots t_M)$ is a subset of Γ.

We will call a subset Γ_c of Γ <u>complete</u> if any given $\gamma \in \Gamma \; \exists \; \gamma \in \Gamma_c$ s.t. $w(\gamma_c) \leq w(\gamma)$.

Obviously, we could restrict our attention to Γ_c as far as obtaining optimal control is concerned.

Now, we can state the main separation result of Yoshikawa and Kobayashi [6]:

<u>Theorem 4.3.1</u>

For a problem satisfying conditions (1-3), the set $\Pi(t_1, t_2, \ldots t_M)$ is complete.

For proof cf [6].

4.3.1 <u>Remarks</u>

(1) This is a very important result. First of all, it reduces significantly the amount of data which needs to be taken into consideration.

Thus the control law π has the advantage over γ in that the past data $I(t_m)$ in $I_j(t)$ can be reduced to $\hat{q}_m$. This becomes especially important where q_m has an equivalent sufficient statistic $\hat{q}_{em}$ [7]. In that case the theorem holds with $\Pi(t_1,t_2,\dots t_M)$ replaced by $\Pi^e(t_1,t_2,\dots t_M)$ which is the set of all $\pi^e = \{\pi^e_{ij}\}$.

$$u_j(t) = \pi^e_{ij}\ (\Delta I_j(t),\hat{q}_m) \tag{4.3.16}$$

(2) If $I_j(t)$ satisfies for some m (≥ 1)

$$I_j(t) = \{y_j(t),u_j(t-1),y_j(t-1),\dots,\\ u_j(t-m+1),y_j(t-m+1),I(t-m)\} \tag{4.3.17}$$

then the system is said to have an m step delay sharing pattern.

If we specialise this to the case m = 1 i.e. we assume that

(a) The system has a one step delay sharing structure

(b) The group of random vectors $\{x(1),v(1)\},v(t)$, $t=2,3,\dots T$ are mutually independent and

(c) The cost function has the form

$$w = \sum_{t=0}^{T} \tilde{w}_t(x(t),x(t+1),u(t))$$

Then by the theorem, the set $\Pi(1,2,\dots,T-1)$ is complete. This result includes the main result of section 4.2 as a special case.

To conclude this section, we have outlined an important result for decentralised discrete time stochastic control problems. For systems with a sequential partition independent random vectors and a separable cost function of the kind given in equation (4.3.10), it is possible to restrict control laws, without any loss, to be functions of the data between the current time t and the most recent time $t_m + 1$, and of the conditional probability of the state at time $t_m + 1$ given the data up to time t.

Having examined these separation properties, let us return to the practical problem of designing the decentralised stochastic controllers. We will begin by constraining the controller to have a fixed structure.

4.4 FIXED STRUCTURE CONTROLLERS

We will first of all derive the equations for the controller if we constrain it to be a linear function of its local measurements. In addition, we will constrain the compensators to be of the same order as the system to be controlled in order to get around the problem of "second guessing". Although this will clearly give a suboptimal solution as we saw in chapter 3, when we were considering Witsenhausen's problem, it will at least give a simple, realisable control which provides a reasonable solution to the problem. We therefore follow Chong and Athans [9] and define the usual two controller models as

$$\dot{x}(t) = A(t)x(t) + B_1(t)u_1(t) + B_2(t)u_2(t) \tag{4.4.1}$$

where u_1, u_2 are the controls applied at stations 1 and 2 and $x \in R^n$.

The measurements at the two stations are given by

$$z_1(t) = C_1(t)x(t) + \theta_1(t) \tag{4.4.2}$$

$$z_2(t) = C_2(t)x(t) + \theta_2(t) \tag{4.4.3}$$

where $Q_1(t), Q_2(t)$ are mutually independent zero mean Gaussian white noise process with covariances Θ_i $i=1,2$.

The information structure is given by Y_1 and Y_2 where

$$Y_1(t) = {}_1(t) \cup \{\text{a priori information}\} \tag{4.4.4}$$

$$Y_2(t) = {}_1(t) \cup \{\text{a priori information}\} \tag{4.4.5}$$

and

$$z_1(t) \triangleq \{z_1(\tau);\ t_o \leqslant \tau < t\} \tag{4.4.6}$$

$$z_2(t) \triangleq \{z_2(\tau);\ t_o \leqslant \tau < t\} \tag{4.4.7}$$

Both the controllers know the statistics of the initial state i.e.

$$E\{x(t_o)\} = \bar{x}_o \tag{4.4.8}$$

$$Cov\{x(t_o);\ x(t_o)\} = \Sigma_o \tag{4.4.9}$$

We are seeking the decentralised control laws

$$u_1(t) - \phi_1(Y_1(t)), \tag{4.4.10}$$

and

$$u_2(t) = \phi_2(Y_2(t)) \tag{4.4.11}$$

which minimise

$$J = E\{x^T(t_f)Fx(t_f) + \int_{t_o}^{t_f} (x^T(t)Q(t)x(t) + u_1^T(t)R_1(t)u_1(t) + u_2^T(t)R_2(t)u_2(t))\,dt\} \tag{4.4.12}$$

As we hinted earlier, we will now constrain the controls to be linear functions of the estimated local states i.e.

$$u_1(t) = D_1(t)\hat{x}_1(t) \tag{4.4.13}$$

$$u_2(t) = D_2(t)\hat{x}_2(t) \tag{4.4.14}$$

where $\hat{x}_1, \hat{x}_2$ are the states of n-dimensional filters which evolve according to

$$\dot{\hat{x}}_1 = E_1(t)\hat{x}_1(t) + G_1(t)z_1(t) + H_1(t)u_1(t) \tag{4.4.15}$$

$$\dot{\hat{x}}_2 = E_2(t)\hat{x}_2(t) + G_2(t)z_2(t) + H_2(t)u_2(t) \tag{4.4.16}$$

Here $D_1(t), D_2(t), E_1(t), E_2(t), G_1(t), G_2(t), H_1(t), H_2(t)$ are matrices of appropriate dimensions which need to be chosen. Figure 4.2 shows the structure of the controllers.

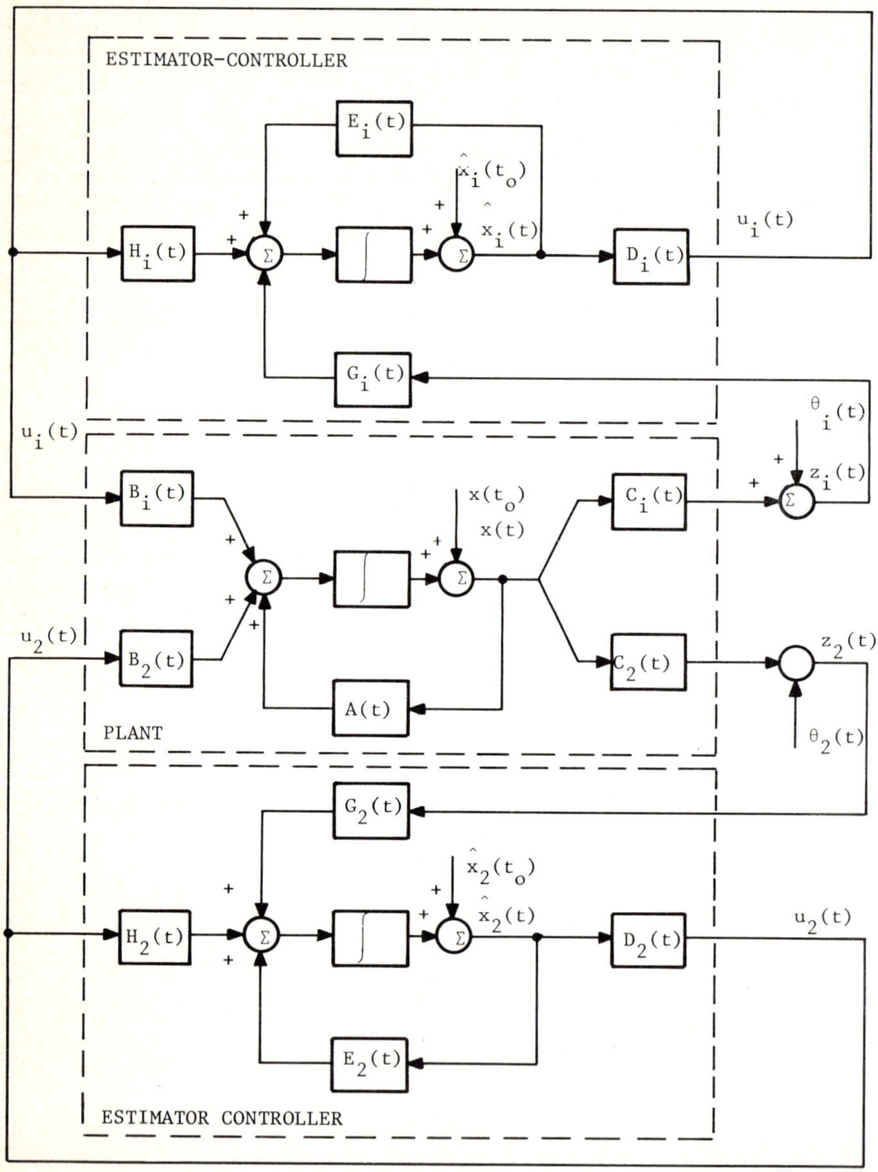

Figure 4.2

On substituting (4.4.2), (4.4.3), (4.4.13) and (4.4.14) into the state and filter equations we obtain

$$\dot{x}(t) = A(t)x(t) + B_1(t)D_1(t)\hat{x}_1(t) + B_2(t)D_2(t)\hat{x}_2(t) \tag{4.4.17}$$

$$\dot{\hat{x}}_1(t) = E_1(t)\hat{x}_1(t) + G_1(t)C_1(t)x(t) + H_1(t)D_1(t)\hat{x}_1(t) + G_1(t)\theta_1(t) \tag{4.4.18}$$

$$\dot{\hat{x}}_2 = E_2(t)\hat{x}_2(t) + G_2(t)C_2(t)x(t) + H_2(t)D_2(t)\hat{x}_2(t) + G_2(t)\theta_2(t) \tag{4.4.19}$$

Subtracting (4.4.18) and (4.4.19) from (4.4.17) we obtain

$$\dot{x} - \dot{\hat{x}}_1 = (A - E_1 + B_2D_2 - G_1C_1)x - B_2D_2(x - \hat{x}_2) + E_1(x - \hat{x}_1) + (B_1 - H_1)D_1\hat{x}_1 - G_1\theta_1 \tag{4.4.20}$$

$$\dot{x} - \dot{\hat{x}}_2 = (A - E_2 + B_1D_1 - G_2C_2)x - B_1D_1(x - \hat{x}_1) + E_2(x - \hat{x}_2) + (B_2 - H_2)D_2x_2 - G_2\theta_2 \tag{4.4.21}$$

Also, in order to ensure that $\hat{x}_1(t)$ and $\hat{x}_2(t)$ are unbiased estimates of $x(t)$ for all $u_1(t)$ and $u_2(t)$ we need to add the additional constraint

$$E\{x(t) - \hat{x}_1(t)|y_j(t)\} = 0 \qquad \forall\ t,\ i=1,2;\quad j=1,2 \tag{4.4.22}$$

This means that

$$E\{\dot{x}(t) - \dot{\hat{x}}_i(t)|y_j(t)\} = 0\ \forall\ t \qquad i=1,2;\quad j=1,2 \tag{4.4.23}$$

On taking the conditional expectations of (4.4.20) and (4.4.21) and applying these conditions, we obtain

$$E_1(t) = A(t) + B_2(t)D_2(t) - G_1(t)C_1(t) \tag{4.4.24}$$

$$E_2(t) = A(t) + B_1(t)D_1(t) - G_2(t)C_2(t) \tag{4.4.25}$$

$$H_1(t) = B_1(t) \tag{4.4.26}$$

$$H_2(t) = B_2(t) \tag{4.4.27}$$

Also, the initial conditions become

$$\hat{x}_1(t_o) = \bar{x}_o \tag{4.4.28}$$

$$\hat{x}_2(t_o) = \bar{x}_o \tag{4.4.29}$$

On defining the combined vector

$$m^T(t) \triangleq [x^T(t), x^T(t) - \hat{x}_1^T(t), x^T(t) - \hat{x}_2^T(t)] \tag{4.4.30}$$

we can rewrite equations (4.4.17), (4.4.20) and (4.4.21) together as

$$\dot{m}(t) = \hat{A}(t)m(t) - \hat{B}(t)Q(t) \tag{4.4.31}$$

where

$$\hat{A}(t) \triangleq \begin{bmatrix} A + B_1D_1 + B_2D_2 & -B_1D_1 & -B_2D_2 \\ 0 & A + B_2D_2 - G_1C_1 & -B_2D_2 \\ 0 & -B_1D_1 & A + B_1D_1 - G_2C_2 \end{bmatrix} \tag{4.4.32}$$

$$\hat{B}(t) \triangleq \begin{bmatrix} 0 & 0 \\ G_1(t) & 0 \\ 0 & G_2(t) \end{bmatrix} \tag{4.4.33}$$

$$\Theta(t) \triangleq \begin{bmatrix} \Theta_1(t) \\ \Theta_2(t) \end{bmatrix} \tag{4.4.34}$$

Let us define the second moment matrix of m i.e.

$$E\{m(t)m^T(t)\} = M(t)$$

$$= \begin{bmatrix} M_{00} & M_{01} & M_{02} \\ M_{10} & M_{11} & M_{12} \\ M_{20} & M_{21} & M_{22} \end{bmatrix} \tag{4.4.35}$$

Then it is easy to show that M(t) satisfies the linear matrix equation

$$\dot{M}(t) = \hat{A}(t)M(t) + M(t)\hat{A}^T(t) + \hat{B}(t)\Theta(t)\hat{B}^T(t) \tag{4.4.36}$$

where

$$\Theta(t) = \begin{bmatrix} \Theta_1(t) & 0 \\ 0 & \Theta_2(t) \end{bmatrix} \tag{4.4.37}$$

Also, it can be shown that the cost function value J is given by the following expression:

$$J = tr[FM_{00}(t_f)] + \int_{t_o}^{t_f} tr[QM_{00}] + tr[D_1^T R_1 D_1$$

$$(M_{00} - M_{10} - M_{01} + M_{11})] + tr[D_2^T R_2 D_2(M_{00} - M_{20} - M_{02} + M_{22})]dt$$

$$= tr[\hat{F}M(t_f)] + \int_{t_o}^{t_f} tr[\hat{Q}(t)M(t)]dt \qquad (4.4.38)$$

$$\text{where } \hat{F} \triangleq \begin{bmatrix} F & 0 & 0 \\ 0 & 0 & 0 \\ 0 & 0 & 0 \end{bmatrix} \qquad (4.4.39)$$

$$\text{and } \hat{Q}(t) \triangleq \begin{bmatrix} D_1^T R_1 D_1 + D_2^T R_2 D_2 + Q & -D_1^T R_1 D_1 & -D_2^T R_2 D_2 \\ -D_1^T R_1 D_1 & D_1^T R_1 D_1 & 0 \\ -D_2^T R_2 D_2 & 0 & D_2^T R_2 D_2 \end{bmatrix} \qquad (4.4.40)$$

Thus, we have reformulated our original problem as the following deterministic minimisation problem i.e.

" Find $D_1^*(t), D_2^*(t), G_1^*(t), G_2^*(t)$ so as to minimise

$$J = tr\ [\hat{F}M(t_f)] + \int_{t_o}^{t_f} tr[\hat{Q}(t)M(t)]dt \qquad (4.4.41)$$

Subject to the dynamic constraints

$$\dot{M}(t) = \hat{A}(t)M(t) + M(t)\hat{A}^T(t) + \hat{B}(t)\Theta(t)\hat{B}^T(t) \qquad (4.4.42)$$

with

$$M(t_o) = \begin{bmatrix} x_o x_o^T + \Sigma_o & \Sigma_o & \Sigma_o \\ \Sigma_o & \Sigma_o & \Sigma_o \\ \Sigma_o & \Sigma_o & \Sigma_o \end{bmatrix} \qquad (4.4.43)$$

"

It is quite straightforward [9] to solve this problem using the matrix minimum principle [10] and the results are summarised as: Defining $I_o^T = [I,0,0]$, $I_1^T = [0,I,0]$, $I_2^T = [0,0,I]$ where I is then nxn identity matrix

$$D_1^*(t) = - R_1^{-1}B_1^T(t) \,[I_oP^*(t)M^*(t)(I_o - I_1) + I_2^TP^*(t)M^*(t)\,(I_2 - I_1)][(I_o - I_1)^TM^*(t)(I_o - I_1)]^{-1} \tag{4.4.44}$$

$$D_2^*(t) = - R_2^{-1}(t)B_2^T(t)\,[I_o^TP^*(t)M^*(t)(I_o - I_2) + I_1^TP^*(t)M^*(t)(I_1 - I_2)][(I_o - I_2)^TM^*(t)(I_o - I_2)]^{-1} \tag{4.4.45}$$

$$G_1^*(t) = (I_1^TP^*(t)I_1)^{-1}[I_1^TP^*(t)M^*(t)I_1]\,C_1^T(t)\,\Theta^{-1}(t) \tag{4.4.46}$$

$$G_2^*(t) = (I_2^TP^*(t)I_2)^{-1}\,[I_2^TP^*(t)I_2]C_2^T(t)\Theta_2^{-1}(t) \tag{4.4.47}$$

where

$$\dot{M}^*(t) = \hat{A}^*(t)M^*(t) + M^*(t)\hat{A}^{*T}(t) + \hat{B}^*(t)\Theta(t)\hat{B}^{*T}(t) \tag{4.4.48}$$

where

$$M^*(t_o) = \begin{pmatrix} \Sigma_o + \bar{x}_o\bar{x}_o^T & \Sigma_o & \Sigma_o \\ \Sigma_o & \Sigma_o & \Sigma_o \\ \Sigma_o & \Sigma_o & \Sigma_o \end{pmatrix} \tag{4.4.49}$$

and the optimal cost is

$$J = \mathrm{tr}\,[\hat{F}M^*(t_f)] + \int_{t_o}^{t_f} \mathrm{tr}[\hat{Q}^*(t)M^*(t)]dt \tag{4.4.50}$$

or

$$J = \mathrm{tr}[P^*(t_o)M^*(t_o)] + \int_{t_o}^{t_f} \mathrm{tr}[P^*(t)\hat{B}^*(t)\Theta(t)\hat{B}^{*T}(t)]dt \tag{4.4.51}$$

4.4.1 Remarks

(1) It must be emphasised that this is the "optimal" solution only in terms of the constrained optimisation problem that we have formulated i.e. where we have constrained the compensators to be linear.

(2) In order to solve this problem numerically, we need to solve a matrix two point boundary value problem. This can be done using either standard techniques [11] (e.g. the gradient method, quasilinearisation etc.) or hierarchical techniques as we do to a deterministic version of this problem in chapter 6.

(3) The filter (4.4.15) with $G_1^*(t)$ is not the optimal Kalman-Bucy filter although it is linear and unbiased.

(4) As we can see from the expressions for the $G_1^*(t)$ and $D_1^*(t)$, the product of the state and costate matrices P(t)M(t) plays an important role. We note that $G_1^*(t)$ depends mainly on $M^*(t)$ while $D_1^*(t)$ depends mainly on $P^*(t)$. In the single controller case, $G_1^*(t)$ depends mainly on $M^*(t)$ and $D_1^*(t)$ depends only on $P^*(t)$. Thus the presence of two controllers has coupled the state and costate together.

(5) Examining the computational implications of the scheme brings home once again the fact that we are paying for limited exchange of on-line information by a considerable increase in the off-line computation. In this sense, informational complexity is posing additional computational problems which do not exist either in the centralised solution procedure or in the hierarchical one [11].

Next, we continue with fixed structure controllers by considering another type of fixed structure i.e. periodic coordination where we allow some information exchange through a coordinator.

4.5 PERIODIC COORDINATION [12]

The idea here is to bring together the notions of hierarchies and decentralisation as we did in chapter 1 except that now we will be considering this within a stochastic context.

Basically, we will be considering a linear stochastic system comprising N interacting subsystems. Each of the subsystems has its own individual controller. The controllers pass some information to the coordinator which is responsible for looking after the interactions and thus ensuring that the team's goals are achieved. The information sent to the coordinator is more frequent than the interventions of the coordinator. The information transmission is neither noiseless nor instantaneous as was the case when we considered the degenerate example of Bismut [13] in the previous chapter.

The dynamic system comprising N interacting subsystems could be described by

$$x(k+1) = Ax(k) + Bu(k) + \xi(k) \qquad k=0,1,\ldots t_f-1 \tag{4.5.1}$$

$$\text{where } x(k) = \begin{bmatrix} x_1(k) \\ x_2(k) \\ \vdots \\ x_N(k) \end{bmatrix} ; \quad u(k) = \begin{bmatrix} u_1(k) \\ u_2(k) \\ \vdots \\ u_N(k) \end{bmatrix}$$

$$\xi(k) = \begin{bmatrix} \xi_1(k) \\ \vdots \\ \xi_N(k) \end{bmatrix} \tag{4.5.2}$$

A is a full matrix and B is a block diagonal matrix with the blocks corresponding to the N subsystems i.e.

$$A = \begin{bmatrix} A_{11} & A_{12} & \cdots & A_{1N} \\ \vdots & \vdots & & \vdots \\ A_{N1} & A_{N2} & \cdots & A_{NN} \end{bmatrix}$$

$$B = \begin{bmatrix} B_1 & 0 & \cdots & 0 \\ 0 & B_2 & \cdots & 0 \\ 0 & \cdots & \ddots & B_N \end{bmatrix} \tag{4.5.3}$$

The measurements y(k) are generated as follows:

$$y(k) = Cx(k) + \theta(k); \quad k=0,\ldots,T-1 \tag{4.5.4}$$

$$\text{where } y(k) = \begin{bmatrix} y_1(k) \\ \vdots \\ y_N(k) \end{bmatrix}; \quad \theta(k) = \begin{bmatrix} \theta_i(k) \\ \vdots \\ \theta_N(k) \end{bmatrix};$$

$$C = \begin{bmatrix} C_1 & \cdots & 0 & \cdots & 0 \\ 0 & \cdots & C_2 & \cdots & 0 \\ \vdots & & & \ddots & \vdots \\ 0 & \cdots & & \cdots & C_N \end{bmatrix} \tag{4.5.5}$$

We notice that C is also a block diagonal matrix i.e. the measurements for subsystem i only relate to that subsystem's state i.e.

$$y_i(k) = C_i x_i(k) + \theta_i(k) \qquad i=1,2,\ldots N$$

$\{x_i(o), \xi_i(k), \theta_i(k); i=1,\dots N, k=0,1,\dots\}$

are assumed to be mutually independent Gaussian random vectors with known means and covariances given by

$$E\{x_i(o)\} = \bar{x}_i(o); \; Cov\{x_i(o)\} = \Sigma_i(o) \tag{4.5.6}$$

$$E\{\xi_i(k)\} = o; \quad Cov\{\xi_i(k)\} = E_i(k) \tag{4.5.7}$$

$$E\{\theta_i(k)\} = o; \; Cov\{\theta_i(k))\} = \Theta_i(k) \tag{4.5.8}$$

The cost function is a seperable quadratic one and it is given by

$$J = \sum_{i=1}^{N} J_i = \sum_{i=1}^{N} E\{x_i^T(t_f)F_i x_i(t_f) + \sum_{k=0}^{t_{f-1}} x_i(k)^T Q_i x_i(k) + u_i(k)^T R_i u_i(k)\} \tag{4.5.9}$$

where F_i, Q_i and R_i are all positive definite

Let us also define

$$\Sigma(o) \triangleq diag\{\Sigma_i(o)\}; \quad E(k) \triangleq diag\{E_i(k)\};$$

$$\Theta(k) \triangleq diag\{\Theta_i(k)\}; \; F \triangleq diag\{F_i\};$$

$$Q \triangleq diag\{Q_i\}; \; R \triangleq diag\{R_i\} \tag{4.5.10}$$

In order to define the information structure for this problem, we form the matrices

$$Y_i(k) \triangleq \{y_i(o), \dots y_i(k)\}$$

$$U_i(k) \triangleq \{u_i(o), \dots u_i(k)\}$$

$$Y(k) \triangleq Y(k) \cup Y_2(k) \dots \cup Y_N(k)$$

$$U(k) \triangleq U_1(k) \cup U_2(k) \dots \cup U_N(k) \tag{4.5.11}$$

Then, the <u>centralised</u>, or <u>classical</u> information structure arises when we pool all the measurements together and compute the controls as some function of these i.e.

$$u(k) = \gamma^k(Y(k), U(k-1)) \tag{4.5.12}$$

The <u>non-classical</u> team information structure arises if we could only use the <u>local</u> measurements to compute the controls i.e.

$$u_i(k) = \gamma_i^k(Y_i(k), U_i(k-1)) \tag{4.5.13}$$

Now, let us consider a two level structure to solve this problem. The two level structure comprises a second level <u>coordinator</u> and independent first level controllers. The information is partitioned among the first level controllers and the coordinator as follows:

(a) Lower level controllers

Each lower level controller possesses the following information:

a priori information:

$$x_i(k+1) = A_{ii}x_i(k) + v_i(k) + B_i \tag{4.5.14}$$

$$y_i(k) = C_i x_i(k) + \theta_i(k) \tag{4.5.15}$$

$$J_i = E\{x_i^T(t_f)F_i x_i(t_f) + \sum_{k=o}^{t_{f-1}} x_i^T(k)$$

$$Q_i x_i(k) + u_i(k)^T R_i u_i(k)\} \tag{4.5.16}$$

The statistics of $x_i(o)$, $\xi_i(k)$, $\theta_i(k)$, $k=0,1,\ldots t_{f-1}$ are also all assumed to be known to each of the lower level controllers. In equation (4.5.14), $v_i(k)$ represents the interaction input from the other subsystems and this is unknown to each subsystem controller.

a posteriori information

On implementation, the lower level untis have available

$Y_i(k)$ $U_i(k-1)$

i.e. their own observation.

(b) The coordinator

The information available to the coordinator comprises:

A priori information

The coordinator is assumed to have complete a priori (structural) information on the system.

A posteriori information

Let $Y_o(k)$ be the information available to the coordinator at time k. Then we will specifically consider two cases i.e.

(1) $Y_o(k) = Y(n\ell) \cup U(n\ell-1)$; $Y_o(o) = \phi$

where ϕ is the null set and n is an integer such that

$$k \geq n\ell, \quad \ell \geq 1 \qquad m\ell = t_f \tag{4.5.17}$$

This means that the coordinator gathers together the measurements from the lower level units every ℓ units of time.

(2) $$Y_o)k) = \phi \tag{4.5.18}$$

In this case the coordinator has no measurement information.

One of the jobs of the coordinator is to provide some information concerning the interactions v_i to the lower level units. This is done by ensuring, on average, that $\hat{v}_i$ is the best estimate of the interactions based on the coordinator's information i.e.

$$E\{v_i(t) - \sum_{j\neq 1} A_{ij}x_j(t)|Y_o(k)\} = 0, \ t \geq k \tag{4.5.19}$$

Also, the coordinator needs to minimise the cost function.

In the way of permitted information transfer, the coordinator is allowed to transmit any portion of his stored data except for the structure of the overall system. For example, he could transmit the values of the matrices A_{ij} without transmitting the interaction relationships. Also, since $v_1(k)$ depends on $Y_o(k)$ we shall also let $u_i(k)$ depend on $Y_o(k)$ i.e. the a posteriori information of the coordinator could be used by the lower level.

Then the first problem of interest which is formulated by Chong and Athans *[12]* is:

<u>P1</u>

Given:

$$x_i(k+1) = A_{ii}x_i(k) + v_i(k) + B_iu_i(k) + \xi_i(k) \tag{4.5.20}$$

$$J = \sum_{i=1}^{N} J_i \tag{4.5.21}$$

$$E\{v_1(t) - \sum_{j\neq 1} A_{ij}x_j(t)| Y_o(k)\} = 0 \qquad t > k \tag{4.5.22}$$

$$u_i(k) = \gamma_i^k(Y_i(k),U_i(k-1),Y_o(k) \tag{4.5.23}$$

$$v_i(k) = \eta_i^k(Y_o(k))$$

Find the optimal strategies γ_i^k and η_i^k $i=1,\ldots,N$, $k=0,\ldots T-1$ such that J is minimised.

4.5.1 <u>Comments</u>

(a) We perform the minimisation both w.r.t. γ_i^k and η_i^k since we need an estimate of the interactions to enable us to compute the controls.

(b) γ_i^k only uses the decoupled system model.

(c) Equation (4.5.22) takes account of the fact that we have a system comprising interconnected subsystems.

(d) In formulating the constraint (4.5.22) we have tacitly assumed a "separation" property for the coordinator. This is justified by the fact that he has a classical information pattern over the ℓ time intervals.

Next we consider a special case of the problem P1 when the coordinator receives the information from the lower level units i.e. $Y_o(k) = \phi$.

P2

The lower level problem here is to find γ_i^k and $v_i(k)$ so as to minimise for $i=1,2,\ldots N$

$$\tilde{J}_i(p) = J_i + E\{\sum_{k=o}^{t_{f-1}} p_i^T(k)v_i(k) - \tilde{p}_i^T(k)x_i(k)\} \tag{4.5.24}$$

where
$$\tilde{p}_i(k) = \sum_{j\neq i} A_{ji}^T p_j(k) \tag{4.5.25}$$

$$x_i(k+1) = A_{ii}x_i(k) + B_iu_i(k) + \xi_i(k) \tag{4.5.26}$$

and
$$u_i(k) = \gamma_i^k(Y_i(k), U_i(k-1)) \tag{4.5.27}$$

with $v_i(k)$ deterministic.

The coordinator problem in this case is to

$$\underset{p}{\text{maximise}} \quad \sum_{i=1}^{N} \tilde{J}^*(p) \tag{4.5.28}$$

In P2, the solution of the lower level problem, which is stochastic, required the knowledge of $\{p(k),\ k=0,1,\ldots t_{f-1},\ A_{ji},\ j\neq i\}$ or equivalently of $\{p_i(k),\ k=0,1,\ldots t_{f-1}\}$ in addition to the local information. The higher level problem is a deterministic static problem. The solution of the lower level problem is given by Chong and Athans [12] to be:

$$u_i^*(k) = -D_i(k+1)(\hat{x}_i(k) - \bar{x}_i(k)) - E_i(k+1)p_i(k) \tag{4.5.29}$$

where

$$D_i(k+1) = T_i^{-1}(k+1)B_i^TK_i(k+1)A_{ii} \tag{4.5.30}$$

$$T_i(k+1) = R_i + B_i^TK_i(k+1)B_i \tag{4.5.31}$$

$$K_i(k) = Q_i + A_{ii}^TK_i(k+1)A_{ii} - A_{ii}^TK_i(k+1) B_iT_i^{-1}(k+1)B_i^TK_i(k+1)A_{ii} \tag{4.5.32}$$

with
$$K_i(t_f) = F_i$$

$$E_i(k+1) = \tfrac{1}{2} R_i^{-1}B_i^T \tag{4.5.33}$$

Also define

$$S_i(k+1) = K_i(k+1) - K_i(k+1)B_iT_i^{-1}(k+1)$$
$$B_i^T K_i(k+1) \quad (4.5.34)$$

The estimates $\hat{x}_i(k)$ evolve according to:

$$\hat{x}_i(k+1) \triangleq E\{x_i(k+1)|Y_i(k+1), U_i(k)\}$$
$$= A_{ii}\hat{x}_i(k) + v_i^*(k) + B_iu_i^*(k)$$
$$+ G_i(k+1)[y_i(k+1) - C_i(A_{ii}\hat{x}_i(k)$$
$$+ v_i^*(k) + B_iu_i(k))] \quad (4.5.35)$$
$$x_i(o) = \bar{x}_i(o)$$

where

$$G_i(k+1) = \Sigma_i(k+1|k+1)C_i^T\Theta_i^{-1}(k+1) \quad (4.5.36)$$
$$\Sigma_i(k+1|k+1) = [C_i^T\Theta_i^{-1}(k+1)C_i + (A_{ii}\Sigma_i(k|k)$$
$$A_{ii}^T + E_i(k))^{-1}]^{-1}$$

with $\Sigma_i(o|o) = [\Sigma_i^{-1}(o) + C_i^T\Theta_i^{-1}(o)C_i]^{-1}$ (4.5.37)

The estimate $\bar{x}_i(k+1)$ can be computed using

$$\bar{x}_i(k+) \triangleq E\{x_i(k+1)\} = -K_i^{-1}(k+1)[r_i(k+1) + \tfrac{1}{2}$$
$$p_i(k)] \quad (4.5.38)$$

where $v_i^*(k) + r_i(k)$ are given by

$$v_i^*(o) = -A_{ii}\bar{x}_i(o) - K_i^{-1}(1)r_i(1)$$
$$- \tfrac{1}{2} S_i^{-1}(1)p_i(o) \quad (4.5.39)$$
$$v_i^*(k) = A_{ii}K_i^{-1}(k)[r_i(k) + \tfrac{1}{2} p_i(k-1)]$$
$$- K_i^{-1}(k+1)r_i(k+1)r_i(k+1) - \tfrac{1}{2} S_i^{-1}(k+1)p_i(k)$$
$$k=1,2,\dots t_{f-1} \quad (4.5.40)$$
$$r_i(o) = -\tfrac{1}{2}\tilde{p}_i(o) - \tfrac{1}{2}A_{ii}^Tp_i(o) - A_{ii}^TS_i(1)$$
$$A_{ii}S_i(1)A_{ii}\bar{x}_i(o) \quad (4.5.41)$$
$$Q_iK_i^{-1}(k)r_i(k) = -\tfrac{1}{2}\tilde{p}_i(k) - \tfrac{1}{2}A_{ii}^Tp_i(k)$$
$$+ \tfrac{1}{2}A_{ii}^TS_i(k+1)A_{ii}K_i^{-1}(k)p_i(k+1)$$
$$k=1,2,\dots t_{f-1} \quad (4.5.42)$$
$$r_i(t_f) = 0 \quad (4.5.43)$$

Computational considerations

Here all the gain matrices can be computed off-line, along with $r_i(k)$ and $v_i^*(k)$ which depends on $p(k)$. $K_i(k)$ is the solution of the independent subsystem Riccati equation while $D_i(k)$ is the optimal gain matrix for each of the local deterministic optimal control problems.

$\bar{x}_i(k)$ is the unconditional mean of $x_i(k)$ given only his a priori information and it can be computed off-line using $r_i(k), p_i(k)$ and $\tilde{p}_i(k)$.

$\hat{x}(k)$ can be generated using the $v_i^*(k)$ calculated off-line and the on-line measurements $u_i^*(k)$ and $y_i(k)$.

The solution of the higher level problem involves solving the following deterministic two point boundary value problem:

$$\begin{aligned} \bar{x}(k+1) &= A\bar{x}(k) - \tfrac{1}{2} BR^{-1}B^T\lambda(k+1) \\ \lambda(k) &= A^T\lambda(k+1) + 2Q\bar{x}(k) \end{aligned} \tag{4.5.44}$$

where $x(o)$ is given and $\lambda(t_f) = 2Fx(t_f)$.

We could either write

$$p^*(k) = -\lambda(k+1) \qquad k=0,2,\ldots t_{f-1} \tag{4.5.45}$$

or

$$p^*(k) = -2K(k+1)\bar{x}(k+1) \tag{4.5.46}$$

where

$$\bar{x}(k+1) = (A - BT^{-1}(k+1)B^TK(k+1)A\bar{x}(k)$$

with given $\bar{x}(o)$ (4.5.47)

and

$$K(k) = Q + A^TK(k+1)\ [I - BT^{-1}(k+1)B^TK(k+1)]A \tag{4.5.48}$$

with $K(t_f) = F$

and $T(k+1) = R + B^TK(k+1)B.$ (4.5.49)

As Chong and Athans [12] pointed out, it is easy to show that if the optimal p^* are used then $\bar{x}_i(k)$, the conditional mean of $x_i(k)$ generated by the ith controller using only its local information and $\{\bar{p}_i, \tilde{p}_i\}$ is the same as that generated by the coordinator.

The control $u_i^*(k)$ which is actually applied by the ith controller has two parts i.e. a closed loop part which depends on the measurements and an open loop part which does not. The open loop part depends only on the coordination p received from the higher level and it is the optimal deterministic control for the whole system assuming no measurements were made. The closed loop part depends on the difference between the a priori and a posteriori estimates of the ith controller about the state of the ith subsystem.

Having considered this special case of no information received by the coordinator, let us move on to see how the control can be improved by providing information to the coordinator.

4.5.2 Open loop periodic coordination

At each coordination instant $k=0,2,2\ell,\ldots T-\ell$

(1) The statistics of the state of the system x(k) are generated from the available data.

(2) If we assume that the coordinator will make no further measurements in the future then the coordinator's problem is the same as P1 except that the system starts from k so that the a priori statistics on x(o) are replaced by the conditional density of x(k) given $Y_o(k)$.

(3) The coordinating parameters $p^*(k),\ldots p^*(t_{f-1})$ can be found and sent to the lower level where γ_i^t, η_i^t, $t=k,\ldots T-1$, $i=1,\ldots N$ can be computed independently by the lower level units. These lower level strategies are applied until $t=k+1-1$ when a new set of data $Y_o(k+\ell)$ becomes available to the coordinator and the whole process is repeated.

Thus the lower level control strategy is given by:

At any instant k, let $n\ell$ be the last updating instant i.e.

$$n\ell \leq k \leq n\ell + \ell \tag{4.5.50}$$

Then

$$u_i^*(k) = - D_i(k+1)(\hat{x}_i(k|k) - \bar{x}_i(k|n\ell)) - E_i(k+1)p_i^*(k,n\ell) \tag{4.5.51}$$

where the gain matrices are the same as those given previously.

$$\hat{x}_i(k|k) \triangleq E\{x_i(k)|Y_i(k),U_i(k-1); I_i,p_i^*(n\ell),\ell)\bar{x}_i(n\ell|n\ell),\Sigma_{ii}(n\ell|n\ell)\} \tag{4.5.52}$$

$$\bar{x}_i(k|n\ell) \triangleq E\{x_i(k)|I_i,p_i^*(n\ell),\tilde{p}_i^*(n\ell), \bar{x}_i(n\ell|n\ell)\} \tag{4.5.53}$$

where I_i is the decentralised a priori information that the ith controller has about his subsystem. $\bar{x}_i$ is the estimate generated by the ith controller based on the coordinating parameters $p^*(n\ell)$ and $\bar{x}_i(n\ell|n\ell)$. $x_i(k|k)$ is the decentralised estimate generated by the ith controller using $\bar{p}^*(n\ell)$, the statistics $\bar{x}_i(n\ell|n\ell)$ and $\Sigma_{ii}(n\ell|n\ell)$ as well as the data $Y_i(k),U_i(k-1)$.

$\hat{x}_i(k|k)$ and $\bar{x}_i(k|n\ell)$ can be computed using:

$$\hat{x}_i(k+1) = A_{ii}x_i(k|k) + v_i^*(k|n\ell) + B_iu_i(k) + G_i(k+1|n\ell)[y_i(k+1) - C_i(A_{ii}x_i(k|k) + v_i^*(k|n\ell) + B_iu_i(k))] \tag{4.5.54}$$

$$\hat{x}_i(n\ell|n\ell) = \bar{x}_i(n\ell|n\ell)$$

$$G_i(k+1|n\ell) = \Sigma_i(k+1;n\ )C_i'\Theta_i^{-1}(k+1) \tag{4.5.55}$$

$$\Sigma_i(k+1|k+1;n\ell) = [C_i'\Theta_i^{-1}(k+1)C_i + (A_{ii}\Sigma_i(k|k;n\ell)A_{ii}' + E_i(k)^{-1})]^{-1} \tag{4.5.56}$$

$$\Sigma_i(n\ell|n\ell;n\ell) = \Sigma_{ii}(n\ell|n\ell)$$

$$\bar{x}_i(k+1|n\ell) = -K_i^{-1}(k+1)[r_i(k+1)|n\ell) + \tfrac{1}{2}p_i^*(k;n\ell)] \tag{4.5.57}$$

$v_i^*(k|n\ell)$ and $r_i(k|n\ell)$ are given by

$$v_i^*(n\ell|n\ell) = -A_{ii}\bar{x}_i(n\ell|n\ell) - K_i^{-1}(n\ell+1)r_i(n\ell+1|n\ell) - \tfrac{1}{2}S_i^{-1}(n\ell+1)P_i^*(n\ell;n\ell) \tag{4.5.58}$$

$$v_i^*(k|n\ell) = A_{ii}K_i^{-1}(k)[r_i(k|n\ell) + \tfrac{1}{2}p_i^*(k-1;n\ell) - K_i^{-1}(k+1)r_i(k+1|n\ell)] - \tfrac{1}{2}S_i^{-1}(k+1)p_i^*(k;n\ell) \tag{4.5.59}$$

$$n\ell < k < n\ell + \ell$$

$$r_i(n\ell|n\ell) = -\tfrac{1}{2}\tilde{p}_i^*(n\ell;n\ell) - \tfrac{1}{2}A_{ii}'p_i^*(n\ell;n\ell) - A_{ii}'S_i(n\ell+1) A_{ii}\tilde{x}_i(n\ell|n\ell) \tag{4.5.60}$$

$$Q_iK_i^{-1}(k)r_i(k|n\ell) = -\tfrac{1}{2}4\tilde{p}_i^*(k;n\ell) - \tfrac{1}{2}A_{ii}'p_i^*(k;n\ell) + \tfrac{1}{2}A_{ii}'S_i(k+1)A_{ii}K_i^{-1}(k)p_i^*(k+1;n\ell) \tag{4.5.61}$$

$$n\ell < k \leq n\ell + \ell$$

$$r_i(t_f/n\ell) = 0 \tag{4.5.62}$$

<u>Higher level</u>

The state estimate for the coordinator is generated using:

$$\bar{x}(k+1|k+1) = A\bar{x}(k|k) + Bu(k) + G(k+1)[y(k+1) - C(A\bar{x}(k|k) + Bu(k))];\bar{x}(o|o) = \bar{x}(o) \tag{4.5.63}$$

where $$G(k+1) = \Sigma(k+1|k+1)C^T\ \Theta^{-1}(k+1) \tag{4.5.64}$$

$$\Sigma(k+1|k+1) = [C^T\Theta^{-1}(k+1)C + A\ \Sigma(k|k)A^T + F(k))^{-1}]^{-1}$$

$$\Sigma(o|o) = \Sigma(o) - \Sigma(o)C^T(C\Sigma(o)C^T + \Theta(o))^{-1}C\Sigma(o) \tag{4.5.6 }$$

and the coordinating parameters $p^*(k;n\ell)$ are given by

$$p^*(k;n\ell) = -\ \lambda^*(k+1;n\ell) = 2K(k+1)\bar{x}(k+1|n\ell) \tag{4.5.6 }$$

where $$\bar{x}(k+1|n\ell) = (I - BT^{-1}(k+1)B^TK(k+1)A\bar{x}(k|n\ell)$$

The procedure therefore involves the coordinator evaluating a state estimate $\bar{x}(n\ell|n\ell)$ and new covariance $\Sigma(n\ell|n\ell)$ using the measurements and thus computing $p^*(k;n\ell)$. Then $x_i(n\ell|n\ell)$, $\Sigma_{ii}(n_{i}|n\ell)$, $p_i^*(k;n\ell)$ and $\bar{p}_i^*(k;n\ell)$, $k = n\ell,\ldots,n\ell+\ \ell-1$ are sent to the ith lower level unit which generates the new a priori estimates $\bar{x}_i(k|n\ell)$ as well as $\hat{x}_i(k|k)$.

It is easy to see that if the frequency of intervention of the coordinations becomes the same as the period k, the coordinator takes control of the whole system and we have the classical centralised solution.

The simplicity of the case that we have treated arose from the fact that we assumed that the coordinator will not get future information. If we assume future information, then we end up with a series of team decision problems. These are treated in detail by Chong and Athans [12].

Next, we examine decentralised state reconstruction since this is important when designing decentralised controllers.

4.6 DECENTRALISED STATE RECONSTRUCTION [15]

Consider a large-scale discrete system S that could be described as an interconnection of N discrete subsystems $S_1, S_2, \ldots, S_N$ by

$$x_i(k+1) = A_ix_i(k) + B_iu_i(k) + h_i(k,x)$$

$$y_i(k) = C_ix_i(k) \quad ; \quad i=1,\ldots,N \tag{4.6.1}$$

where

$x_i(k): R \to R^{n_i}$ is the state vector of the ith subsystem S_i,

$u_i(k): R \to R^{m_i}$ is the input vector controlling the ith subsystem S_i,

$y_i(k): R \to R^{p_i}$ is its output vector,

A_i, B_i, C_i, are constant matrices of appropriate dimensions,

$x(k): R \to R^n$ is the integrated state vector

$[x_1^T x_2^T \ldots x_N^T]^T$, and

$h_i(k,x): R \times R^{n_i}$ is a vector function that characterises

the couplings between S_i and the remaining subsystems.

It will be assumed that the N-pairs $\{A_i C_i\}$ are completely observable. The large-scale discrete system S may be correspondingly described by the dynamic equations

$$x(k+1) = Ax(k) + Bu(k) + h(t,x)$$

$$y(k) = Cx(k) \qquad (4.6.2)$$

where $u(k): R \to R^m$, $y(k): R \to R^p$, and $h(k,x): R \times R^n \to R^n$ are the composite vectors $u(k) = [u_1^T u_2^T(k) \ldots u_N^T(k)]^T$, $y(k) = [y_1^T(k) y_2^T(k) \ldots y_N^T(k)]^T$ and $h(k,x) = [h_1^T(k,x) h_2^T(k,x) \ldots h_N^T(k,x)]^T$, respectively. Also, $A = \text{diag}\ [A_1 A_2 \ldots A_N]$, $B = \text{diag}\ [B_1 B_2 \ldots B_N]$ and $C = \text{diag}\ [C_1 C_2 \ldots C_N]$. The vector function $h(k,x)$ represents the totality of interconnections among the subsystems.

Instead of building up an integrated state reconstructor, we will consider the design problem of N-independent state reconstructors based on the subsystem dynamics. This obviously has the merit of distributing the computational efforts, and reducing the associated wirings in the implementation.

We will design full-order state reconstruction schemes for interconnected discrete-time systems in two basic steps. First, we will consider the subsystems of the integrated system to be decoupled, by suppressing the interaction terms $h(k,x)$, and we will build a state reconstructor for each of them. This gives a set of independent state-reconstructors of lower dimension. Second, we will provide conditions under which the set of local state-reconstructors with appropriate modifications will perform satisfactorily for the original system.

Under the decoupling condition, each subsystem is a typical discrete-time control system whose dynamics are represented by:

$$x_i(k+1) = A_i x_i(k) + B_i u_i(k)$$

$$y_i(k) = C_i x_i(k) \quad ; \qquad i=1,\ldots,N \qquad (4.6.3)$$

Depending on the observation records available for reconstructing the state $x_i(k)$, two different problems could arise:

Problem (a)

The observation records consists of the measurements $\{y_i(k-1), y_i(k-2)\ldots y_i(0)\}$; $i=1,\ldots,N$.

Problem (b)

The observation records consist of the measurements $\{y_i(k), y_i(k-1),\ldots, y_i(0)\}$; $i=1,\ldots,N$.

Whereas problem (a) entails a one-step delay in the measurement pattern on the subsystem level, problem (b) considers the situation whereby the subsystem has an instantaneous measurement.

Solution of Problem (a)

Since the pair $\{A_i, C_i\}$ are completely observable, it is fairly straightforward to design a state-reconstructor of the form:

$$\hat{x}_i(k+1) = A_i\hat{x}_i(k) + B_iu_i(k) + K_i\{y_i(k) - C_i\hat{x}_i(k)\}$$
$$i=1,\ldots,N \tag{4.6.4}$$

where $\hat{x}_i(k): R \to R^{n_i}$ is the estimate of $x_i(k)$ and $K_i \in R^{n_i x p_i}$ are the design parameters that may be suitably selected to ensure any desired degree of convergence of the state reconstruction scheme.

Defining the estimation error $\hat{e}_i(k) - \hat{x}_i(k)$, its dynamics can be obtained from (4.6.3) and (4.6.4) as:

$$\hat{e}_i(k+1) = [A_i - K_iC_i]\hat{e}_i(k) \tag{4.6.5}$$

It is well known that $\hat{e}_i \to 0$ as $k \to \infty$ provided that the eigenvalues of the matrix $[A_i - K_iC_i]$ are located within the unit circle of the complex plane.

Theorem 4.6.1

If the interaction pattern is a linear function of the state variables, $h_i(k,x) = \sum_{j=1}^{N} H_{ij}x_j(k)$ with the matrix $H = \{H_{ij}\}$ satisfying the rank qualification:

$$\text{rank}\begin{pmatrix} C \\ H \end{pmatrix} = \text{rank}\begin{pmatrix} C \end{pmatrix} = p = \sum_{i=1}^{N} p_i \tag{4.6.6}$$

Then the addition of the quantity $\sum_{j=1}^{N} M_{ij}y_j(k)$ to each local state reconstructor (4.6.4) with $M = \{M_{ij}\} = HC^T[CC^T]^{-1}$ ensures the asymptotic convergence of the integrated state reconstruction scheme.

Proof:

Summing up the state reconstruction scheme (4.6.4) over $i=1,\dots,N$ yields:

$$\hat{x}(k+1) = A - KC\,\hat{x}(k) + Bu(k) + Ky(k) \qquad (4.6.7)$$

where $\hat{x}(k) = [\hat{x}_1^T(k)\hat{x}_2^T(k)\dots\hat{x}_N^T(k)]^T$ and $K = \text{diag}\ K_1K_2\dots K_N$ has an arbitrary degree of convergence. The validity of equation (4.6.6) implies that

$$H = MC \qquad (4.6.8)$$

or equivalently $M = HC^T[CC^T]^{-1}$, and due to the observability condition we arrive at

$$\text{rank}\ \{A,C\} = \text{rank}\{(A + H),C\} \qquad (4.6.9)$$

Moreover, because A and C are block-diagonal matrices and the pairs $\{A_iC_i\}$ are completely observable, then $\{A,C\}$ has full rank. Consequently, $\{(A + H),C\}$ is completely observable. In this way the state of the integrated system

$$x(k+1) = [A + H]x(k) + Bu(k) \qquad (4.6.10)$$

can be reconstructed by

$$\hat{x}(k+1) = [A + H - WC]\hat{x}(k) + Bu(k) + Wy(k) \qquad (4.6.11)$$

The choice of $W = K + M$ would eventually yield a similar eigenvalue distribution as $[A + H - WC]$ and $[A - KC]$, hence ensuring the prescribed convergence for the reconstruction scheme (4.6.11). However, by the same choice of W, the scheme (4.6.11) becomes:

$$\hat{x}(k+1) = [A - KC]\hat{x}(k) + Bu(k) + Ky(k) + My(k) \qquad (4.6.12)$$

which is obviously the scheme (4.6.7) plus the modification terms

$My(k) = \sum_{j=1}^{N} M_{ij}y_j(k)$ and the proof is completed.

In Figure 4.3 the wiring diagram of this observation scheme is given.

4.6.1 Remarks

1) An interesting feature of this state reconstruction scheme is that all the computations are performed on the subsystem level which is of relatively small dimension. Not only does this scheme yield a substantial saving in the computational effort, but it also provides more flexibility in allocating the required eigenvalues.

2) Only the observability condition of the decoupled subsystems is needed for the present analysis. Indeed, this simplifies the task of control design by avoiding the need to check the overall system.

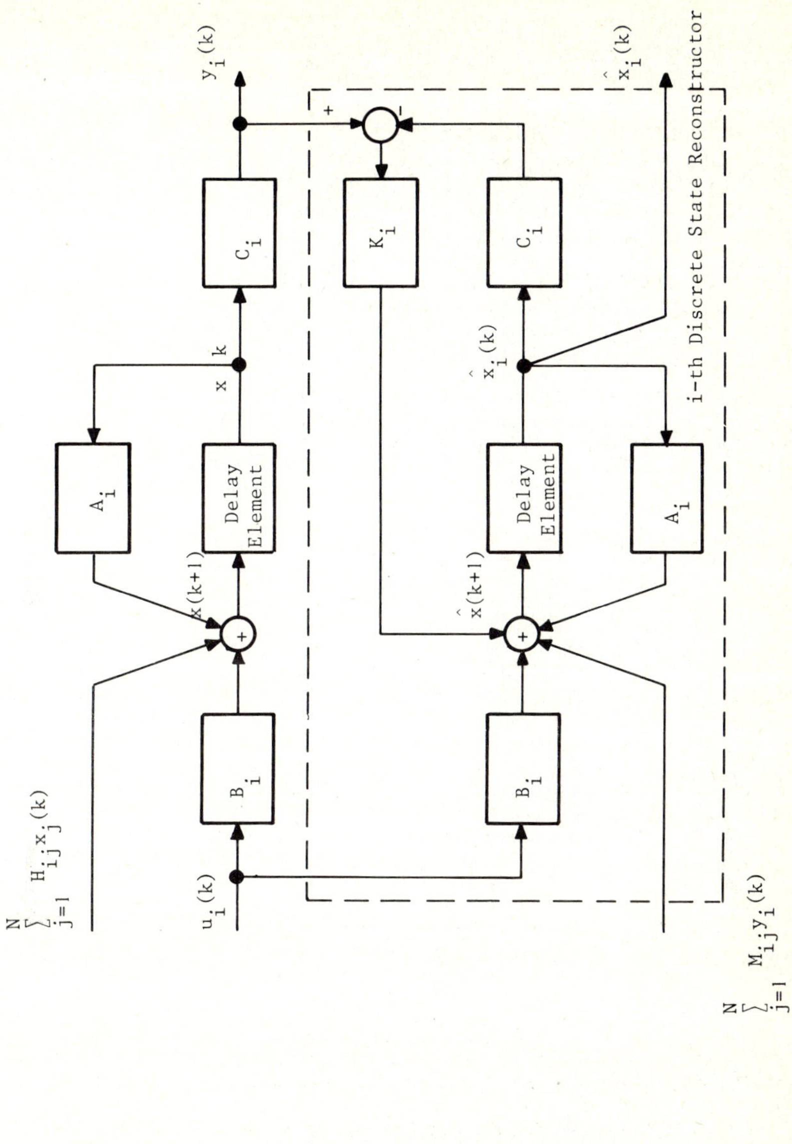

i-th Discrete State Reconstructor
Delay Element
Delay Element

Figure 4.3

3) The condition imposed on the interconnections is not restrictive, but it does identify a class of patterns that will guarantee the asymptotic behaviour of the state reconstruction scheme.

Solution of Problem (b)

In this case, the state reconstruction design is based on the present output measurement. The only modification to the scheme (4.6.4) is to replace the last term in (4.6.4) by the difference between the present output record $y_i(k+1)$ and its estimate $C_i[A_u\tilde{x}_i(k) + B_iu_i(k)]$. As a result, the corresponding state reconstruction scheme takes the form:

$$\tilde{x}_i(k+1) = A_i\tilde{x}_i(k) + B_iu_i(k)$$
$$+ F_i[y_i(k+1) - C_iA_i\tilde{x}_i(k) - C_iB_iu_i(k)]$$

rearranging

$$\tilde{x}_i(k+1) = [A_i - F_iC_iA_iA_i]\tilde{x}_i(k) + [B_i - F_iC_iB_i]u_i(k) + F_iy_i(k+1) \tag{4.6.13}$$

Comparing equations (4.6.3) and (4.6.13), it can be easily seen that the estimation error $\tilde{e}_i(k) = x_i(k) - \tilde{x}_i(k)$ evolves as

$$\tilde{e}_i(k+1) = [A_i - F_iC_iA_i]\tilde{e}_i(k) \tag{4.6.14}$$

where F_i is a matrix of design parameters. By similar arguments, the error $\tilde{e}_i(k)$ dies out as K increases indefinitely provided that the eigenvalues of the matrix $[A_i - F_iC_iA_i]$ are positioned within the unit circle on the complex plane. It should be noted that the possibility of arbitrarily locating these eigenvalues depends on the observability of the pair $\{A_i, C_iA_i\}$. If the matrix A_i is nonsingular, then the observability of the pair $\{A_i, C_i\}$ is equivalent to the observability of the pair $\{A_i, C_iA_i\}$. On the other hand, if A_i is singular and the pair $\{A_i, C_i\}$ is observable then it can be shown that the pair $\{A_i, C_iA_i\}$ is both detectable and reconstructable [3].

Lemma 1 [14]

The state reconstruction schemes (4.6.4) and (4.6.13) are identical under the condition

$$K_i = A_iF_i \tag{4.6.15}$$

Proof:

The link between the schemes (4.6.4) and (4.6.13) could be established by reconstructing the state $x_i(k+1)$, from the records $\{y_i(k), \ldots, y_i(o)\}$ in two steps. First, the state $x_i(k)$ is estimated using the scheme (4.6.13):

$$\tilde{x}_i(k) = [A_i - F_iC_iA_i]\tilde{x}_i(k-1) + [B_i - F_iC_iB_i]u_i(k-1) + F_iy_i(k) \tag{4.6.16}$$

Then the estimate $\tilde{x}_i(k+1)$ using the dynamic system (4.6.3) is derived.

$$\hat{x}_i(k+1) = A_i\tilde{x}_i(k) + B_iu_i(k) \tag{4.6.17}$$

Substituting (4.6.16) into (4.6.17) yields:

$$x_i(k+1) = [A_i^2 - A_iF_iC_iA_i]x_i(k-1) + [A_iB_i - A_iF_iC_iB_i]u_i(k+1)$$
$$+ A_iF_iy_i(k) + B_iu_i(k) \tag{4.6.18}$$

But

$$\hat{x}_i(k) = A_i\hat{x}_i(k-1) + B_iu_i(k-1) \tag{4.6.19}$$

Finally from (4.6.18) and (4.6.9)

$$\hat{x}_i(k+1) = [A_i - A_i - A_iF_iC_i]\hat{x}_i(k) + B_iu_i(k) + A_iF_iy_i(k) \tag{4.6.20}$$

Equation (4.6.20) is obviously identical to (4.6.4) if one chooses

$$A_iF_i = K_i$$

which proves the lemma.

Theorem 4.6.2

The modified state-reconstruction scheme

$$\tilde{x}_i(k+1) = [A_i - F_iC_iA_i]\tilde{x}_i(k) + [B_i - F_iC_iB_i]u_i(k) + F_iy_i(k+1)$$
$$+ \sum_{j=1}^{N} S_{ij}y_j(k)$$

ensures the asymptotic convergence of the integrated state vector x(t) if the condition

1) A_i is nonsingular

2) $\text{rank}\begin{bmatrix} C \\ H \end{bmatrix} = \text{rank}\,[C]$

are satisfied where $S = \{S_{ij}\} = HC^T[CC^T]^{-1}$ with $F = \text{diag}\,[F_1,F_2,\ldots F_N]$

Proof:

The theorem can be easily proved using the results of theorem 4.6.1 and lemma 4.6.1 and by noting that S replaces M in equation (4.6.8).

It should be noted that theorem 4.6.2 is based on the same rank qualification as for theorem 4.6.1. In addition, the nonsingularity of the subsystem matrices A_i is required. Although this further restricts the selection of the design parameters, we need to use less information from the output records.

4.6.2 Reduced-Order Design

Design of reduced-order state reconstruction schemes are required in several practical applications where only some of the state variables can be directly computed from the available measurement posts. The remaining state variables need to be reconstructed. For interconnected systems, the state reconstructor design on the subsystem level is facilitated by a suitable re-arrangement of the corresponding model variables. One way of achieving this is through the use of the following coordinate transformation.

$$x(k) = T \begin{bmatrix} w_i(k) \\ y_i(k) \end{bmatrix} \tag{4.6.21}$$

such that the output matrix is brought into the form

$$CT^{-1} = [0 \quad I_{p_i}] \tag{4.6.22}$$

Therefore, under the decoupling condition the ith subsystem dynamics becomes equivalent to:

$$\begin{bmatrix} w_i(k+1) \\ y_i(k+1) \end{bmatrix} = \begin{bmatrix} A_{1i} & A_{2i} \\ A_{3i} & A_{4i} \end{bmatrix} \begin{bmatrix} w_i(k) \\ y_i(k) \end{bmatrix} + \begin{bmatrix} B_{1i} \\ B_{2i} \end{bmatrix} u_i(k) \tag{4.6.23}$$

where $y_i \in R^{p_i}$ is the available output vector and $w_i(k) \in R^{n_i - p_i}$ is the unmeasurable part of the state vector.

It should be remarked that equation (4.6.21) for the interconnected system implies that the original model has been initially converted into one of the canonical forms.

Consider first the case in which the measurement pattern consists of $\{y_i(k), \ldots y_i(0)\}$ which is similar to problem (b) above. The objective is to design a state-reconstructor of order $(n_i - p_i)$. Notice that the available information of the output records together with the input pattern motivates writing the second part of equation (4.6.23) as:

$$A_{3i} w_i(k) = y_i(k+1) - A_{4i} y_i(k) - B_{2i} u_i(k) \tag{4.6.24}$$

Equivalently stated, the quantity $A_{3i} w_i(k)$ can be treated as the output record for the first part of equation (4.6.23).

$$w_i(k+1) = A_{1i} w_i(k) + A_{2i} y_i(k) + B_{1i} u_i(k) \tag{4.6.25}$$

Therefore the state-reconstruction design is similar to that for problem (b) i.e.

$$\hat{w}_i(k+1) = A_{1i} \hat{w}_i(k) + A_{2i} y_i(k) + B_{1i} u_i(k) + P_i [A_{3i} \hat{w}_i(k) - A_{3i} w_i(k)]$$

(4.6.24)

Combining equations (4.6.24) and (4.6.26) yields

$$\hat{w}_i(k+1) = [A_{1i} - P_iA_{3i}]\hat{w}_i(k) + [B_{1i} - P_iB_{2i}]u_i(k) + [A_{2i} - P_iA_{4i}]y_i(k) + P_iy_i(k+1) \tag{4.6.27}$$

Define $r_i(k) = \hat{w}_i(k) - P_iy_i(k)$, equation (4.6.27) then reduces to

$$r_i(k+1) = [A_{1i} - P_1A_{3i}]r_i(k) + [B_{1i} - P_iB_{2i}]u_i(k) + [A_{2i} - P_iA_{4i} + A_{1i}P_i - P_iA_{3i}P_i]u_i(k) \tag{4.6.28}$$

as the observer equation and

$$\hat{w}_i(k) = r_i(k) + P_iy_i(k) \tag{4.6.29}$$

as its output equation. Finally, the reconstructed state vector takes the form

$$x_i(k) = T\begin{bmatrix} r_i(k) + P_iy_i(k) \\ y_i(k) \end{bmatrix} \tag{4.6.30}$$

where P_i is the matrix of design parameters. The estimation errors evolves as

$$\tilde{e}_i(k+1) = [A_{1i} - P_iA_{3i}]\tilde{e}_i(k) \tag{4.6.31}$$

The eigenvalues of the reduced-order state reconstruction scheme can be arbitrarily selected since the observability of $\{A_i, C_i\}$ implies the observability of $\{A_{1i}, A_{3i}\}$.

<u>Theorem 4.6.3.</u>

The modified reduced-order state-reconstructor

$$\hat{w}_i(k+1) = A_{1i} - P_iA_{3i}\hat{w}_i(k) + B_{1i} - P_iB_{2i}u_i(k) + A_{2i} - P_iA_{4i}y_i(k) + P_iy_i(k+1) + \sum_{j=1}^{N} G_{ij}y_i(k) \tag{4.6.32}$$

where $G = \{G_{ii}\} = HC^T\{CC^T\}^{-1}$, guarantees the asymptotic convergence of the state reconstruction of the interconnected system provided that

$$\text{rank}\begin{bmatrix} C \\ H \end{bmatrix} = \text{rank } [C]$$

Proof:

In terms of the state vector x(k), equation (4.6.23) can be rewritten as:

$$\begin{bmatrix} \hat{w}_i(k+1) \\ \hat{y}_i(k+1) \end{bmatrix} = \begin{bmatrix} A_{1i} - P_iA_{3i} & A_{2i} - P_iA_{4i} \\ 0 & 0 \end{bmatrix} \begin{bmatrix} \hat{w}_i(k) \\ \hat{y}_i(k) \end{bmatrix}$$

$$+ \begin{bmatrix} B_{1i} - P_iB_{2i} \\ 0 \end{bmatrix} u_i(k) + \begin{bmatrix} P_i \\ I_{p_i} \end{bmatrix} y_i(k+1) \qquad (4.6.33)$$

Define $F_i = \begin{bmatrix} P_i \\ I_{p_i} \end{bmatrix}$ and using equation (4.6.32); equation (4.6.34) can be manipulated to read:

$$\hat{x}_i(k+1) = [A_i - F_iC_iA_i]\hat{x}_i(k) + [B_i - F_iC_iB_i]u_i(k) \; A \; F_iy_i(k+1) \quad (4.6.34)$$

Now, the state-reconstruction scheme (4.6.34) is identical to that of (4.6.13). In the light of theorems 4.6.1 and 4.6.2, the remaining procedure of the proof is straightforward. It should be noted that the implementation of the schemes developed here requires essentially a pole-placement algorithm and some matrix operations. The schemes have been simulated for a number of examples. In what follows, we provide some of these simulation results.

4.6.3 Examples

To demonstrate the application of the design procedure for state reconstruction, two system examples are considered and the results obtained are given below.

Example 1:

The interconnected system is described by

$$A = \left[\begin{array}{cc|cc|cc} 0.7 & 0.2 & -6 & 3 & 0 & 0 \\ 0.1 & 0.8 & 0 & 0 & -3 & 3 \\ \hline 0 & 0 & 0.5 & 0.15 & 2 & -2 \\ -2 & -2 & 0 & 0.4 & 0 & 0 \\ \hline 1 & 1 & 0 & 0 & 0.6 & 0 \\ 0 & 0 & 4 & -2 & 0.3 & 0.1 \end{array}\right]$$

Because the rank qualification is met, the additional term for the local state reconstructors is:

$$M = \begin{bmatrix} 0.0 & -3.0 & 0.0 \\ 0.0 & 0.0 & 3.0 \\ 0.0 & 0.0 & -2.0 \\ -2.0 & 0.0 & 0.0 \\ 1.0 & 0.0 & 0.0 \\ 0.0 & 2.0 & 0.0 \end{bmatrix}$$

If the measurement records were instantaenous, and because of the non-singularity of subsystem matrices, problem (b) could be solved to give

$$F_1 = \begin{bmatrix} 0.749 \\ -0.094 \end{bmatrix} ; \quad F_2 = \begin{bmatrix} 0.05 \\ 0.0 \end{bmatrix}$$

$$F_3 = \begin{bmatrix} 0.183 \\ -0.55 \end{bmatrix}$$

Example 2:

The interconnected system is of order 6 and consists of three subsystems as follows

$$A_1 = \begin{bmatrix} 0.7 & 0.2 \\ .1 & .8 \end{bmatrix} ; \quad B_1 = \begin{bmatrix} 0 \\ 1 \end{bmatrix}$$

$$C_1 = \begin{bmatrix} 1 & 0 \\ 0 & 1 \end{bmatrix}$$

$$A_2 = \begin{bmatrix} 0.5 & 0.15 \\ 0 & 0.4 \end{bmatrix} ; \quad B_2 = \begin{bmatrix} 1 \\ 0 \end{bmatrix}$$

$$C_2 = \begin{bmatrix} 2 & 0 \\ 0 & -1 \end{bmatrix}$$

$$B = \left(\begin{array}{c} 0 \\ 1 \\ \hline 1 \\ 0 \\ \hline 1 \end{array}\right)$$

$$C = \left(\begin{array}{cc|cc|cc} 1 & 1 & 0 & 0 & 0 & 0 \\ \hline 0 & 0 & 2 & -1 & 0 & 0 \\ \hline 0 & 0 & 0 & 0 & -1 & 0 \end{array}\right)$$

Thus the system comprises 3-2nd order subsystems. This means that we have three state-reconstructor design problems, as given below.

$$A_1 = \begin{pmatrix} 0.7 & 0.2 \\ 0.1 & 0.8 \end{pmatrix} \quad ; \quad B_1 = \begin{pmatrix} 0 \\ 1 \end{pmatrix}$$

$$C_1 = [1 \quad 1]$$

$$A_2 = \begin{pmatrix} 0.5 & 0.15 \\ 0 & 0.4 \end{pmatrix} \quad ; \quad B_2 = \begin{pmatrix} 1 \\ 0 \end{pmatrix}$$

$$C_2 = [2 \quad -1]$$

$$A_3 = \begin{pmatrix} 0.6 & 0 \\ 0.3 & 0.1 \end{pmatrix} \quad ; \quad B_3 = \begin{pmatrix} 1 \\ 1 \end{pmatrix}$$

$$C_3 = [-1 \quad 1]$$

Observe that A_1, A_2, A_3 are nonsingular, rank $(C_i) = 1$; $i=1,\ldots,3$ and $\{A_1,C_1\}$, $\{A_2,C_2\}$, $\{A_3,C_3\}$ are observable. In case of one-step-delay measurement records, three eigenvalues could be assigned. They are $z_1 = 0.25$, $z_2 = 0.45$, $z_3 = 0.65$. The result of computer simulation gives

$$K_1 = \begin{pmatrix} 0.506 \\ 0.0 \end{pmatrix} \quad ; \quad K_2 = \begin{pmatrix} .025 \\ 0.0 \end{pmatrix} \quad ; \quad K_3 = \begin{pmatrix} 0.11 \\ 0.0 \end{pmatrix}$$

$$A_3 = \begin{bmatrix} 0.6 & 0 \\ 0.3 & 0.1 \end{bmatrix}; \quad B_3 = \begin{bmatrix} 1 \\ 1 \end{bmatrix}$$

$$C_3 = \begin{bmatrix} -1 & 0 \\ 0 & 1 \end{bmatrix}$$

along with the interconnection matrix

$$H = \begin{bmatrix} 0 & 0 & -6 & 0 & 0 & 0 \\ 0 & 0 & 0 & 0 & 0 & 3 \\ 0 & 0 & 0 & 0 & 0 & 0 \\ 0 & -2 & 0 & 0 & 2 & 0 \\ 1 & 0 & 0 & 0 & 0 & 0 \\ 0 & 0 & 0 & -2 & 0 & 0 \end{bmatrix}$$

Observe that rank (C_1) = rank (C_2) = rank (C_3) = 2 and the pairs $\{A_1, C_1\}$, $\{A_2, C_2\}$ and $\{A_3, C_3\}$ are observable. Also rank $\begin{bmatrix} C \\ H \end{bmatrix} = 2$. The computer simulation results for problem (a), (one-step-delay) is

$$K_1 = \begin{bmatrix} 0.583 & 0.083 \\ 0.217 & 1.517 \end{bmatrix} \quad \text{at } z_{11} = .1, \; z_{12} = -.7$$

$$K_2 = \begin{bmatrix} 0.1 & 0.3 \\ 0 & 0 \end{bmatrix} \quad \text{at } z_{21} = .3, \; z_{22} = .4$$

$$K_3 = \begin{bmatrix} -0.8 & 0 \\ -0.7 & 0.7 \end{bmatrix} \quad \text{at } z_{31} = -.6, \; z_{32} = -.2$$

and the modification matrix M is

$$M = \begin{pmatrix} 0 & 0 & -3 & 0 & 0 & 0 \\ 0 & 0 & 0 & 0 & 0 & 3 \\ 0 & 0 & 0 & 0 & -2 & 0 \\ 0 & -2 & 0 & 0 & 0 & 0 \\ 1 & 0 & 0 & 0 & 0 & 0 \\ 0 & 0 & 0 & 2 & 0 & 0 \end{pmatrix}$$

If instantaneous measurement records are available (problem b), the additional F matrices are obtained to be:

$$F_1 = \begin{pmatrix} 0.784 & -0.438 \\ 0.173 & 1.951 \end{pmatrix}$$

$$F_2 = \begin{pmatrix} 0.2 & 0.6 \\ 0 & 0 \end{pmatrix}$$

$$F_3 = \begin{pmatrix} -1.33 & 0 \\ -3.0 & 7.0 \end{pmatrix}$$

In conclusion we have developed a decentralised design procedure which consists of two phases. In the first phase, a set of local state-reconstructors is designed to estimate the local state vectors on a decoupled basis. Then, an appropriate modification is introduced in the second phase in order to ensure that the modified local state-reconstructors provide an asymptotically convergent estimate of the overall state vector. The modification depends on the structural coupling matrix. Both full-order and reduced-order state-reconstruction schemes are considered with respect to the cases of instantaneous and one-step-delay output measurements.

Finally, in this chapter we consider the structural informational needs of decentralised stochastic control systems by explicitly measuring the informational needs of the subsystems using a cost function.

4.7 FEASIBLY DECENTRALISED CONTROL SYSTEMS *[16]*

The idea here is to add explicitly a term in the subsystem cost function to take into account the informational needs. For example, instead of the standard quadratic cost function

$$J = \sum_{o}^{\infty} (x^T Q x + \sum_{i=1}^{N} u_i^T R_i u_i) dt \tag{4.7.1}$$

which we have considered so far, let us modify the cost function to

$$J^1 = J + \sum_{i=1}^{N} m_1^2 \tag{4.7.2}$$

where the terms m_i, i=1,2,...N penalise the non-local information used in the local control stations. The measure m_i of the non-local information is chosen so as to include a set of weighting factor γ_{ij} j=1,...N, j≠i which penalises the exchange of information from the jth to the ith control station. By increasing any particular weighting factor, we can decrease the influence of the information exchanged. If we let all the γ_{ij} j=1,2,...N go to infinity, we have a completely decentralised control.

There are of course various ways of defining a measure of non-local information. One way is to split the control u_i of the ith station into two parts i.e. u_{ie} which uses the locally available information for agent i and u_{im} which needs information available at other control stations. Then we could use a weighted vector function norm of the non-local parts of the control u_i as the measure of non-local information i.e.

$$m_i = ||\gamma_{ij} u_{im}|| \tag{4.7.3}$$

Let us next apply this idea to the stochastic control problem.

The dynamic behaviour of the system under consideration is given by

$$\dot{x} = Ax + \sum_{i=1}^{N} B_i u_i + v \tag{4.7.4}$$

$$y_i = C_i x + w_i \tag{4.7.5}$$

$$i=1,2,...N$$

$$E\{v\} = 0, \; E\{w_i\} = 0, \; i=1,2,...N$$

$$E\{v(t)v^T(s)\} = \Sigma\delta(t-s) \tag{4.7.7}$$

$$E\{w_i(t)w_i^T(s)\} = \theta_i \, \delta(t-s),$$

$$i=1,2,...N \tag{4.7.8}$$

The initial state x_o of the plant is also assumed to be a zero-mean Gaussian random variable with finite variance i.e.

$$E\{x_o\} = 0, \; E\{x_o x_o^T\} = x_o \tag{4.7.9}$$

Also, we assume that

$$E\{vw_i^T\} = 0 \qquad i=1,...N \tag{4.7.10}$$

$$E\{vx_o^T\} = 0 \tag{4.7.11}$$

$$E\{w_i w_j^T\} = 0 \qquad i=j=1,\ldots N_j \; j\neq i \tag{4.7.12}$$

$$E\{x_o w_i^T\} = 0 \qquad i=1,\ldots N \tag{4.7.13}$$

Let us define an augmented system by concatenating the state vectors of the plant and of the local dynamic compensators as

$$\dot{x}^* = A^* x^* + \sum_{i=1}^{N} B_i^* u_i^* + v^* \tag{4.7.14}$$

$$y_i^* = C_i^* x^* + w_i^* \tag{4.7.15}$$

The linear dynamic compensator is described by a linear output feedback of the augmented system

$$u_i = D_i y \tag{4.7.16}$$

where y denotes the concatenation of all local measurement vectors y_i of the local control stations.

Now we know that the standard quadratic criterion given below is unbounded for all non-zero matrices R_i, $i=1,\ldots N$.

$$J = E\left[\int_o^\infty (x^T Q x + \sum_{i=1}^{N} u_i^T R_i u_i)dt\right] \tag{4.7.17}$$

To avoid this, let us weight the <u>deterministic</u> part of the control in the quadratic criterion.

The system will never reach a final state because of the disturbance. Hence the integral over an infinite time horizon will be unbounded. To avoid this, we use the "average" of the criterion i.e.

$$J = \lim_{T\to\infty} \frac{1}{T} E \int_o^T (x^T Q x + \sum_{i=1}^{N} u_{d_i}^T R_i u_{d_i})dt \tag{4.7.18}$$

where u_{d_i} the deterministic part of the control is given by

$$u_{d_i} = D_i C x \tag{4.7.19}$$

The local control u_i is given by

$$u_i = D_i y = D_i (Cx+w) \tag{4.7.20}$$

The final cost function which includes the non-local information effects is given by

$$J^1 = \lim_{T\to\infty} (\tfrac{1}{T})\, E \{\int (x^T Q x + \sum_{i=1}^{N} u_{d_i}^T R_i u_{d_i})dt + \lim_{T\to\infty} (\tfrac{1}{T}) \sum_{i=1}^{N} ||u_{d_{im}}||^2 \tag{4.7.21}$$

where $u_{d_{im}}$ is the non-local part of the deterministic part of the control. The matrix norm $||.||$ is defined as

$$||.||^2 = \mathrm{tr}\int_0^T (.^T.)dt \tag{4.7.22}$$

Thus the optimisation problem is to find the feedback matrices $D_{*1},\ldots D_{*N}$ such that

$$J^1(D_{*1},\ldots,D_{*N}) \leqslant J^T(D_1,\ldots D_N) \tag{4.7.23}$$

for all permissible feedback matrices $D_1,\ldots,D_N$.

The solution is given by the following theorem:

Theorem 4.7.1

The optimal solution of the above problem is given by

$$R_iD_iC{*}XC{*} + D_i\Gamma_iC{*}XC{*}\Gamma_i + B{*}_i^TPXC^T + B{*}_i^TPB{*}_i$$

$$D_iQ_i = 0$$

$$i=1,2,\ldots N$$

$$A_o^TP + PA_o + Q + C{*}^T\sum_{i=1}^{N} D_i^TR_iD_i + \Gamma_iD_i^T\Gamma_i)C{*} = 0$$

$$A_oX + XA_o^T + \Sigma + \sum_{i=1}^{N}(B_iD_iQ_iB_i^T) = 0$$

where $A_o = A{*} + \sum_{i=1}^{N} B{*}_iD_iC{*}$

and $X = \lim_{T\to\infty}\frac{1}{T}\int_0^T E(x{*}x{*}^T)dt$

For proof cf Senning [16]. The basis of Senning's development is to write an expression for the gradient of J^1 w.r.t. the feedback matrices D_i and to set this to zero. We will see a similar development for the determinis- tic case in Chapter 6.

4.7.1 A Power Systems example [16]

To illustrate the procedure let us consider a power systems example. The system is a multimachine power system comprising one thermal and 2 hydro plants with an infinite bus. Figure 4.4 shows the structure.

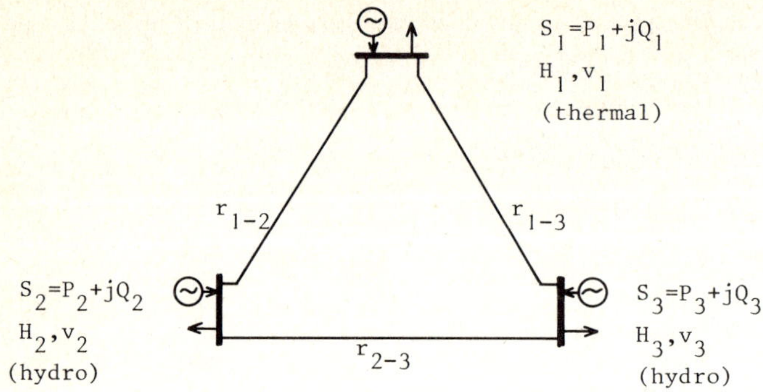

Figure 4.4

Each plant is modelled as a third order system, the states being the flux linkage, the torque angle and the angular velocity. The example considered by Senning assumes a fixed bus voltage.

The system matrices are given below:

$$A = \begin{pmatrix} A_{11} & A_{12} & A_{12} \\ A_{21} & A_{22} & A_{23} \\ A_{31} & A_{32} & A_{33} \end{pmatrix} \; ; \; B_1 = \begin{pmatrix} B_{11} \\ 0 \\ 0 \end{pmatrix} \; ; \; B = \begin{pmatrix} 0 \\ B_{22} \\ 0 \end{pmatrix}$$

$$B_3 = \begin{pmatrix} 0 \\ 0 \\ B_{33} \end{pmatrix}$$

$$C_1 = [C_{11} \quad 0 \quad 0] \; ; \; C_2 = [0 \quad C_{22} \quad 0] \; ;$$

$$C_3 = [0 \quad 0 \quad C_{33}]$$

The synchronous generators in the power system are described by the following matrices

$$A_{11} = \begin{pmatrix} -1.91 & -0.75 & -0.022 \\ 0.0 & 0.0 & 1.0 \\ -4.95 & -55.5 & -0.039 \end{pmatrix}$$

$$B_{11} = \begin{bmatrix} 12.98 \\ 0.0 \\ 0.0 \end{bmatrix}$$

$$C_{11} = [0.0 \quad 0.0 \quad 1.0]$$

$$A_{22} = \begin{bmatrix} -1.26 & 3.56 & -0.071 \\ 0.0 & 0.0 & 1.0 \\ -3.1 & -56.0 & -0.32 \end{bmatrix}$$

$$B_{22} = \begin{bmatrix} 43.83 \\ 0.0 \\ 0.0 \end{bmatrix} \quad ; \quad C_{22} = [0.0 \quad 0.0 \quad 1.0]$$

$$A_{33} = \begin{bmatrix} -2.92 & 2.3 & -0.12 \\ 0.0 & 0.0 & 1.0 \\ -3.4 & -21.0 & -0.017 \end{bmatrix}$$

$$B_{33} = \begin{bmatrix} 50.0 \\ 0.0 \\ 0.0 \end{bmatrix} \quad ; \quad C_{33} = [0.0 \quad 0.0 \quad 1.0]$$

The coupling between the first and the two other synchronous generators is described by

$$A_{12} = \begin{bmatrix} -0.032 & 0.31 & -0.002 \\ 0.0 & 0.0 & 0.0 \\ 0.22 & 8.17 & 0.004 \end{bmatrix}$$

$$A_{13} = \begin{bmatrix} -0.093 & 0.75 & -0.004 \\ 0.0 & 0.0 & 0.0 \\ 0.92 & 17.5 & 0.02 \end{bmatrix}$$

The network between the second synchronous generator and the rest of the system is given by:

$$A_{21} = \begin{pmatrix} -0.59 & -0.78 & -0.005 \\ 0.0 & 0.0 & 0.0 \\ -2.43 & 1.37 & 0.034 \end{pmatrix}$$

$$A_{23} = \begin{pmatrix} -0.49 & 1.28 & -0.02 \\ 0.0 & 0.0 & 0.0 \\ 0.12 & 29.8 & -0.028 \end{pmatrix}$$

Finally, the couplings between the third and the two other synchronous generators is given by:

$$A_{31} = \begin{pmatrix} -0.34 & -0.42 & -0.004 \\ 0.0 & 0.0 & 0.0 \\ -1.24 & 0.5 & -0.017 \end{pmatrix}$$

$$A_{32} = \begin{pmatrix} -0.094 & 0.305 & -0.006 \\ 0.0 & 0.0 & 0.0 \\ -0.07 & 6.37 & -0.011 \end{pmatrix}$$

Further we assume

$$\Sigma = \text{diag}\ (10.0)$$

$$Q_i = \text{diag}\ (5.0) \qquad i=1,2,3$$

The weighting matrices for the plant and for the dynamic compensators were chosen to be

$$Q_{plant} = \text{diag}\ (100.0)$$

$$Q_{comp} = \text{diag}\ (.0)$$

The weightings for the input to the plant and to the compensator were taken to be

$$R_i = \text{diag}\ (1.0)$$

$$i=1,2,3)$$

The non-local information was weighted equally for all the subsystems using a weighting factor of 100.

Results

The compensator for this case turns out to be completely decentralised and it is given by

$$u_1 = 0.387y_1 + 0.579z_1$$

$$\dot{z}_1 = -7.338z_1 - 6.262y_1$$

$$u_2 = 0.151y_2 + 0.226z_2$$

$$\dot{z}_2 = -7.857z_2 - 6.542y_2$$

$$u_3 = 0.118y_3 + 0.069z_3$$

$$\dot{z}_3 = -4.098z_3 - 6.111y_3$$

4.8 CONCLUDING REMARKS

In this chapter we have attempted to give the current "state of the art" of decentralised stochastic control. We started by examining the special case of one step delay sharing patterns where a kind of separation property between estimation and control exists. We subsequently considered this whole question of separation of estimation and control in the context of stochastic decentralised control. Next we moved on to the practical problem of controller design and here we limited ourselves to fixed structure linear controllers. We also considered periodic coordination schemes where some information can be transferred using a hierarchy. We subsequently considered suboptimal decentralised state reconstruction. Finally, we explored the idea of using a measure of information transfer in the cost function itself. In the next chapter we will examine the decentralised control problem from a completely different point of view i.e. using pole placement.

4.9 ACKNOWLEDGEMENTS

I have relied heavily on the pioneering papers of Kurtaran, Yoshikawa, Chong and Athans and the thesis of Senning in writing this chapter.

4.10 REFERENCES

Specific References

1. Kurtaran B. and Sivan R. "L-Q-G Control with one-step-delay sharing pattern", IEEE Trans AC, October 1974, pp. 571-574.

2. Witsenhausen H. "A counter example in stochastic optimal control", SIAM J. Contr. 6, pp. 131-147, 1968.

3. Kwakernaak H. and Sivan R. "Linear optimal control systems", John Wley, New York, 1972.

4. Astrom K. "Introduction to stochastic control theory", Academic Press, 1970.

5. Kailath T. "An innovations approach to least-squares estimation - Part 1: linear filtering with additive white noise", IEEE Trans AC 13, pp. 646-655, 1968.

6. Yoshikawa T. and Hirodki K. "Separation of estimation and control for decentralised stochastic control systems", Automatica, 14, pp. 623-628, 1978.

7. Streibel C. "Sufficient statistics in the optimum control of stochastic systems", J Math Analysis Applic, 12, pp. 576-592, 1965.

8. Aoki M. "On decentralised linear stochastic control problems with quadratic cost", IEEE Trans AC 18, pp. 243-258, 1973.

9. Chong C. and Athans M. "On the stochastic control of linear systems with different information sets", IEEE Trans AC 16, 5, pp. 423-430, 1971.

10. Athans M. "The matrix minimum principle", Inform. Contr, 11, 5 and 6, pp. 592-606, 1967.

11. Singh M.G. and Titli A. "Systems: decomposition, optimisation and control", Pergamon Press, Oxford, 1978.

12. Chong C. and Athans M. "On the periodic coordination of linear stochastic systems", Automatica 12, pp. 321-335, 1978.

13. Bismut J. "An example of interaction between information and control: the transparency of a game", IEEE Trans AC 18, pp. 518-522, 1973.

14. Willems J. "Design of state observers for linear discrete-time systems", Int.J.Syst.Sci., 11, pp. 139-147, 1980.

15. Mahmoud M.S. and Singh M.G. "Decentralised state reconstruction of interconnected discrete systems", CSC Report No. 488, UMIST, Manchester, 1980. To appear in LSS, 2, 1981.

16. Senning M.F. "Feasibly decentralised control", Doctor of Technical Science thesis, ETH, Zurich, 1979.

General references

1. Kurtaran B. "Dynamic two person two objective control problems with delayed sharing information pattern", IEEE Trans AC 22, 4, pp. 659-661, 1977.

2. Ho Y.C. and Chu K. "Information structure in dynamic multi-person control problems", Automatica 10, pp. 341-351, 1974.

3. Varaiya P. and Walrand J. "On delayed sharing patterns", IEEE Trans AC 23, 3, pp. 443-445, 1978.

4. Yoshikawa T. "Dynamic programming approach to decentralised stochastic control problems", IEEE Trans AC, pp. 796-797, December, 1975.

5. Sandell Nl and Athans M. "Solution of some non-classical LQG stochastic decision problems", IEEE Trans AC 19, 2, pp. 108-116, 1974.

6. Kobayashi H., Hanatusa H. and Yoshikawa T. "Controllability under decentralised information structure", IEEE Trans AC 23, 2, pp. 182-188, 1978.

7. Witsenhausen H. "Separation of estimation and control for discrete time systems", Proc. IEEE 9, pp. 1557-1566, 1971.

CHAPTER 5

DECENTRALISED STABILISATION AND POLE PLACEMENT

5.1 INTRODUCTION

In the previous chapter we studied the decentralised control problem from the point of view of stochastic control. In this chapter, we concentrate on stability issues.

A fundamental result in control theory which is due to Wonham [1] is that the poles of a closed-loop system may be placed arbitrarily using state feedback provided the system is controllable. Thus controllable systems can be stabilised using a centralised controller which places the poles in the left half plane. However, all the states of a system may not be measurable. Brasch and Pearson [2] have extended Wonham's [1] result to the case of output feedback. Their basic result is that arbitrary pole placement may be obtained by adding a compensator whose order does not exceed $(p_c-1,\ p_o-1)$ where p_c is the controllability index and p_o is the observability index of the plant. This result is obtained using a theorem which shows that given any controllable-observable linear time invariant system

$$\dot{x} = Ax + Bu \tag{5.1.1}$$

$$y = Cx \tag{5.1.2}$$

there exists a matrix K and a vector g such that the pair (A+BKC,Bg) is controllable.

In a basic sense, this result is very similar to the use of an observer for estimating the unmeasurable states so as to control the system using complete state feedback. Essentially, in the case of full state feedback, a <u>static</u> compensator is adequate to stabilise the system whilst output feedback requires <u>dynamic</u> compensation. At the present time no necessary and sufficient conditions exist for achieving stabilisation using <u>static</u> <u>output</u> feedback although in certain special cases it is clearly possible to do so.

Before we go on to considering decentralised stabilisation, let us examine some further fundamental notions in the case of centralised control.

5.1.1 Minimal Realisation

Let us consider once again the system given by equations (5.1.1) and (5.1.2) and let us denote this system by the triple R = (C,A,B). It is possible to isolate all the controllable and observable states from the representation

$R = (C,A,B)$.

Next, we state the well known minimal realisation theorem:

Theorem: $\tilde{R}$ is a minimal realisation if and only if $\tilde{R}$ is controllable and observable.

By a minimal realisation of a system we mean the minimal order of the state space to adequately describe the system. Thus given an impulse response matrix, we would seek a minimal realisation. Since it is possible to isolate all controllable and observable states, we could use the above theorem to obtain a minimal realisation.

5.1.2 Canonical structure theorem

Consider once again the system $R = (C,A,B)$ which is not completely controllable and observable. Then there exists a real non-singular matrix T such that

$$\tilde{A} \triangleq TAT^{-1} = \begin{bmatrix} \tilde{A}_{11} & 0 & \tilde{A}_{13} & 0 \\ \tilde{A}_{21} & \tilde{A}_{22} & \tilde{A}_{23} & \tilde{A}_{24} \\ 0 & 0 & \tilde{A}_{33} & 0 \\ 0 & 0 & \tilde{A}_{43} & \tilde{A}_{44} \end{bmatrix} \qquad (5.1.3)$$

$$\tilde{B} \triangleq TB = \begin{bmatrix} \tilde{B}_1 \\ \tilde{B}_2 \\ 0 \\ 0 \end{bmatrix} \qquad (5.1.4)$$

$$\tilde{C} \triangleq CT^{-1} = [\tilde{C}_1 \quad 0 \quad \tilde{C}_3 \quad 0] \qquad (5.1.5)$$

where $\left\{ \begin{bmatrix} \tilde{A}_{11} & 0 \\ \tilde{A}_{21} & A_{22} \end{bmatrix}, \begin{bmatrix} \tilde{B}_1 \\ \tilde{B}_2 \end{bmatrix} \right\}$ is a completely controllable pair and

$\left\{ [\tilde{C}_{11} \quad \tilde{C}_3], \begin{bmatrix} \tilde{A}_{11} & \tilde{A}_{13} \\ 0 & \tilde{A}_{33} \end{bmatrix} \right\}$ is a completely observable pair.

Due to the existence of such a T matrix, it can be shown that both the representations R and $\tilde{R}$ have the same impulse response matrix and are, thus, dynamically equivalent.

Next, let us write the dynamical representation in terms of its controllable and observable subsystems.

The subsystem

$$\tilde{R}_{12} = \left\{ [\tilde{C}_1 \quad 0] \;,\; \begin{bmatrix} \tilde{A}_{11} & 0 \\ \tilde{A}_{21} & A_{22} \end{bmatrix} ,\; \begin{bmatrix} \tilde{B}_1 \\ \tilde{B}_{22} \end{bmatrix} \right\} \tag{5.1.6}$$

is completely controllable and for this subsystem we have

$$\begin{aligned} \dot{x}_1(t) &= \tilde{A}_{11}\, x_1(t) + \tilde{B}_1\, u(t) \\ \dot{x}_2(t) &= \tilde{A}_{21}\, x_1(t) + \tilde{A}_{22}\, x_2(t) + B_2\, u(t) \end{aligned} \tag{5.1.7}$$

The subsystem representation

$$\tilde{R}_{24} = \left\{ [0_1 \quad 0] \;,\; \begin{bmatrix} \tilde{A}_{22} & \tilde{A}_{24} \\ 0 & \tilde{A}_{44} \end{bmatrix} ,\; \begin{bmatrix} \tilde{B}_2 \\ 0 \end{bmatrix} \right\}$$

is completely unobservable and here we have

$$\begin{aligned} \dot{x}_2(t) &= \tilde{A}_{22}\, x_2(t) + \tilde{A}_{24}\, x_4(t) + \tilde{B}_2\, u(t) \\ \dot{x}_4(t) &= \tilde{A}_{44}\, x_4(t) \\ y_{24} &= 0 \end{aligned}$$

The subsystem representation

$$\tilde{R}_4 = \{0, \quad \tilde{A}_{44}, \quad 0\}$$

is both uncontrollable and unobservable.

Thus the only subsystem that is both controllable and observable is subsystem 1.

Next, we will use these preliminary ideas to examine the question of stabilisation of decentralised control systems.

5.2 STABILISATION OF DECENTRALISED CONTROL SYSTEMS

In the case of decentralised control, as we have seen previously, certain constraints are imposed on the structure of the controller. Typically, instead of controlling the system using one centralised controller, the control is done at different control stations which may be distributed in space. Thus the system could be described by equations of the form:

$$\dot{x}(t) = A\, x(t) + \sum_{i=1}^{\nu} B_i\, u_i(t) \tag{5.2.1}$$

where $x(t) \in R^n$, $u_i(t) \in R^{n_i}$, and A, B_i are constant matrices.

Equation (5.2.1) describes a system with ν control stations.

One way of providing decentralised control in this case could be by static state feedback i.e. controls of the form

$$u_i = K_i x_i \tag{5.2.2}$$

could be used. Thus the overall gain matrix

$$K = \begin{bmatrix} K_1 & & \bigcirc \\ & K_2 & \\ \bigcirc & & K_\nu \end{bmatrix} \tag{5.2.3}$$

has a block diagonal structure.

Unfortunately, once we impose constraints on the overall gain matrix K, the nice stability properties of the centralised control no longer apply. Thus even if (A,B) is a controllable pair and each subsystem is controllable, it is not necessary for the system to be stabilisable by decentralised control. The following example of Wang [4] illustrates this. In this example, we consider a system with 3 local stations and the pairs (A,B), (A_{ii},B_{ii}) (i = 1,2,3) are all controllable. Yet the system can not be stabilised by local state feedback. The system is described by

$$\begin{bmatrix} \dot{x}_{11} \\ \dot{x}_{12} \\ \dot{x}_{21} \\ \dot{x}_{22} \\ \dot{x}_{31} \\ \dot{x}_{32} \end{bmatrix} = \left[\begin{array}{cc|cc|cc} 0 & 1 & \multicolumn{2}{c|}{\bigcirc} & \multicolumn{2}{c}{\bigcirc} \\ 0 & 0 & & & & \\ \hline 1 & 0 & 0 & 1 & 0 & 1 \\ 0 & 0 & 0 & 0 & 0 & 0 \\ \hline \multicolumn{2}{c|}{\bigcirc} & 0 & 1 & 0 & 1 \\ & & 0 & 0 & 0 & 0 \end{array}\right] \begin{bmatrix} x_{11} \\ x_{12} \\ x_{21} \\ x_{22} \\ x_{31} \\ x_{32} \end{bmatrix}$$

$$+ \left[\begin{array}{c|c|c} 0 & \multicolumn{2}{c}{\bigcirc} \\ 1 & & \\ \hline & 0 & \\ & 1 & \\ \hline \multicolumn{2}{c|}{\bigcirc} & 0 \\ & & 1 \end{array}\right] \begin{bmatrix} u_1 \\ u_2 \\ u_3 \end{bmatrix}$$

where u_i is the control input of station i (i = 1,2,3). The set of decentralised local state feedback is of the form

$$u_i = [k_{i1}, k_{i2}] \begin{bmatrix} x_{i1} \\ x_{i2} \end{bmatrix} \qquad (i = 1,2,3)$$

Applying this set of feedbacks, the overall system matrix becomes:

$$\left[\begin{array}{cc|cc|cc} 0 & 1 & & \bigcirc & & \bigcirc \\ k_{11} & k_{12} & & & & \\ \hline 1 & 0 & 0 & 1 & 0 & 1 \\ 0 & 0 & k_{21} & k_{22} & 0 & 0 \\ \hline 0 & 0 & 0 & 1 & 0 & 1 \\ 0 & 0 & 0 & 0 & k_{31} & k_{32} \end{array}\right]$$

It can be easily checked that the first, second, third and fifth rows of the above matrix are linearly dependent for arbitrary k_{ij}s. As we will see later in this chapter, this has arisen because the system has an unstable "fixed" mode at the origin and hence the closed loop system can not be asymptotically stable. This notion of fixed modes will be discussed in great detail in this chapter.

5.3 STABILISATION BY OUTPUT FEEDBACK

In practice, we usually seek to control systems using output feedback since state feedback often requires an excessive number of feedback links especially for large scale systems. In this section we will examine the conditions which will allow us to implement decentralised output feedback. It must be emphasised that at the present time necessary and sufficient conditions do not exist for <u>static</u> output feedback even for centralised systems and a search for such conditions in special cases is an active area of research. We would therefore not expect to see spectacular results for decentralised static output feedback. Here, we will concentrate on dynamic output feedback for decentralised control.

We begin our analysis by examining the fundamental results of Wang and Davison *[5]*.

Consider the multivariable linear time-invariant system with ν control stations described by:

$$\left.\begin{aligned} \dot{x}(t) &= Ax(t) + \sum_{i=1}^{\nu} B_i\, u_i(t) \\ y_i(t) &= C_i\, x(t) \qquad i = 1,2,\ldots\nu \end{aligned}\right] \tag{5.3.1}$$

where $\quad x(t) \in R^n,\ u_i(t) \in R^{m_i},\ y_i(t) \in R^{p_i}$

and $\quad A, B_i, C_i$ are real constant matrices.

The problem is to find ν local output feedback control laws with dynamic compensation in order to stabilise the whole system.

The set of local feedback laws are assumed to be generated by the following feedback controllers:

$$\left.\begin{aligned} \dot{Z}_i(t) &= S_i\, Z_i(t) + B_i\, y_i(t) \\ u_i(t) &= Q_i\, Z_i(t) + K_i\, y_i(t) + v_i(t) \\ & \qquad i = 1,2,\ldots\nu \end{aligned}\right] \tag{5.3.2}$$

where $Z_i(t) \in R^{n_{oi}}$, $v_i(t)$ which is an external input $\in R^{m_i}$.

We can rewrite equation (5.3.2) in a more compact form as:

$$\left.\begin{aligned} \dot{Z}(t) &= S\, Z(t) + R\, y(t) \\ u(t) &= Q\, Z(t) + K\, y(t) + v(t) \end{aligned}\right\} \tag{5.3.3}$$

where
$$S \triangleq \text{block diag } [S_1, S_2, \ldots S_\nu]$$
$$Q \triangleq \text{block diag } [Q_1, Q_2, \ldots Q_\nu]$$
$$R \triangleq \text{block diag } [R_1, R_2, \ldots R_\nu]$$
$$K \triangleq \text{block diag } [K_1, K_2, \ldots K_\nu]$$

If we apply the feedback law (5.3.3) to the system (5.3.1), then the closed loop system can be represented by

$$\begin{pmatrix} \dot{x}(t) \\ \dot{Z}(t) \end{pmatrix} = \left(\begin{array}{c|c} A + BKC & BQ \\ \hline RC & S \end{array}\right) \begin{pmatrix} x(t) \\ Z(t) \end{pmatrix} + \begin{pmatrix} B \\ 0 \end{pmatrix} v(t) \tag{5.3.4}$$

$$\text{where } C = \begin{pmatrix} C_1 \\ C_2 \\ \vdots \\ C_\nu \end{pmatrix} ; \quad B = [B_1, B_2, \ldots B_\nu] \tag{5.3.4a}$$

The problem is to choose local feedback control laws (5.3.3) such that the overall system (5.3.4) is stable i.e. all its poles are in the open-left complex plane C'.

Definition

Let $\bar{K}$ be the set of block diagonal matrices

$$\bar{K} = \{K \mid K = \text{block diag } [K_1, \ldots K_\nu]$$

where $K^i \in R^{m_i x p_i}$

Then the greatest common divisor (g.c.d) of the set of characteristic polynomials of (A+BKC), for all $K \in \bar{K}$ is called the fixed polynomial of the triple (C,A,B) with respect to $\bar{K}$ and it is denoted by $\psi(\lambda; C,A,B,\bar{K})$

$$\psi(\lambda; C,A,B,\bar{K}) = \underset{K \in \bar{K}}{\text{g.c.d}} \{\det(\lambda I - A - BKC)\}$$

Comment

It can be shown that there exists a unique monic polynomial (i.e. a polynomial whose leading coefficient is unity) ψ such that ψ divides all the polynomials generated by $\bar{K}$. Such a polynomial is called the greatest common divisor (g.c.d) of the set of polynomials generated by $\bar{K}$.

Definition

For a given triple (C,A,B) and the given set $\bar{K}$ of block diagonal matrices in (5.3.5), the set of fixed modes of (C,A,B) w.r.t. $\bar{K}$ is defined as:

$$\Lambda(C,A,B,\bar{K}) = \bigcap_{K \in K} \sigma\,(A+BKC)$$

where $\sigma(A+BKC)$ denotes the set of eigenvalues of (A+BKC) i.e. of the matrix whose stability is of interest.

Alternatively, we can define $\Lambda(C,A,B,\bar{K})$ as

$$\Lambda(C,A,B,\bar{K}) = \{\lambda \mid \lambda \in \not{C} \text{ and } \psi(\lambda;C,A,B,\bar{K}) = 0\}$$

This definition can be interpreted as follows:

Since ψ divides all polynomials, it has all the common eigenvalues shared by these polynomials.

Remark

We next see why $\Lambda(C,A,B,\bar{K})$ represents the fixed modes of the system and how these fixed modes can be computed.

Let $K \in \bar{K}$ and let $k(i,r,s)$ denote the $(r,s)^{th}$ element of K_i $(i = 1,2,\ldots\nu)$.

Since $\bar{K}$ contains all block diagonal matrices, it contains the null matrix.

Hence $\qquad \Lambda(C,A,B,\bar{K}) \subset \sigma(A) \qquad (5.3.5)$

Figure 5.1 illustrates the relationship (5.3.5)

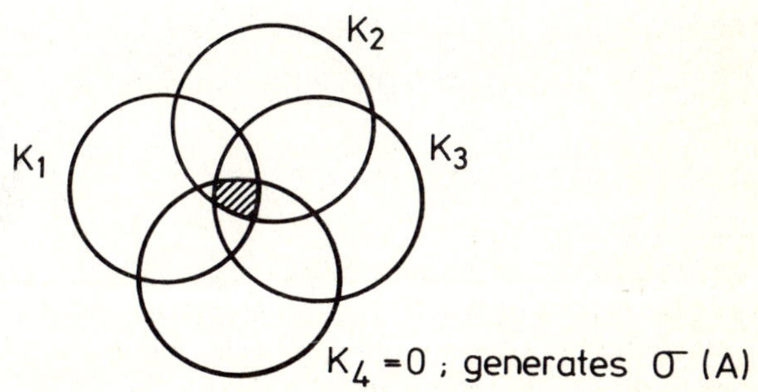

Figure 5.1

For each $\lambda_i \in \sigma(A)$, $\det(\lambda_i I-A-BKC)$ is a polynomial in $k(i,r,s)$. If $\det(\lambda_i I-A-BKC) = 0$, then $\lambda_i \in \Lambda(C,A,B,\bar{K})$. On the other hand if $\det(\lambda_i I-A-BKC) \neq 0$, then there exists a $\hat{K} \in K$ (at least one as shown in Fig. 5.1) for which $\det(\lambda_i I-A-BKC) \neq 0$. Hence $\lambda_i \in \Lambda(C,A,B,\bar{K})$.

This is one way of determining the set of fixed modes of a system i.e. we compute all the eigenvalues of A and we then apply the above test.

The reason for the name <u>fixed modes</u> is now clear, since these modes do not depend on the values assumed by $K \in \bar{K}$ but depend on the structure of $\bar{K}$ (block diagonal in our case).

Suppose we have found the eigenvalues of A that belong to A and these are λ_i, $i = 1,2,\ldots k$. Then we can write the g.c.d. as

$$\psi(\lambda,A,B,\bar{K}) = (\lambda-\lambda_1)^{\mu_i} (\lambda-\lambda_2)^{\mu_2} \ldots\ldots (\lambda-\lambda_k)^{\mu_k}$$

where μ_i is the multiplicity of ν_i amongst the eigenvalues of A.

With these preliminary definitions out of the way, we are now in a position to examine the main result of Wang and Davison [5].

5.3.1 Main result of Wang and Davison [5]

Theorem 5.3.1

Consider the system (C,A,B). Let $\bar{K}$ be the set of block diagonal matrices. Then a necessary and sufficient condition for the existence of a set of local feedback control laws (5.3.3) such that the closed loop system (5.3.4) is asymptotically stable is

$$\Lambda(C,A,B,K) \subset \bar{C}$$

where $\bar{C}$ denotes the open left half complex plane.

The following lemma relates the uncontrollable modes and unobservable modes of a triple (C,A,B) to its fixed modes.

Lemma 5.3.1 [5]

Consider a given triple (C,A,B) and its Kalman canonical structure form $(\tilde{C},\tilde{A},\tilde{B})$. Then

$$\psi(\lambda,C,A,B,R^{mxp}) = \prod_{i=2}^{4} \det(\lambda I-A_{ii})$$

where $m = \sum_{i=1}^{\nu} m_i \qquad p = \sum_{i=1}^{\nu} p_i$

This lemma shows that the notion of fixed modes of a system corresponds to the idea of uncontrollable and unobservable modes of a triple (C,A,B). Recall that the subsystem 1 in the Kalman canonical structure form is the only subsystem which contains controllable and observable modes.

We note that our interest in controllable and observable systems arises from the fact that at least for the centralised case, we can freely assign the poles by dynamic output feedback.

Lemma 5.3.2

Consider a given triple (C,A,B) which is not necessarily completely controllable or completely observable and let us assume that this triple is transformed into its Kalman canonical structure form $(\tilde{C},\tilde{A},\tilde{B})$. Assume that the submatrix $\tilde{A}_{11}$ is of dimension $n_1 x n_1$ $(0 \leqslant n_1 \leqslant n)$, where n is the dimension of the state vector.

Then there exists an integer $\tilde{n}$ $(0 \leqslant \tilde{n} \leqslant n_1)$ such that for any polynomial $\phi(\lambda) = \lambda^{\tilde{n}+n_1} + a_{\tilde{n}+n_1-1} + \ldots + \ldots + a_o$ of degree $\tilde{n}+n_1$ with real coefficients, there exists a real constant K of discussion $(m+\tilde{n})$ x $(p+\tilde{n})$ such that $\det(\lambda I-A^{\ell}-B^{\ell}K^{\ell}C^{\ell}) = \phi(\lambda)\prod_{i=2}^{4} \det(\lambda I-\tilde{A}_{ii})$

where

$$A^{\ell} = \begin{bmatrix} A & 0 \\ 0 & 0 \end{bmatrix} \begin{matrix} \}\, n \\ \}\, \tilde{n} \end{matrix} \quad (\text{columns: } n,\ \tilde{n}) \qquad B^{\ell} = \begin{bmatrix} B & 0 \\ 0 & I \end{bmatrix} \begin{matrix} \}\, n \\ \}\, \tilde{n} \end{matrix} \quad (\text{columns: } m,\ \tilde{n})$$

$$C^{\ell} = \begin{bmatrix} C & 0 \\ 0 & I \end{bmatrix} \begin{matrix} \}\, n \\ \}\, n \end{matrix} \quad (\text{columns: } n,\ \tilde{n})$$

Comment

Lemma 5.3.2 states that with a suitable number of integrators, the controllable and observable modes of (C,A,B) can be arbitrarily assigned via output feedback. It can also be shown that the remaining uncontrollable and unobservable modes of (C,A,B) are invariant under output feedback even with dynamic compensation.

It should be noted that since the controllable and observable modes can be placed in any location, the problem of stabilisation reduces to the suitable placement in the open left-half complex plane of the controllable and observable modes. We note also that we can not effect the uncontrollable and unobservable modes. That is the reason why theorem 5.3.1 stipulates that these modes must remain in $\bar{C}$.

5.3.2 Restatement of the problem

In the light of Lemma 5.3.2, we can restate our problem as: consider the system (C,A,B) where B,C are of the form given in equation (5.3.5). The problem of stabilising this system using local output feedback with dynamic compensation is to find a set of integers $n_{o_1} \geqslant 0, \ldots n_{o_\nu} \geqslant 0$ and a real constant matrix K^{ℓ} of dimension $(m+n_o)$ x $(p+n_o)$ where

$$n_o \triangleq \sum_{i=1}^{\nu} n_{oi} \ , \ m \triangleq \sum_{i=1}^{\nu} \quad , \ p = \sum_{i=1}^{\nu} p_i$$

such that all the zeroes of the polynomial

$$\det(\lambda I - A^{\ell} - B^{\ell} K^{\ell} C^{\ell})$$

are in the open left-half complex plane

where

$$A^{\ell} = \underbrace{\begin{bmatrix} A & 0 \\ 0 & 0 \end{bmatrix}}_{n \quad n_o} \begin{matrix} \}\ n \\ \}\ n_o \end{matrix} \quad ; \quad B^{\ell} = \underbrace{\begin{bmatrix} B & 0 \\ 0 & I \end{bmatrix}}_{m \quad n_o} \begin{matrix} \}\ n \\ \}\ n_o \end{matrix}$$

$$C^{\ell} = \underbrace{\begin{bmatrix} C & 0 \\ 0 & I \end{bmatrix}}_{n \quad n_o} \begin{matrix} \}\ p \\ \}\ n_o \end{matrix}$$

and

$$K^{\ell} = \left[\begin{array}{ccc|ccc} K_1 & & & Q_1 & & \\ & K_2 & & & Q_2 & \\ & & K_\nu & & & Q_\nu \\ \hline R_1 & & & S_1 & & \\ & R_2 & & & S_1 & \\ & & R_\nu & & & S_\nu \end{array}\right] \begin{matrix} \}\ m_1 \\ \}\ m_2 \\ \}\ m_\nu \\ \}\ n_{o_1} \\ \}\ n_{o_2} \\ \}\ n_{o_\nu} \end{matrix}$$

$$\quad p_1 \ p_2 \quad p_{\nu_1} \quad n_{o_1} n_{o_2} \quad n_{o_\nu}$$

where K_i, Q_i, R_i and S_i $(i=1,\ldots\nu)$ are arbitrary constant matrices. We note that the structure of K^{ℓ} determines the controller given in (5.3.2).

This leads us to proposition 1.

Proposition 1 *[5]*

Consider a given triple (C,A,B). Then for any given set of integers $n_{o_1} \geqslant 0, \ldots n_{o_\nu} \geqslant 0$

$$\psi(\lambda;C,A,B,\bar{K}) = \Psi(\lambda;C^{\ell},A^{\ell},B^{\ell},\bar{K})$$

Comment

It is easy to follow the proof of this proposition in *[5]* and the result is interesting. It states that the fixed polynomials with respect to $\bar{K}$ and $\bar{K}$ are equal, or in other words, that the uncontrollable and unobservable modes are not affected by output feedback with dynamic compensation.

We are now in a position to examine the proof of theorem 5.3.1 i.e. the main result of Wang and Davison *[5]*.

Proof of the necessity part of theorem 5.3.1

Suppose that System (5.3.1) can be stabilised using local output feedback with dynamic compensation. Then there exists a matrix K^{ℓ} such that the polynomial $\det(\lambda I-A^{\ell}-B^{\ell}K^{\ell}C^{\ell})$ has only stable zeroes. By definition, $\psi(\lambda;C^{\ell},A^{\ell},B^{\ell},\bar{K}^{\ell})$ is a divisor of $\det(\lambda I-A^{\ell}-B^{\ell}K^{\ell}C^{\ell})$. But from proposition (1)

$$\psi(\lambda;C,A,B,\bar{K}) = \psi(\lambda;C^{\ell},A^{\ell},B^{\ell},\bar{K}^{\ell}).$$

Hence $\psi(\lambda,C,A,B,\bar{K})$ has only stable zeroes i.e.

$$\Lambda(C,A,B,\bar{K}) \subset \bar{C}$$

Q.E.D.

Lemma 5.3.3

Consider a given triple (C,A,B) and the set $\bar{K}$ of block diagonal matrices. For each $K \in \bar{K}$, the characteristics polynomial of (A+BKC) can be factored as

$$\det(\lambda I-A-BKC) = \psi(\lambda;C,A,B,\bar{K}) \quad \phi(\lambda;K)$$

This follows from the definition of ψ. Now let us write $\phi(\lambda;0)$ as:

$$\phi(\lambda;0) = (\lambda-\lambda_1)^{\mu_1} \quad (\lambda-\lambda_2)^{\mu_2} \quad \ldots \quad (\lambda-\lambda_k)^{\mu_k}$$

where λ_j, (j=1,...k) are distinct zeroes of $\phi(\lambda;0)$ with multiplicity μ_j. Among the set of zeroes λ_j (j=1,...k) assume that Re $\lambda_j \geq 0 \;\forall\; j \in \{\alpha,\beta,\ldots\delta\}$ and the remaining zeroes have negative real parts. Under these conditions we have:

Part 1

$$\exists\; \varepsilon > 0 \text{ such that}$$

$$B(\lambda_j;\varepsilon) \triangleq \{\lambda | \lambda \in C, |\lambda-\lambda \quad| < \varepsilon\} \subset \bar{C} \;\forall\; j \in \quad \{1,\ldots k\}/\{\alpha,\beta,\ldots j\}$$

Comment

All we are saying here is that there exists a ball of radius ε such that all the stable modes remain in $\bar{C}$.

Part II

For all $\varepsilon > 0$, there exists $\gamma > 0$ such that for all $K \in \bar{K}$ with $||K|| < \gamma$, the polynomials $\phi(\lambda;0)$ and $\phi(\lambda;K)$ have the same number of zeroes in $B(\lambda^*;\varepsilon') \subset B(\lambda^*,\varepsilon)$, $0 < \varepsilon' \leq \varepsilon$. When we apply this argument to $\lambda^* = \lambda_j$ $(j=1,\ldots k)$ we see that there are exactly u_j zeroes of $\phi(\lambda;K)$ in $B(\lambda_j, \varepsilon')$ $\forall j \in \{1,\ldots k\}$.

Figure 5.2 illustrates this.

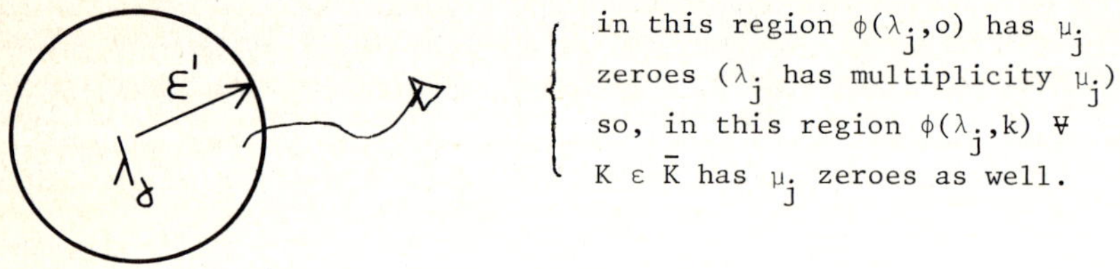

Figure 5.2

Comment

We need Part II in order to determine the range of ε' i.e. we wish to state results in such a way that we do not prejudice the modes that are already stable.

Part III

$\forall \gamma > 0 \exists K \in \bar{K}$ with $||K|| < \gamma$ such that $\phi(\lambda;K) \neq 0 \; \forall j \in \{1,\ldots K\}$.

Proof of the sufficiency part of theorem 1

Suppose that $(C,A,B,\bar{K}) \subset \bar{C}$ i.e. all the uncontrollable and unobservable modes are stable.

Let us write

$$\det(\lambda I-A-BKC) = \psi(\lambda;C,A,B,\bar{K}) \quad \phi(\lambda;K). \quad K \in \bar{K}$$

and $$\phi(\lambda;o) = (\lambda-\lambda_1)^{\mu_1} (\lambda-\lambda_2)^{\mu_2} \ldots (\lambda-\lambda_k)^{\mu_k}$$

The set of unstable zeroes of $\phi(\lambda;0)$ is $\{\lambda_\alpha,\lambda_\beta,\ldots\lambda_\delta\}$. From Lemma 5.3.3, $\exists \gamma > 0$ and $\hat{K} \triangleq \text{block diag } |\hat{K}_1\ldots K_\nu| \in K$ with $||K|| < \gamma$ such that the following holds

I : $\phi(\lambda_j;K) \neq 0 \; \forall j \in \{1,\ldots K\}$

Comment

This relationship is very important since it states that there exists a matrix $\hat{K}$ such that $\det(\lambda_j I - A - B\hat{K}C) \neq 0$ $j=1,\ldots k$. This is equivalent to saying that the unstable poles can be stabilised.

II : $\forall\, K \in \bar{K}$ with $||K|| < \gamma$, the number of unstable zeroes of $\phi(\lambda;K)$ is no greater than the number of unstable zeroes of $\phi(\lambda;0)$. Figure 5.3 illustrates this

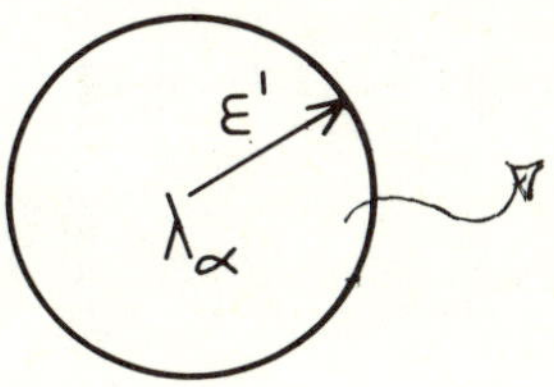

In this region $\phi(\lambda_j o)$ has μ_α unstable zeroes (we assume that λ_α is unstable with multiplicity α). So, in this region, by Lemma 5.3.3 $(\lambda_i K)$ has μ_α zeroes and hence at most μ_α zeroes are unstable.

Figure 5.3

The relationship II implies that the number of unstable eigenvalues of (A+BKC) is no greater than the number of unstable eigenvalues of A (provided these are counted by taking into consideration their multiplicity).

5.4 PROCEDURE FOR STABILISATION

Wang and Davison suggest the following procedure for stabilisation:

From the matrix $\hat{K} \triangleq$ block diag $[\hat{K}_1 \ldots \hat{K}_\nu]$ define the following set of matrices

$$\hat{K}(i) \triangleq \text{block diag } [0,\ldots \hat{K}_i,\ldots \hat{K}_\nu] \in \bar{K}$$

$$i=1,\ldots \nu$$

and $\hat{K}(\nu+1) = 0$

Let $\Pi \in \{1\ldots\nu\}$ be an integer with the following properties

$$\phi(\lambda_j;\hat{K}(\Pi) \neq 0 \;\forall\; j \in \{\alpha,\beta,\ldots\delta\}$$

and $\phi(\lambda_j;\hat{K}(\Pi+1) = 0$ for some $j \in \{\alpha,\beta,\ldots\delta\}$

The existence of Π follows from the fact that

$$\phi(\lambda_j;\hat{K}(1)) = \phi(\lambda_j;\hat{K}) \neq 0 \text{ (Lemma 5.3.3)}$$

and
$$\phi(\lambda_j;\hat{K}(\nu+1)) = \phi(\lambda_j;0) = 0 \;\forall\; j \in \{\alpha,\beta,\ldots\delta\} \qquad (5.4.1)$$

Let $\hat{A}$ denote $A+B\hat{K}(\Pi+1)C$. Then (5.4.1) is the same as A having an unstable eigenvalue (since $\det(\lambda_j - A - B\hat{K}(\Pi+1)C) = 0$) and there exists $\hat{K}\Pi$ such that

$\lambda_j \notin \sigma\,(\hat{A}+B_\Pi\hat{K}_\Pi C_\Pi)$.

The example below illustrates why we write $\hat{A}+B_\Pi\hat{K}_\Pi C_\Pi$. Assume that $\Pi = 2$ and $\nu = 3$. Then

$$A + [B_1 \quad B_2 \quad B_3] \begin{pmatrix} 0 & 0 & 0 \\ 0 & 0 & 0 \\ 0 & 0 & K_3 \end{pmatrix} \begin{pmatrix} C_1 \\ C_2 \\ C_3 \end{pmatrix}$$

$$= A + B_3K_3C_3 = \hat{A}$$

Now

$$A + [B_1 \quad B_2 \quad B_3] \begin{pmatrix} 0 & 0 & 0 \\ 0 & K_2 & 0 \\ 0 & 0 & K_3 \end{pmatrix} \begin{pmatrix} C_1 \\ C_2 \\ C_3 \end{pmatrix}$$

$$= A + B_2K_2C_2 + B_3K_3C_3$$

$$= \hat{A} + B_2K_2C_2$$

Hence, as $\lambda_j \notin \sigma[\hat{A}+B_\Pi\hat{K}_\Pi C_\Pi]$

i.e. $\det(\lambda I-A-B_\Pi\hat{K}_\Pi C_\Pi) \neq 0$, then λ_j is a controllable and observable mode of the triple $(C_\Pi,\hat{A},B_\Pi)$. Then from Lemma 5.3.2 the controllable and observable mode λ_j can be stabilised via output feedback with dynamic compensation (at the Πth control station). Hence, the number of unstable eigenvalues of A can be reduced by at least one via the above local feedback. Since there are only a finite number of unstable eigenvalues $(\lambda_\alpha,\lambda_\beta,\ldots\lambda_j)$ of the given $\hat{A}$ matrix, the system (C,A,B) can therefore be stabilised by a repeated application of the above procedure. As a design procedure, this has its weaknesses since one is not assigning poles at each application - one's merely moving them and, of course, the previously shifted poles don't remain fixed after each new application.

The other fundamental aspect in trying to stabilise by decentralised control is the calculation of the fixed modes of the system. Next we describe the approach suggested by Davison for computing these.

5.5 CALCULATION OF THE FIXED MODES [6]

Recall the main result of Wang and Davison [5]:

Consider the system

$$\underline{\dot{x}} = A\underline{x}(t) + \sum_{i=1}^{\nu} B_i\underline{u}_i(t) \qquad (5.5.1)$$

$$\underline{y}_i(t) = c_i\underline{x}(t) \qquad i=1,2,\ldots\nu \qquad (5.5.2)$$

Let $\bar{K}$ be a set of block diagonal matrices. Then a necessary and sufficient condition for the existence of a set of dynamical local feedback control laws such that the closed loop system is stable is

$$\Lambda(C,A,B,\bar{K}) \subset \bar{C} \tag{5.5.3}$$

where $\bar{C}$ denotes the open left-half complex plane and

$$\Lambda(C,A,B,\bar{K}) = \bigcap_{K \in \bar{K}} \sigma(A+BKC) \tag{5.5.4}$$

or $$\Lambda(C,A,B,\bar{K}) = \{\lambda \mid \lambda \in C \text{ and } \psi(\lambda;C,A,B,\bar{K}) = 0\} \tag{5.5.5}$$

where $$\psi(\lambda,C,A,B,\bar{K}) = \underset{K \in \bar{K}}{\text{g.c.d.}}\{\det(\lambda I-A-BKC)\} \tag{5.5.6}$$

Since $\bar{K}$ contains all block diagonal matrices, it contains the null matrix. Hence $\Lambda(C,A,B,\bar{K}) \subset \sigma(A)$ (cf Fig. 5.4)

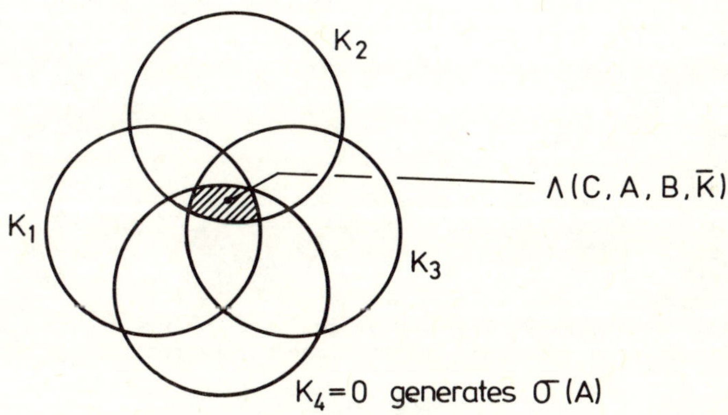

Figure 5.4

For each $\lambda_i \in \sigma(A)$, $\det(\lambda_i I-A-BKC)$ is a polynomial of $R(i,r,s)$ where i denotes the ith matrix in the diagonal of K and r,s denotes rows and columns in this ith matrix.

If $\det(\lambda_i I-A-BKC) = 0$, then $\lambda_i \in \Lambda(C,A,B,\bar{K})$. On the other hand, if $\det(\lambda_i I-A-BKC) \neq 0$, then there exists $\hat{K} \in \bar{K}$ such that $\det(\lambda_i I-A-B\hat{K}C) \neq 0$ hence $\lambda_i \in \Lambda(C,A,B,\bar{K})$.

<u>Example</u>

Consider the system

$$\dot{\underline{x}} = \begin{pmatrix} 1 & 0 & 0 \\ 0 & 2 & 0 \\ 0 & 0 & -1 \end{pmatrix} \underline{x} + \begin{pmatrix} 1 & 0 \\ 0 & 1 \\ 0 & 1 \end{pmatrix} \begin{pmatrix} u_1 \\ u_2 \end{pmatrix}$$

$$\begin{pmatrix} y_1 \\ y_2 \end{pmatrix} = \begin{pmatrix} 1 & 1 & 0 \\ 0 & 0 & 1 \end{pmatrix} \underline{x}$$

with $\bar{K} = \left\{ \begin{pmatrix} k_1 & 0 \\ 0 & k_2 \end{pmatrix} \middle| \; k_1, k_2, \varepsilon \quad R \right\}$

Then $\det(\lambda I-A-BKC) = \begin{vmatrix} \lambda-1-k_1 & -k_1 & 0 \\ 0 & 1-2 & k_2 \\ 0 & 0 & \lambda+k_2 \end{vmatrix}$

Thus we see that this system has a fixed mode $\lambda=2$ and it can not therefore be stabilised using decentralised control.

Theorem 5.5.1 *[6]*

If the system

$$\dot{\underline{x}}(t) = A\underline{x}(t) + B\underline{u}(t) \qquad u \; \varepsilon \; R^m \tag{5.5.7}$$

$$\underline{y}(t) = C\underline{x}(t) \qquad y \; \varepsilon \; R^p \tag{5.5.8}$$

is completely controllable and observable, then for any finite number of points $p \triangleq \{p_1,\ldots p_j\}$ on the complex plane C, the set

$$K^* = \{K | K \; \varepsilon \; R^{mxp} \text{ and } \sigma[A+BKC] \cap P \neq \phi\}$$

is either an empty set or a hypersurface in R^{mxp}.

Corollary 5.5.1

Consider the system

$$\dot{\underline{x}}(t) = A\underline{x}(t) + \sum_{i=1}^{\nu} B_i\underline{u}_i(t) \tag{5.5.9}$$

$$\underline{y}_i(t) = C_i\underline{x}(t) \qquad i=1,\ldots\nu \tag{5.5.10}$$

Then for any finite number of points

$p \triangleq \{p_1,\ldots p_j\} \supset \Lambda(C,A,B,\bar{K})$ on the complex plane C, the set

$$\bar{K}^* = \{K | K \; \varepsilon \; \bar{K} \text{ and } \sigma[A+BKC] \cap P \neq \Lambda[C,A,B,\bar{K}]\}$$

is either an empty set or a hypersurface in $\bar{K}$.

We can use this corollary to modify Fig. 5.4 (cf Fig. 5.5).

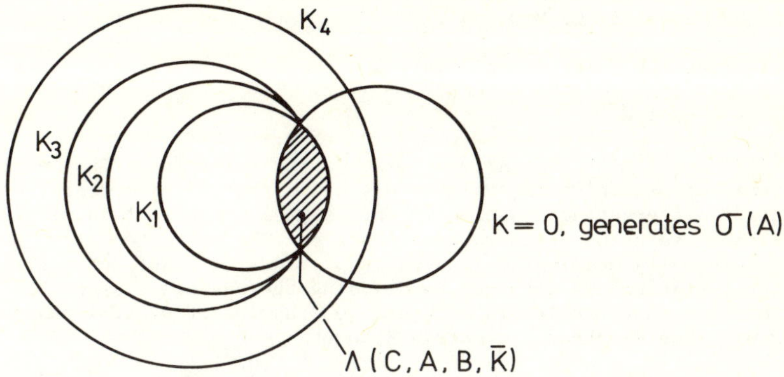

Figure 5.5

If we select K at random, it is very unlikely that we will obtain a K such as the K_4 shown in Fig. 5.4 since that would mean that we have somehow selected a K that satisfies the hypersurface equation.

Based on this corollary 5.5.1 Davison *[6]* proposes the following algorithm for the determination of the fixed modes.

Step 1: Find the eigenvalues of A

Step 2: Select an arbitrary matrix $K \in \bar{K}$ and scale it such that $||A|| \div ||BKC||$ in order to avoid computational difficulties in the next step.

Step 3: Find the eigenvalues of A+BKC.

Step 4: The fixed modes of (C,A,B) with respect to $\bar{K}$ are contained in those eigenvalues of (A+BKC) which are common with those of A (if there are any). Also, using corollary 5.5.1, the fixed modes will be identical to those eigenvalues of A+BK which are common with the eigenvalues of A.

Next we consider a realistic example to illustrate the calculation of the fixed modes.

5.6 FIXED MODES FOR THE SHIP EXAMPLE

The system under consideration is a 20th order system which can be broken up into three interconnected subsystems. These subsystems have the property that each one of the subsystems is controllable. The system model is a part of the model for the energy cycle of a big container ship. This model was identified by Bardiaux *[7]* using non-linear filtering techniques (cf. *[8]* Chapter 1). The particular state space model used here was provided to the author by C. Moog *[9]*.

Table 5.1 gives the A matrix and Table 5.2 the B matrix for the system.

$$\dot{x} = Ax + Bu \tag{5.6.1}$$

where $x \in R^{20}$, $u \in R^{5}$. Thus A is a 20x20 matrix and B is a 20x5 matrix. We note that on examining the values in A and B that (5.6.1) represents a set of stiff differential equations. This system is in fact quite difficult to stabilise using decentralised control. The subsystem structure which is physically meaningful (and which provides at least one control at each control station) is obtained by dividing the overall system into 3 subsystems such that the 1st is of order 10, the 2nd of order 4 and the 3rd of order 6. Thus we chose the matrix K to be

$$K = \begin{array}{l} 2\{ \\ \\ 2\{ \\ \\ 1\{ \end{array} \left[\begin{array}{cccccccccc|cccc|cccccc} -1 & -1 & -1 & -1 & -1 & -1 & -1 & -1 & -1 & -1 & & & & & & & & & & \\ -1 & -1 & -1 & -1 & -1 & -1 & -1 & -1 & -1 & -1 & & & O & & & & O & & & \\ \hline & & & & & & & & & & -1 & -1 & -1 & -1 & & & & & & \\ & & & & O & & & & & & -1 & -1 & -1 & -1 & & & O & & & \\ \hline & & & & O & & & & & & & & O & & -1 & -1 & -1 & -1 & -1 & -1 \end{array} \right]$$

$$\underbrace{\qquad\qquad\qquad}_{10} \qquad \underbrace{\qquad}_{4} \qquad \underbrace{\qquad\quad}_{6}$$

We found that there was no intersection between the eigenvalues of the matrix A and those of (A+BK). Table 5.3 gives the eigenvalue of A and Table 5.4 those of (A+BK). Thus the system has no unstable fixed modes and it can therefore be stabilised by decentralised control.

In Chapter 6, we will examine some of the more efficient methods for stabilising such systems using decentralised control.

TABLE 5.1

MATRIX A

-0.1250E-02	0.0000E+00	0.0000E+00	0.0000E+00	0.0000E+00
0.0000E+00	0.0000E+00	0.0000E+00	0.0000E+00	0.0000E+00
0.0000E+00	0.0000E+00	0.0000E+00	0.0000E+00	0.0000E+00
0.0000E+00	0.0000E+00	0.0000E+00	0.0000E+00	0.0000E+00
-0.2300E-01	0.0000E+00	0.0000E+00	0.0000E+00	0.0000E+00
0.0000E+00	-0.2530E-10	0.0000E+00	0.0000E+00	-0.6940E-01
0.0000E+00	-0.9070E-01	0.0000E+00	0.8000E-01	0.0000E+00
0.0000E+00	0.1120E+00	0.0000E+00	0.0000E+00	0.0000E+00
-0.2390E-02	0.1000E+01	0.0000E+00	0.0000E+00	0.0000E+00
0.0000E+00	-0.9870E-08	0.0000E+00	0.0000E+00	-0.6760E-02
0.0000E+00	-0.9450E-02	0.0000E+00	0.7790E-02	0.0000E+00
0.0000E+00	0.1090E-01	0.0000E+00	0.0000E+00	0.0000E+00
-0.9590E-04	0.0000E+00	0.1000E+01	0.0000E+00	0.0000E+00
0.0000E+00	-0.1540E-05	0.0000E+00	0.0000E+00	-0.2490E-03
0.0000E+00	-0.6030E-04	0.0000E+00	0.2860E-03	0.0000E+00
0.0000E+00	0.4010E-03	0.0000E+00	0.0000E+00	0.0000E+00
-0.1840E-05	0.0000E+00	0.0000E+00	0.1000E+01	0.0000E+00
0.0000E+00	-0.1220E-03	0.0000E+00	0.0000E+00	-0.4300E-05
0.0000E+00	-0.1100E-05	0.0000E+00	0.4950E-05	0.0000E+00
0.0000E+00	0.6930E 05	0.0000E+00	0.0000E+00	0.0000E+00
-0.1690E-07	0.0000E+00	0.0000E+00	0.0000E+00	0.1000E+01
0.0000E+00	-0.5220E-02	0.0000E+00	0.0000E+00	-0.3470E-07
0.0000E+00	-0.9690E-00	0.0000E+00	0.4000E-07	0.0000E+00
0.0000E+00	0.5600E-07	0.0000E+00	0.0000E+00	0.0000E+00
-0.5820E-10	0.0000E+00	0.0000E+00	0.0000E+00	0.0000E+00
0.1000E+01	-0.1140E+00	0.0000E+00	0.0000E+00	-0.1050E-09
0.0000E+00	-0.3260E-10	0.0000E+00	0.1210E-09	0.0000E+00
0.0000E+00	0.1690E-09	0.0000E+00	0.0000E+00	0.0000E+00
0.2760E-03	0.0000E+00	0.0000E+00	0.0000E+00	0.0000E+00
0.0000E+00	0.1200E-03	0.0000E+00	0.0000E+00	0.9920E-03
0.0000E+00	0.1390E-03	0.0000E+00	0.0000E+00	0.0000E+00
0.0000E+00	-0.1600E-02	0.0000E+00	0.0000E+00	0.0000E+00

TABLE 5.1 (Contd)

0.0000E+00	0.0000E+00	0.0000E+00	0.0000E+00	0.0000E+00
0.0000E+00	0.0000E+00	0.4170E-02	0.0000E+00	-0.6670E-02
0.0000E+00	0.1460E+00	0.0000E+00	0.0000E+00	0.0000E+00
0.0000E+00	0.0000E+00	0.0000E+00	0.0000E+00	0.0000E+00
0.0000E+00	0.0000E+00	0.0000E+00	0.0000E+00	0.0000E+00
0.0000E+00	0.0000E+00	0.8340E-03	0.1000E+01	-0.2330E+00
0.0000E+00	0.4870E-02	0.0000E+00	0.0000E+00	0.0000E+00
0.0000E+00	0.0000E+00	0.0000E+00	0.0000E+00	0.0000E+00
0.0000E+00	0.0000E+00	0.0000E+00	0.0000E+00	0.0000E+00
0.0000E+00	-0.8340E-02	0.0000E+00	0.0000E+00	0.0000E+00
0.0000E+00	-0.5710E-03	0.0000E+00	0.0000E+00	0.0000E+00
0.0000E+00	0.0000E+00	0.0000E+00	0.0000E+00	0.0000E+00
0.0000E+00	0.0000E+00	0.0000E+00	0.0000E+00	0.0000E+00
0.0000E+00	-0.8340E-03	0.0000E+00	0.0000E+00	0.0000E+00
0.1000E+01	-0.1060E+00	0.0000E+00	0.0000E+00	0.0000E+00
0.0000E+00	0.0000E+00	0.0000E+00	0.0000E+00	0.0000E+00
0.0000E+00	0.0000E+00	0.0000E+00	0.0000E+00	0.0000E+00
0.0000E+00	-0.1340E-03	0.0000E+00	0.0000E+00	0.0000E+00
0.0000E+00	0.0000E+00	0.0000E+00	0.0000E+00	0.0000E+00
0.0000E+00	0.0000E+00	0.0000E+00	0.0000E+00	0.0000E+00
0.0000E+00	0.0000E+00	0.0000E+00	0.0000E+00	0.0000E+00
0.0000E+00	-0.5360E-06	0.0000E+00	0.0000E+00	0.0000E+00
0.0000E+00	0.0000E+00	0.1000E+01	-0.4000E-02	0.0000E+00
0.0000E+00	0.0000E+00	0.0000E+00	0.0000E+00	0.0000E+00
0.0000E+00	0.0000E+00	0.0000E+00	0.0000E+00	0.0000E+00
0.0000E+00	0.0000E+00	0.0000E+00	0.0000E+00	0.1240E-01
0.0000E+00	0.0000E+00	0.0000E+00	0.0000E+00	0.0000E+00
0.0000E+00	-0.1110E-04	0.0000E+00	0.0000E+00	0.0000E+00
0.0000E+00	0.0000E+00	0.0000E+00	0.0000E+00	0.0000E+00
0.0000E+00	0.0000E+00	0.0000E+00	0.0000E+00	0.4480E-03
0.0000E+00	0.0000E+00	0.0000E+00	0.0000E+00	0.1000E+01
0.0000E+00	-0.1720E-01	0.0000E+00	0.0000E+00	0.0000E+00

TABLE 5.1 (Contd)

0.0000E+00	0.0000E+00	0.0000E+00	0.0000E+00	0.0000E+00
0.0000E+00	0.0000E+00	0.0000E+00	0.0000E+00	0.3430E-05
0.0000E+00	0.0000E+00	0.0000E+00	0.0000E+00	0.0000E+00
0.1000E+01	-0.7610E-01	0.0000E+00	0.0000E+00	0.0000E+00
0.0000E+00	0.0000E+00	0.0000E+00	0.0000E+00	0.0000E+00
0.0000E+00	0.0000E+00	0.0000E+00	0.0000E+00	-0.4300E-02
0.0000E+00	0.0000E+00	0.0000E+00	0.0000E+00	0.0000E+00
0.0000E+00	-0.9200E-02	-0.6670E-01	0.0000E+00	0.0000E+00
0.0000E+00	0.0000E+00	0.0000E+00	0.0000E+00	0.0000E+00
0.0000E+00	0.0000E+00	0.0000E+00	0.0000E+00	-0.4360E-03
0.0000E+00	0.0000E+00	0.0000E+00	0.0000E+00	0.0000E+00
0.0000E+00	0.7040E-03	0.1930E+00	0.0000E+00	-0.4000E-02
0.0000E+00	0.0000E+00	0.0000E+00	0.0000E+00	0.0000E+00
0.0000E+00	0.0000E+00	0.0000E+00	0.0000E+00	-0.4360E-05
0.0000E+00	0.0000E+00	0.0000E+00	0.6000E+00	0.0000E+00

TABLE 5.2

MATRIX B

-0.1250E-02	0.0000E+00	0.0000E+00	0.0000E+00	0.0000E+00
-0.2670E-02	-0.2250E-02	0.0000E+00	0.0000E+00	0.3090E+00
-0.2160E-03	-0.2000E-03	0.0000E+00	0.0000E+00	0.3010E-01
-0.6730E-05	-0.6700E-05	0.0000E+00	0.0000E+00	0.1110E-02
-0.1010E-06	-0.1060E-06	0.0000E+00	0.0000E+00	0.1910E-04
-0.7290E-09	-0.7960E-09	0.0000E+00	0.0000E+00	0.1550E-06
-0.2030E-11	-0.2270E-11	0.0000E+00	0.0000E+00	0.4660E-09
0.1450E-02	0.1280E-02	0.0000E+00	0.0000E+00	-0.4420E-02
0.0000E+00	0.0000E+00	0.0000E+00	0.0000E+00	0.0000E+00
0.0000E+00	0.0000E+00	0.0000E+00	0.0000E+00	0.0000E+00
0.0000E+00	0.0000E+00	-0.1370E-02	0.3310E-02	0.2300E-01
0.0000E+00	0.0000E+00	-0.1370E-02	0.3310E-03	0.1310E-03
0.0000E+00	0.0000E+00	-0.4500E-03	0.1600E-03	0.0000E+00
0.0000E+00	0.0000E+00	-0.1800E-05	0.6400E-06	0.0000E+00
0.0000E+00	0.0000E+00	0.0000E+00	0.0000E+00	-0.2660E-01
0.0000E+00	0.0000E+00	0.0000E+00	0.0000E+00	-0.1360E-02
0.0000E+00	0.0000E+00	0.0000E+00	0.0000E+00	-0.1180E-04
0.0000E+00	0.0000E+00	0.0000E+00	0.0000E+00	0.7610E-02
0.0000E+00	0.0000E+00	0.0000E+00	0.0000E+00	0.1940E-02
0.0000E+00	0.0000E+00	0.0000E+00	0.0000E+00	0.1940E-03

TABLE 5.3

EIGENVALUES OF MATRIX A

0.3986E+00	0.0000E+00	0.2854E+00	0.2880E+00	0.2854E+00
-0.3466E+00	-0.2483E+00	0.2249E-01	0.3563E+00	0.2249E-01
-0.2208E+00	0.2239E+00	-0.2208E+00	-0.22393+00	0.1987E+00
-0.2880E+00	-0.4706E+00	0.0000E+00	-0.3466E+00	0.2483E+00
-0.3563E+00	-0.6839E-01	0.3402E+00	-0.6839E-01	-0.3402E+00
0.0000E-03	-0.1000E+00	0.0000E+00	-0.6670E-01	0.0000E+00

TABLE 5.4

EIGENVALUES OF MATRIX (A+BK)

-0.4837E+00	0.0000E+00	0.4804E+00	0.0000E+00	0.1847E+00
-0.3282E+00	-0.2720E+00	-0.2143E+00	0.3415E+00	-0.2143E+00
0.1934E+00	0.8633E-01	0.1934E+00	-0.8633E-01	-0.2449E+00
0.1080E+00	0.0000E+00	-0.1938E-01	0.0000E+00	-0.7688E-03
0.3336E+00	0.1847E+00	-0.3336E+00	-0.3282E+00	0.2720E+00
-0.3415E+00	0.2150E-01	0.3297E+00	0.2150E-01	-0.3297E+00
0.0000E+00	-0.1377E+00	0.1334E+00	-0.1377E+00	-0.1334E+00
0.1356E-02	-0.7688E-03	-0.1356E-02	0.1498E-01	0.0000E+00

Now we go on to examining a different approach to the analysis of decentralised controllers. This approach was developed by Corfmat and Morse [10, 11]. The new element in this approach arises from the fact that aside from stabilisation by decentralised control, there exist more degrees of freedom which could be used to achieve good dynamic response. This is done by Corfmat and Morse by assigning the poles of the closed loop system under decentralised control.

In order to understand the work of Corfmat and Morse, we first of all group together some preliminary mathematical notions on invariant subspaces, quotient spaces, induced linear transformations, reachability and controllability.

5.7 SOME PRELIMINARY MATHEMATICAL NOTIONS

5.7.1 Invariant Subspaces

Let T be a linear operator such that $T : V \rightarrow V$ where V is a vector space. A subspace W of V is said to be invariant under T (or T-invariant) if T maps W into itself, i.e. if $v \in W$ implies $T v \in W$. In this case T restricted to W (T/w) defines a linear operator on W, i.e. T induces a linear operator $T/w : W \quad W$ defined by $T/w = Tw$ for every $w \in W$.

Theorem 5.7.1

Suppose W is an invariant subspace of $T : V \rightarrow V$. Then T has a block matrix representation $\begin{pmatrix} A & B \\ 0 & C \end{pmatrix}$ where A is a matrix representation of the restriction of T to W, T/W.

Proof:

We choose a basis $\{w_1 \ldots w_r\}$ of W and extend it to a basis

$$\{w_1, w_2 \ldots w_r, \quad v_1 \ldots v_s\} \text{ of } V.$$

Then

$$[T_{w1} \ldots T_{wr} \quad T_{v1} \ldots T_{vs}] = [T/w1 \ldots T/wr \quad Tv1 \ldots Tvs]$$

$$= [w_1 \ldots w_r \quad v_1 \ldots v_s] \begin{matrix} \\ r\left\{\begin{matrix} \\ \\ \\ \end{matrix}\right. \\ \\ s\left\{\begin{matrix} \\ \\ \end{matrix}\right. \end{matrix} \left[\begin{matrix} \overbrace{\begin{matrix} a_{11} \ldots a_{1r} \\ a_{22} \ldots a_{2r} \\ \vdots \\ a_{r1} \ldots a_{xs} \\ \vdots \end{matrix}}^{r} & \overbrace{\begin{matrix} b_{11} \ldots b_{1s} \\ b_{21} \ldots b_{2s} \\ \vdots \\ b_{x1} \ldots b_{xs} \\ \vdots \end{matrix}}^{s} \\ \begin{matrix} 0 - - - 0 \\ \vdots \\ 0 - - - 0 \end{matrix} & \begin{matrix} c_{11} \ldots c_{s1} \\ \vdots \\ c_{s1} \ldots c_{ss} \end{matrix} \end{matrix}\right] \tag{5.7.1}$$

$$= \begin{pmatrix} A_{rxr} & B_{rxs} \\ 0 & C_{sxs} \end{pmatrix}$$

The interpretation of C will be clear in the sequel.

5.7.2 Quotient Spaces

Let V be a vector space over a field K and let W be a subspace of V. If v is any vector in V, we can write the v+w for the set of sums v+w with $w \varepsilon W$

$$v + W = \{v+w : w \varepsilon W\}$$

These sets are called the cosets of W in V.

Example:

Let W be the subspace of R^2 defined by W = {(a,b) : a = b} i.e. W is the line given by the equation x.y = 0. We can view v+W as a translation of the line obtained by adding the vector v to each point in W. v+W is also a line and is parallel to W. Thus the cosets of W in R^2 are all the lines parallel to W.

In the next theorem we use the cosets of a subspace W of a vector space V to define a new vector space; it is called the quotient space of V by W and is denoted by V/W.

Theorem 5.7.2

Let W be a subspace of a vector space over a field K. Then the cosets of W in V form a vector space over K with the following operations of addition and scalar multiplication:

i) $(u+W) + (v+W) = (v+v) + W$

ii) $k(u+W) = k\ u+W; \quad k \varepsilon K$

Theorem 5.7.3

Let V be a vector space and W a subspace of V. Then the natural map $\eta : V/W$ defined by $\eta(v) = v+w$ is linear.

Proof:

For any u, $v \varepsilon V$ and any $k \varepsilon K$ we have

$$\eta(u+v) = u+v+W = u+W+v+W$$

$$= \eta(u) + \eta(v)$$

$$\eta(kv) = kv+W = k(v+w) = k\ \eta(v)$$

Note that the linear operator η projects a vector $v \varepsilon V$ into a coset. But if $v \varepsilon W$, then $\eta(v) = 0$.

Theorem 5.7.4

Let W be a subspace of a vector space V. Suppose $\{w_1 \ldots w_r\}$ is a basis of W and the set of cosets $\{\bar{v}_1 \ldots \bar{v}_s\}$ where $\bar{v}_j = v_j + W$ is a basis of the quotient space V/W. Then $B = \{v_1 \ldots v_s, w_1 \ldots w_r\}$ is a basis of V. Thus $\dim V = \dim W + \dim(V/W)$.

5.7.3 Induced Linear Transformations

Consider a T-invariant subspace W. We will now show that in a certain sense T also induces linear transformations in the quotient space $\bar{V} = V/W$. We recall that a vector $\bar{v}$ of $\bar{V}$ is a coset consisting of the vector of the form v+w where v is fixed and w ranges over W.

Now

$$T(v+w) = Tv + Tw = Tv+w' \text{ where}$$

$$w' = Tw \in W$$

Hence the image of any vector in the coset $\bar{v}$ is a vector in the coset $\bar{T}v$ determined by Tv.

Thus the mapping $\bar{T}$ that associates with $\bar{v}$ the vector $\bar{T}v$ of $\bar{V}$ is single valued. We call $\bar{T}$ the transformation induced by T in $\bar{V}$.

Consider the mapping $P = \eta : \underline{x} \to \bar{x} = x+W$. Now, we have defined $\bar{T}$ by the rule $\bar{T}\,\bar{v} = \overline{Tv}$. Thus we have $\bar{T}\,P = PT$ or in other words there is a unique map $\bar{T} : V \to V$ such that

$$\bar{T}\,P = PT \tag{5.7.2}$$

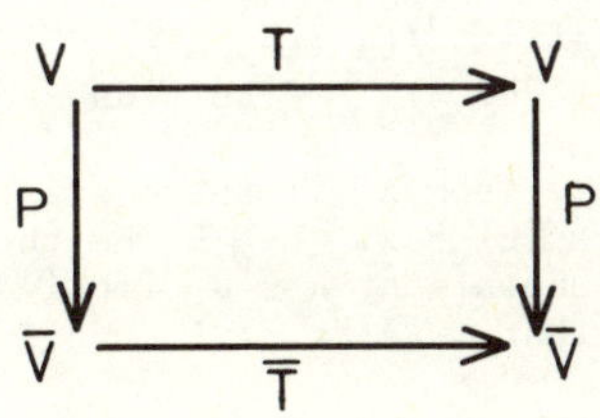

Figure 5.6

Let $\{\bar{v}_i\}$ i = 1,...s be a basis for $\bar{V}$ where $\bar{v}_i = Pv_i$. Then

$$\bar{T}\,\bar{x}_i = \bar{T}\,Px_i = PTx_i \qquad i=1,\ldots s.$$

On the other hand $TW \subset W$ and $PW = 0$. This shows that the relationship (5.7.2) is unambiguous.

Now assume that $\{w_1 \ldots w_r\}$ is a basis for w. Then for $\bar{v}_k \in \bar{V}$ we have

$$\bar{T}\,\bar{v}_k = \overline{T\,v_k} = \sum_{i=1}^{r} \overline{b_{ik}w_i} + \sum_{i=1}^{s} \overline{C_{ik}v_i}$$

$$= \sum_{i=1}^{s} C_{ik}\bar{v}_i$$

since $\bar{T} : \bar{V} \quad \bar{V}$

We know that $\bar{v}_i$, i=1...s form a basis for $\bar{V}$. Hence these relations show that the matrix of $\bar{T}$ relative to this basis is the matrix C in (5.7.1) i.e. in

$$\begin{pmatrix} A_{rxr} & B_{rxs} \\ 0 & \boxed{C_{rxs}} \end{pmatrix}$$

5.7.4 Reachability and controllability

It is reasonable to call a dynamical system "controllable" if by a suitable manipulation of its inputs, the system outputs can be made to behave in some desirable way. Consider the linear dynamical system

$$\dot{x}(t) = Ax(t) + Bu(t);\ t \geq 0 \tag{5.7.3}$$

we begin by examining these states which, roughly speaking, the control u(.) is able to influence.

Let $\bar{U}$ denote the linear space of piecewise continuous controls $t \to u(t) \in U$ where U is a linear space and is defined for $t \geq 0$. Denote by $\phi(t,x_0,u)$ the solution of (5.7.3) for $x(0) = x_0$ i.e.

$$\phi(t;x_0,u) = e^{tA}\,x_0 + \int_0^t e^{(t-s)A}\,Bu(s)ds \tag{5.7.4}$$

A state $\underline{x} \in X$ is reachable from x_0 if there exists a t and u with $0 < t < \infty$ and $u \in \bar{U}$ such that $\phi(t;x_0,u) = x$. Let R_o be the set of states reachable from $x_0 = 0$. It can be shown that $x \in R_o$ implies that for every $t > 0$ there exists $u \in \bar{U}$ such that

$$x = \phi(t;o,u)$$

It is easy to see from (5.7.4) and the admissability of piecewise constant controls that R_o is a linear subspace of X. Let $\beta = I_m B$ and

$$< A|B > = \beta + A\beta + \ldots + A^{n-1}\beta$$

Then

$$R_o = < A|\beta >$$

From (5.7.4) it follows that $x \varepsilon X$ is reachable from x_o if and only if $x - e^{tA} x_0 \varepsilon R_o$ for some t, $0 < t < \infty$

The subspace $R_o = < A|\beta > \subset X$ is the controllable subspace of the pair (A,B). It is easy to show using the Cayley-Hamilton theorem that R_o is A-invariant. It can also be shown using the Cayley-Hamilton theorem that R_o is the smallest A-invariant subspace containing β. Now let $\bar{X} = X R_o$, $P : X \to \bar{X}$ be the canonical projection, $\bar{A}$ the map induced in $\bar{X}$ by A and write $\bar{x} = Px$. As $\beta \subset R_o$ and $\beta = I_m B$, we have

$$PB = 0. \quad \text{Then}$$

$$\dot{\bar{x}}(t) = \bar{A}\,\bar{x}(t)$$

Thus the control u(.) has no influence on the coset of x and R_o.

In this notation, all states can be reached from zero when $\bar{X} = 0$ i.e. $R_o = X$.

Definition

The pair (A,B) is controllable if its controllable subspace is the whole space i.e. $< A|\beta > = X$.

5.7.5 Controllability and Feedback

Consider once again the system

$$\dot{x}(t) = Ax(t) + Bu(t), \qquad t \geq 0 \tag{5.7.5}$$

Assume that we choose

$$u(t) = Fx(t) + v(t), \qquad t \geq 0 \tag{5.7.6}$$

where v is a new external input, and $F : X \to U$ is an arbitrary map. We refer to F as state feedback. Substituting (5.7.6) into (5.7.5) we change the pair (A,B) into the pair (A+BF,B). We will now show that controllability is not affected by state feedback.

Lemma 5.7.1

For any state feedback $F : X \to U$

$$< A+BF|\beta > = < A|\beta >$$

In particular, if (A,B) is controllable, so is (A+BF,B).

Proof:

We note that

$$\beta + (A+BF)R = \beta + AR \tag{5.7.7}$$

$\forall R \subset X$ and $F : X \to U$ i.e. $BFR = \beta$

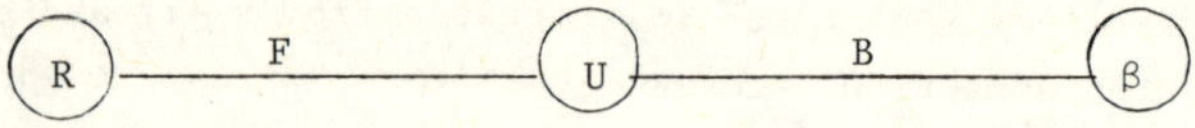

Writing $\hat{A} = A+BF$, we have $< A+BF, \beta > \; = \beta + \hat{A}\beta + \ldots + \hat{A}^{n-1}\beta$

$= \beta + \hat{A}\,(\beta + A\,(\ldots(\beta+\hat{A}\beta))\ldots)$

But $\beta+\hat{A}\beta \; = \; \beta+A\beta$ by (5.7.7)

Proceeding in this way, we obtain $< A+BF, \beta > \; = \; \beta+A\beta + \ldots + A^{n-1}\beta = < A|\beta >$

Now, we are in a position to examine the analysis of Corfmat and Morse [13,14] on the effect of decentralised feedback on the closed loop properties of k-channel, jointly controllable, jointly observable linear systems.

5.8 CONTROL THROUGH SPECIFIED INPUT CHANNELS [10,11]

The model considered is of a k-channel linear system $\sum \equiv \{C_i, A, B_i, \bar{k}\}$ described by the equations

$$\dot{x}(t) = Ax(t) + \sum_{i=1}^{k} B_i u_i(t) \tag{5.8.1}$$

$$y_i(t) = C_i x(t) \qquad\qquad i \in \bar{k} \equiv \{1,2,\ldots k\}$$

where $u_i(t)$ and $v_i(t)$ are vectors of control input and measured output variables associated with channel i; the time functions $u_i(.)$, $x(.)$ and $y_i(.)$ take values in

$$U_i \; (\equiv R^{m_i}), \; X \; (\equiv R^n) \text{ and } Y_i \; (\equiv R^p).$$

We assume that $\sum$ is a jointly controllable, jointly observable system i.e.

$$< A| \sum_{i=1}^{k} \beta_i > \text{ and } X \text{ and } \bigcap_{i=1}^{k} [C_i|A] = 0$$

where

$$\beta_i = \text{image } B_i \text{ and } < A|\beta_i > = \beta_i + A\beta_i + \ldots + A^{n-1}\beta_i \tag{5.8.2}$$

$$[C_i|A] = \bigcap_{i=1}^{n} \text{Kernel } CA^{i-1}$$

Note that this assumption does not imply that the $\sum$ is controllable or observable through a single channel. For example joint controllability of $\sum$ does not imply controllability of, for example (A,B_1). In other words, it is possible that the vector of control inputs u_1 associated with channel 1 does not influence some of the state variables. Also, we may not be able to estimate all the states from the output y_1.

In terms of the necessary and sufficient conditions for a system to be controllable, consider the 2 channel case (i.e. k=2). Then, since the system is completely controllable we have

$$\text{Range } [B_1, AB_1 \ldots A^{n-1} B_1, B_2, AB_2 \ldots A_{n-1}B_2] = R^n$$

The above matrix is of dimension $n \times (m_1+m_2)$. If we can find n linearly independent columns among

$$B_1, AB_1 \ldots A^{n-1}B_1, \text{ then the pair } (A,B_1) \text{ is controllable.}$$

A feedback control for $\sum$ is decentralised if each local control $\underline{u}_i$ depends only on local measurements $\underline{y}_i$.

A local controller for the ith channel of $\sum$ is a linear dynamical system $\sum_i$ forced by $\underline{y}_i$ and possibly a reference input $\underline{v}_i$; the output of $\sum_i$ then serves as the control input u_i to the channel i. Thus $\sum_i$ has a model of the form:

$$\begin{aligned} \dot{\underline{z}}_i(t) &= H_i\underline{z}_i(t) + L_i\underline{y}(t) + R_iv_i(t) \\ \underline{u}_i(t) &= M_iz_i(t) + F_iy_i(t) + G_iv_i(t) \end{aligned} \tag{5.8.3}$$

where $z_i(.)$ takes values in Z_i $(\equiv R^{n_i})$. Note that the non-dynamic local feedback law $\underline{y}_i(t) = F_i\underline{y}_i(t) + \underline{v}_i(t)$ is a special case of (5.8.3).

Applying the local control (5.8.3) to the system given by equations (5.8.1) we have the closed loop system of dimension

$$n + \sum_{i=1}^{k} n_i$$

$$\begin{pmatrix} \dot{x} \\ \dot{z} \end{pmatrix} = \left(\begin{array}{c|c} A+BFC & BM \\ \hline LC & H \end{array}\right) \begin{pmatrix} x(t) \\ z(t) \end{pmatrix} + \begin{pmatrix} BG \\ R \end{pmatrix} v(t)$$

The matrices H,L,R,M,F,G are all block diagonal.

The spectrum (i.e. the eigenvalue set) of the closed loop system depends on the properties of the local controllers $\sum_i$.

The main contribution of the work of Corfmat and Morse *[10,11]* was to develop explicit conditions in terms of the matrices which characterise $\sum$ for determining when the closed-loop spectrum of $\sum$ is freely assignable using decentralised control.

One possible approach to the spectrum assignment problem is first to consider local static controls of the form

$$u_i(t) = F_iy_i(t) + v_i(t) \qquad i \,\varepsilon\, \bar{k} \tag{5.8.4}$$

The object is to select the gains F_i such that the resulting closed loop system

$$\begin{aligned} \dot{x}(t) &= (A + \sum_{j=1}^{k} B_jF_jC_j)\, x(t) + \sum_{i=1}^{k} B_iv_i(t) \\ y_i(t) &= C_ix(t), \qquad i \,\varepsilon\, \bar{k} \end{aligned} \tag{5.8.5}$$

is <u>both controllable and observable through a single channel</u>, say j. If this could be accomplished, then standard techniques could be used to construct a second local control system.

$$\dot{z}_j(t) = H_j z_j(t) + L_j v_j(t)$$
$$v_j(t) = M_j z_j(t) + N_j v_j(t)$$

which when applied to channel j of (5.8.5) results in a closed loop system with a prescribed spectrum.

To simplify the notation, we will henceforth write F for the matrix : diag $[F_1 \ldots F_k]$, $\mathcal{F}$ for the vector space of all such matrices, A_F for $A + \sum_{i=1}^{k} B_i F_i C_i$ and $\sum_F = \{C_i, A_F, B_i \bar{k}\}$ for the k-channel system represented by (5.8.5).

As we have seen in the earlier part of this chapter, $\sum$ can be stabilised using decentralised control only if the fixed polynomial

$$\alpha_{\sum}(\lambda) \equiv \underset{F \in \mathcal{F}}{\Lambda} \det [\lambda I - A_F] \tag{5.8.6}$$

has all its roots in the open left-half complex plane.

The necessity of this condition is a direct consequence of the following proposition which can be directly deduced from the results of Wang and Davison [5] described in an earlier part of this chapter.

<u>Proposition 5.8.1</u>

Let $\Lambda(\sum_i, \bar{k})$ denote the spectrum of the closed loop system described by (5.8.1) and (5.8.3) and write $\Lambda_{\sum}$ for the set of roots $\alpha_{\sum}(\lambda)$. Then

$$\Lambda_{\sum} \subset \Lambda(\sum_i, \bar{k})$$

for all linear controls $\sum_i$, $i \in \bar{k}$

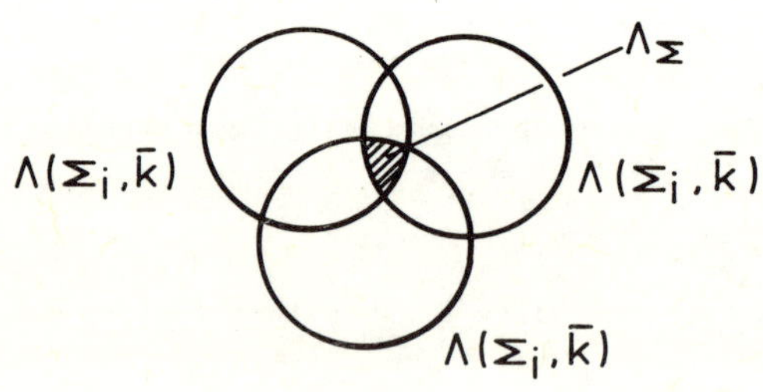

Figure 5.7

5.8.1 <u>Controllable single channel subsystems</u>

The principal problem being considered is as follows: for given $\sum$ and fixed

$j \in \bar{k}$, find conditions for the existence of an $F \in \mathcal{F}$ such that (C_j, A_F, B_j) is a controllable and observable system. If an F with the required property exists for each and every $j \in \bar{k}$, we can say that $\sum$ can be made controllable and observable through a single channel with non-dynamic decentralised control.

We begin by stating some results for the case k=2 obtained by Corfmat and Morse *[10]*.

Consider the problem of selecting for the two-input, controllable, linear system

$$\dot{x}(t) = Ax(t) + Bu(t) + Dv(t)$$
$$y = Cx(t) \tag{5.8.7}$$

an output feedback law v = Fy so that the resulting closed loop system

$$\dot{x} = (A+DFC)\, x(t) + Bu(t) \tag{5.8.8}$$

is controllable with u(.).

We could consider (5.8.7) to be the composite model of two non-interacting subsystems, one, $\sum_1 = (C,A,B)$ with input u and output y and the other $\sum_2$ with input v and output

$$z(t) = Lx(t) \tag{5.8.9}$$

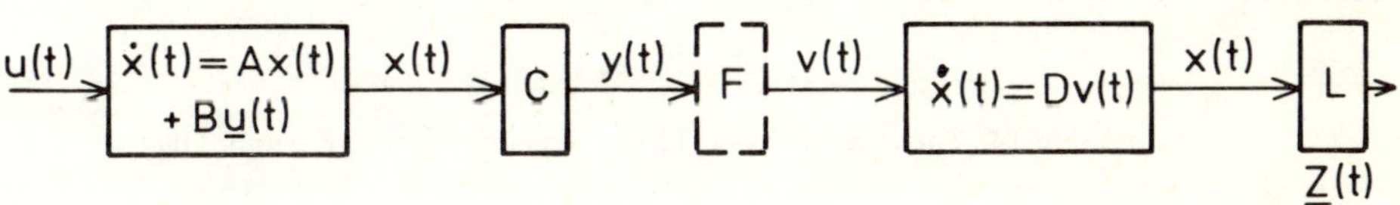

Figure 5.8

we have equation (5.8.8) if $\sum_1$ is connected in cascade with $\sum_2$ by means of the control v = Fv. Note that the subsystem $\sum_1 = (C,A,B)$ may not be necessarily controllable or observable, even if (5.8.7) together with (5.8.9) is. By studying the conditions under which F can be chosen to make (5.8.8) controllable with u (and observable with z), one can determine the requirements for stabilisation or for the control of the dynamic response with feedback from z to u.

5.8.2 Main Problem

Let A,B,C,D be fixed with $< A|\beta + D > = X$. Find condition for the existence of a map $F : Y \to V$ such that

$$< A + DFC|\beta > = X$$

In the sequel, it will be assumed without loss of generality that the transfer matrix

$$C(\lambda I-A)^{-1} B \neq 0 \qquad (5.8.10)$$

for if (5.8.10) were false, $y(t) = v(t) = 0$ and $\langle A+DFC|\beta \rangle$ would be independent of F.

Let F^* denote the class of all maps F for which the dimension $\langle A+DFC|\beta \rangle$ is as large as possible.

Let $\{r_i(z),\ i \in \bar{t}\}$ denote the list of transmission polynomials (t.p) of (C,A,B) ordered such that r_i/r_{i-1}, $i=2\ldots t$ (r_i divides r_{i-1}). Write r for the rank of the non-zero transfer matrix $C(\lambda I-A)^{-1}B$ over the field of rational functions in λ, and define

$$n^* = \begin{cases} n & \text{if } \quad r \geq t \\ n - \deg \left(\prod_{i=r+1}^{t} r_i \right) & \text{if } r > t \end{cases}$$

Theorem 5.8.1

Let A,B,C,D be fixed with $\langle A+\beta+D \rangle = X$ and $C(\lambda I-A)^{-1}B \neq 0$. Then

$$\max_F d(\langle A+DFC|\beta \rangle) = n^*$$

This theorem says that if F is selected so that the dimension of $\langle A+DFC|\beta \rangle$ is as large as possible, then this dimension must equal the dimension of X less the degree of the product of the last t−r polynomials in the list of transmission polynomials of (C,A,B).

Corollary 5.8.1

Under the hypothesis of Theorem 5.8.1, there exists a map F such that

$$\langle A+DFC|\beta \rangle = X$$

if and only if the number of transmission polynomials of (C,A,B) is no greater than the rank of the transfer matrix of (C,A,B).

Remark:

We call (C,A,B) complete if $t \leq \Omega$ where Ω is the rank of the transfer function matrix. Since the rank of a transfer matrix is invariant under transposition, and since the t.ps of (C,A,B) are the same as those of the dual system (B',A',C') it follows that (C,A,B) is complete if and only if its dual system is complete.

Corollary 5.8.1 implies that if (A,B) is controllable, $\langle A|\beta \rangle = X$, then (C,A,B) is complete. This and duality allows us to assert that either the controllability of (A,B) or the observability of (C,A) implies completeness of (C,A,B) i.e.

$$(A,B) \text{ controllable} \Rightarrow (C,A,B) \text{ complete}$$

$$\Leftrightarrow (B',A',C') \text{ complete}$$

$$\Uparrow$$

$$(C,A) \text{ observable}$$

From this and corollary 5.8.1 we have:

Corollary 5.8.2

If (C,A) is observable and $< A|\beta + D > = X$, there exists a map F such that (A+DFC,B) is controllable.

If (C,A,B) is incomplete (r<t) then corollary (5.8.1) implies that (A+DFC,B) can not be made observable. In this case, the following theorem provides a characterisation of the list $\{r_{r+1} \ldots r_t\}$ (i.e. the remanent polynomials of (C,A,B)).

Theorem 5.8.2

Suppose (C,A,B) is incomplete (t>r) and for arbitrary F let $\{\alpha_i^F, \ i \in \underline{q}^F\}$ denote the list of invariant factors of the map induced by (A+DFC) in $X| < A+DFC|\beta >$. Under the hypotheses of theorem 5.8.1:

$$q^F \geq t-r \text{ and } r_{r+i} \,|\, \alpha_i^F; \quad i \in \{1,2,\ldots t-r\} \qquad \forall F$$

In addition, if $d(< A+DFC|\beta >) = n^*$, then

$$q^F = t-r \text{ and } r_{r+i} = \alpha_i^F; \quad i \in \{1,2,\ldots t-r\}$$

Let Λ_o denote the symmetric set of complex roots of the polynomial $\prod_{i=r+1}^{t}$ repeated according to multiplicity (i.e. the remanent zeroes) of (C,A,B). Theorem 5.8.2 implies that Λ_o is a fixed subset of the spectrum of (A+DFC) and $< A+DFC|\beta >$ (i.e. the uncontrollable spectrum of (A+DFC) whenever $d(< A+DFC|\beta >)$ is as large as possible.

Corollary 5.8.3

Under the hypotheses of theorem 5.8.2, there exists a map F such that the pair (A+DFC,B) is stabilisable if and only if Λ_o is a stable set.

With these notions in mind, let us go back to the paper of Corfmat and Morse [10] recalling that the remanant polynomial of a (single-channel) linear system (C,A,B) is defined as

$$\rho(C,A,B) = \begin{cases} 1 & \text{if } r \geq t \\ \prod_{i=\Omega+1}^{t} r_i(\lambda) & \text{if } r < t \end{cases}$$

(C,A,B) is a complete (single-channel) system if its transfer matrix is non-zero and if $\rho(C,A,B) = 1$.

Proposition 5.8.2 (theorem 5.8.2)

Let (C,A,B) and $D : V \to X$ be fixed with $< A|\beta > + D > = X$ and $C(\lambda I-A)^{-1}B \neq 0$. For all $K : Y \to V$, $\rho(C,A,B)$ divides the uncontrollable polynomial of

(A+DKC,B). The word "divides" could be replaced by "equals" for the case K is selected such that $d(< A+DKC|\beta >)$ is as large as possible.

Lemma 5.8.1

For the two-channel system $\sum \equiv \{C_i, A_i, B_i, 2\}$ and any decentralised feedback matrix F = diag

$$(F_1, F_2) \qquad < A_F|\beta_1 > = < A+B_2F_2C_2|\beta_1 > \text{ and}$$

$$\text{u.p. } (A_{F_1}, B_1) = \text{u.p. } (A+B_2F_2C_2, B_1)$$

Just to get a feel for what is happening here, let us show how

$$< A_F|\beta > = < A+B_2F_2C_2|\beta_1 >.$$

$$A_F = A + B_1F_1C_1 + B_2F_2C_2$$

$$< A_F|\beta_1 > \beta_1 + A_{F\ 1} + A_F^2\beta_1$$

$$= \beta_1 + A_F(\beta_1 + A_F\beta_1) + (A+B_2F_2C_2)\beta_1)$$

$$= \beta_1 + (A+B_2F_2C_2)\ \beta_1 + A_F(A+B_2F_2C_2)\beta_1$$

$$= \beta_1 + (A+B_2+F_2C_2)\beta_1 + (A+B_2F_2C_2)^2\beta_1$$

$$+ B_1F_1C_1\ (A+B_2F_2C_2)\beta_1$$

But $(A+B_2F_2C_2)\beta_1 \subset X \Rightarrow B_1F_1C_1\ (A+B_2F_2C_2)\beta_1 = \beta$

and $< A_F|\beta_1 > = \beta_1 + (A+B_2F_2C_2)\beta_1 + (A+B_2F_2C_2')^2$

Hence the result.

Lemma 5.8.1 and proposition 5.8.2 imply:

Proposition 5.8.3

If $\sum = \{C_i, A, B_i, 2\}$ is jointly controllable and if $C_2(\lambda I-A)^{-1}B_1 \neq 0$, then $\forall$ $F \in \mathcal{F}$ $\rho(C_2, A, B_1)$ divides the uncontrollable polynomial of (A_F, B_1) the word "divides being replaceable by "equals" for the case where $F \in \mathcal{F}$ is chosen such that $d(< A_F|\beta >)$ is as large as possible.

From this proposition it follows that if $\sum$ is jointly controllable and if $C_2(\lambda I-A)^{-1}B_1 \neq 0$ then F can be selected such that $< A_F|\beta > = X$ if and only if (C_2, A, B_1) is complete.

In order to extend these results to k-channel systems it is necessary to take into account which transfer matrices within $\sum$ are zero and which ones are not.

We define the graph of $\sum$ denoted by G to be the pair $(\bar{k}, \Gamma)$ comprising k nodes lablelled 1,2,...k and node i representing

$$\Gamma(j) = \{i : H_{ij} \neq 0\}; \quad \text{where } H_{ij} \equiv C_i(\lambda I-A)^{-1}B_j$$

is the transfer matrix from channel input u_j of $\sum$ to channel output y_i.

The graph may be represented by a diagram comprising a set of k nodes labelled 1,2,...k with a directed arc drawn from node j to node if u_i influences y_i in the sense of $H_{ij} \neq 0$. The graph of $\sum$ thus represents the open loop pattern of signal flow between the channels of $\sum$.

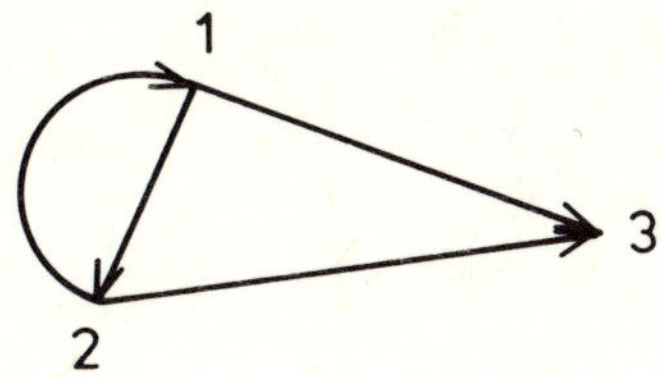

Figure 5.9

It is reasonable to expect that if (A_F, β_j) is controllable from some F, then each node of G must be reachable from node j along a directed path in G .

A path from node i to node $G_\sum$ can be interpreted as a sequence of nodes $i_1,,,i_p$ satisfying $i_p \in \Gamma(i_{q-1})$, $q \in \{2,3,...p\}$ i.e. a path is an arc progression where no arc is repeated and which joins i_1 to i_p.

<u>Proposition 5.8.4</u>

If for some $j \in \bar{k}$ and $F \in \mathcal{F}$, (A_F, β_j) is controllable, then for each $i \in \bar{k}$, with $i \neq j$, there exists a path in G from node j to node i.

Figure 5.9 shows node i corresponding to a controllable (A_F, B_1)

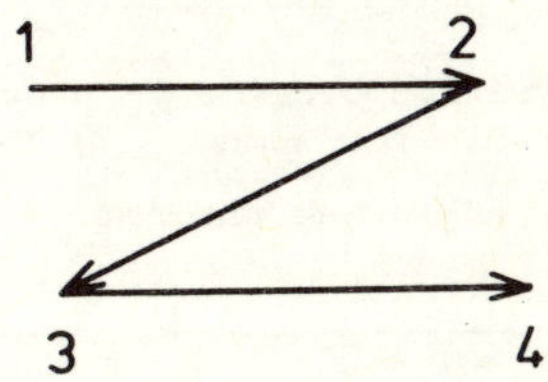

Figure 5.10

While this property of G_{Σ} is necessary for (A_F, β_j) to be controllable for some F, it is not sufficient; as we have already noted, when k=2, completeness of a certain subsystem of Σ is also required.

To describe analogous subsystems for k-channel systems, we introduce the following notation: if s is a non-empty subset of $\bar{k}$ with elements $i_1, i_2, \ldots i_p$ ordered such that $i_1 < i_2 < \ldots < i_p$, then we define B_s and C to be:

$$C_s = \begin{bmatrix} C_{i1} \\ C_{i2} \\ \vdots \\ C_{iq} \end{bmatrix} \qquad B_s = [B_{i1}, B_{i2} \ldots B_{iq}]$$

Thus (C_{s1}, A, B_{s2}) is a subsystem of Σ which models the relationship between channel inputs u_1, $i \in s_2$ and channel outputs y_j, $j \in s_1$. We call (C_{s1}, A, B_{s2}) a complimentary subsystem of Σ if s_2 is a proper subset* of $\bar{k}$ and if $s_1 = \bar{k} - s_2$; a complimentary subsystems (C_{k-s}, A, B_s) is said to contain the input channel j if j is an element of s.

Corfmat and Morse [11] have shown that there exists a path in G_{Σ} from node j to node i if and only if

$$C_{k-s} (\lambda I - A)^{-1} B_o \neq 0 \qquad s \in \bar{k}(j)$$

$\bar{k}(j)$ denotes the class of all proper subsets of $\bar{k}$ which contain j.

Since $\{(C_{k-s}, A, B_s) : s \in k(j)\}$ is a set of all complimentary susbsytems of Σ containing the input channel j, we can state the lemma

<u>Lemma 5.8.2</u>

The transfer matrices of all complimentary subsystems of Σ which contain input channel j, are non-zero if and only if for each $i \in k$, with $i \neq j$, there exists a path in the graph of Σ from the node j to the node i.

Proposition 5.8.4 is proved using Lemma 5.8.2.

As noted previously, proposition 5.8.4 only provides necessary conditions for (A_F, B_j) to be controllable for some F. A complete solution to the problem is provided by the following theorem.

---

* A proper subset of $\bar{k}$ is a set that does not contain all the integers in $\bar{k}$.

Theorem 5.8.3

Let $\sum \equiv \{C_i, A, B_i, \bar{k}\}$ and $j \in \bar{k}$ be fixed with $\sum$ jointly controllable. There exists an $F \in \mathcal{F}$ such that

$$< A_F | \beta_j > = X$$

if and only if each complimentary subsystem of $\sum$ which contains the input channel is complete.

The next theorem further characterises the pair (A_F, B_j) for the case where the conditions of theorem 5.8.3 fail to hold.

Theorem 5.8.4

Let $\sum$ and j satisfy the hypotheses of theorem 5.8.3 and write $\alpha_F(\lambda)$ for the uncontrollable polynomial of (A_F, B_j). Then

(i) $\forall F \in \mathcal{F}$

$$\text{LCM } \rho(C_{\overline{k-s}}, A, B_s) \text{ divides } L_F(\lambda)$$

(LCM denotes the least common multiple)

(ii) If in addition for each $i \in \bar{k}$ with $i \neq j$ there is a path in $G_\sum$ from node j to node i, then $\alpha_F(\lambda)$ is independent of $F \in \underline{\mathcal{F}}_j$ and

$$\alpha_F(\lambda) \text{ divides } \prod_{s \in k(j)} \alpha(C_{\overline{k-s}}, A, B_s)$$

where $\underline{\mathcal{F}}_j$ denotes the set of all $F \in \mathcal{F}$ for which $d(< A_F | \beta_j >)$ is as large as possible.

Theorem 5.8.4 implies that the uncontrollable spectrum of (A_F, B_j), namely the root set of $\alpha_F(\lambda)$ contains the root set of $\text{LCM}_{s \in \bar{k}(j)} \rho(C_{\overline{k-s}}, A, B_s)$ as a fixed subset $\forall F \in \mathcal{F}$. The theorem also implies that if node j of $G_\sum$ is connected by a directed path to every other node of $G_\sum$, then the uncontrollable spectrum of (A_F, B_j) is contained in another fixed subset, namely the root set of $\prod_{s \in \bar{k}(j)} \rho(C_{\overline{k-s}}, A, B_s)$, whenever F is selected such that $d(< A_F | \beta_j >)$ is as large as possible.

From the relationship between the two subsets bounding the root set of $\alpha_F(\lambda)$ it follows that if node j of $G_\sum$ is connected to every other node of $G_\sum$, then F can be selected such that (A_F, B_j) is stabilisable for the case where each polynomial $\rho(C_{\overline{k-s}}, A, B_s)$, $s \in \bar{k}(j)$ is stable.

5.8.3 Controllable and observable single-channel subsystems

By use of theorem 5.8.3 and of duality, it is possible to derive conditions for the existence of an $F \in \mathcal{F}$ which makes the jth channel of $\sum_F$ both controllable and observable.

We call a k-channel system complete whenever all its complementary subsystems are complete.

Theorem 5.8.5

Let $\sum \equiv \{C_i, A, B_i, k$ be a jointly controllable observable system and let $j \in k$ be fixed. There exists an $F \in F$ such that

$$< A_F | \beta_j > = X$$

and $[C_j \; A_F] = 0$

if and only if $\sum$ is complete.

Since completeness is a property which does not depend on j (since we ask for all the complementary subsystems to be complete), if $\sum$ is complete then for any $j \in \bar{k}$, F_j can be selected such that (C_j, A_F, B_j) is controllable and observable.

Corollary 5.8.4

If $\sum$ is a complete, jointly controllable, jointly observable system, then there exists an $F \in F$ such that for all $j \in \bar{k}$ (C_j, A_F, B_j) is controllable and observable.

In order for $\sum$ to be complete, it is necessary for the transfer matrices of each of its complementary subsystems to be non-zero. But from Lemma 5.8.2, it is obvious that:

Lemma 5.8.3: The transfer matrices of all complementary subsystems of $\sum$ are non-zero if and only if the graph of $\sum$ is strongly connected.

A directed graph is said to be strongly connected if for every pair of distinct vertices i and j, there exists a path from i to j as well as from j to it as illustrated in Fig. 5.11.

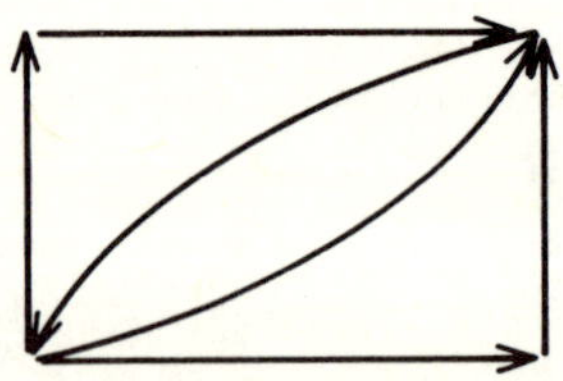

Figure 5.11

This figure shows a strongly connected graph. In our context, such graphs mean that each controller influences all the outputs or each output is influenced by all the controllers.

5.8.4 Strongly connected systems

By using all the preceding results, it is now possible to obtain complete and constructive solutions to the spectrum assignment and stabilisation problems for strongly connected systems.

It is well known that the closed loop spectrum of a linear system (C,A,B), resulting from the application of a centralised linear control law consists of two subsets: one which is controller dependent and freely assignable using dynamic controllers and the other called the unassignable spectrum of (C,A,B) which is controller independent and is therefore a fixed subset of the closed loop system's spectrum.

It is also known that if $\alpha(\lambda)$ is the characteristic polynomial of A mod $< A|\beta >$ (uncontrollable polynomial) and if $\beta(\lambda)$ is the characteristic polynomial of A restricted to $< A|\beta > \cap [C|A]$ (the controllable but unobservable polynomial), then the root set of $\alpha(\lambda)\ \ \beta(\lambda)$ coincides with the <u>unassignable polynomial</u> of (C,A,B).

<u>Theorem 5.8.6</u>

Let $\sum \equiv \{C_i, A, B_i \bar{k}\}$ be a jointly controllable, jointly observable strongly connected system. Let $u_F(\lambda)$ denote the unassignable polynomial of (C_j, A_F, B_j) and write $\alpha_\sum(\lambda)$ for the fixed polynomial of $\sum$. Then

i) $\underset{s\,\varepsilon\,k^*}{\text{LCM}}\ \rho(C_{k-s}, A, B_s)$ divides $u_F^j(\lambda)$ $\quad j\ \varepsilon\ \bar{k},\ F\ \varepsilon\ \mathcal{F}$

ii) $u_F^j(\lambda)$ divides $\prod_{s\,\varepsilon\,k^*} \rho(C_{k-s}, A, B_s)$ $\quad j\ \varepsilon\ \bar{k},\ F\ \varepsilon\ \mathcal{F}^*$

iii) $u_F^j(\lambda) = \alpha_\sum(\lambda) \quad\quad j\ \varepsilon\ \bar{k},\ F\ \varepsilon\ \mathcal{F}^*$

where k* denotes the set of all proper subsets of k and $\mathcal{F}^*$ is a robust set of $\mathcal{F}$.

Note that condition iii) asserts that $u_F^j(\lambda)$ is independent of $F\ \varepsilon\ \mathcal{F}^*$ and and $j\ \varepsilon\ \bar{k}$. Thus for fixed $F\ \varepsilon\ \mathcal{F}^*$, the set of possible closed loop spectra which we can achieve by applying dynamic feedback to a single channel of $\sum_F$ is the same for all channels of $\sum_F$.

Combining iii), Lemma 5.8.3 and theorem 5.8.5, we have:

<u>Theorem 5.8.7</u>

Let $\sum \equiv \{C_i, A, B_i, \bar{k}\}$ be a jointly controllable, jointly observable, strongly connected system. Then, under decentralised control:

i) The spectrum of $\sum$ can be freely assigned if and only if $\sum$ is complete

ii) can be stabilised if and only if for each $s \in k^*$, $\rho(C_{k-s}, A, B_s)$ is a stable polynomial.

5.8.5 k-channel systems

We next examine the spectrum assignment and stabilisation problems for systems which are not necessarily strongly connected. Here, theorem 5.8.5 is not applicable i.e. it is not possible to find an F which will make a single channel of $\sum_F$ controllable and observable.

However, it may still be possible to assign or stabilise the system's spectrum with decentralised control.

Using graph theoretic techniques, we can decompose $G_{\sum}$ into a number (say p) of strongly connected components $G_i \equiv (q_i, \Gamma_i)$; the subsets q_i form a partition of $\bar{k}$ (i.e. $\bar{k}$ can be written as the disjoint union of the q,s); Γ_i is a function mapping q_i according to the rule

$$\Gamma_i(j) \equiv \{r : r \in (\Gamma(j)) \cap \bar{q}_i\}$$

we recall that

$$\Gamma(j) = \{i : H_{ij} \neq 0\}; \quad H_{ij} = C_i(\lambda I - A)^{-1} B_j$$

$\Gamma_i(j)$ can be considered as an internal mapping to $\bar{q}_i$.

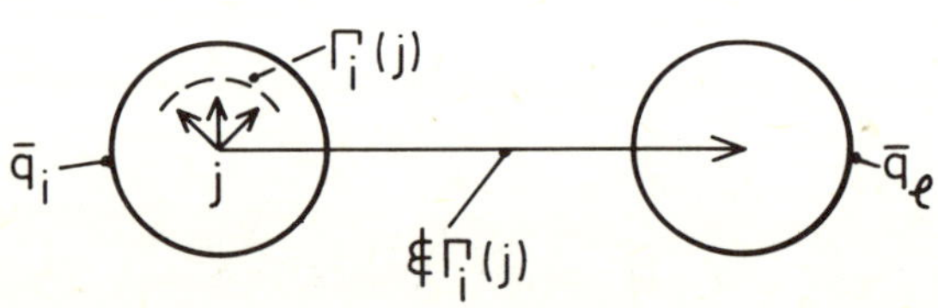

Figure 5.12

The required graph of $\sum$, written G^R, is the pair $(\bar{q}^R, \Gamma^k)$ where $\bar{q}^R \equiv \{(q_1, q_2, \ldots q_p\}$ and Γ^R is the mapping from $\bar{q}^k$ (an external mapping), defined in such a way that $\bar{q}_i \in \Gamma^R(q_j)$ if and only if there exist nodes $i_0 \in e\, \bar{q}_i$ and $j_0 \in q_j$ such that $i_0 \in \Gamma(j_0)$.

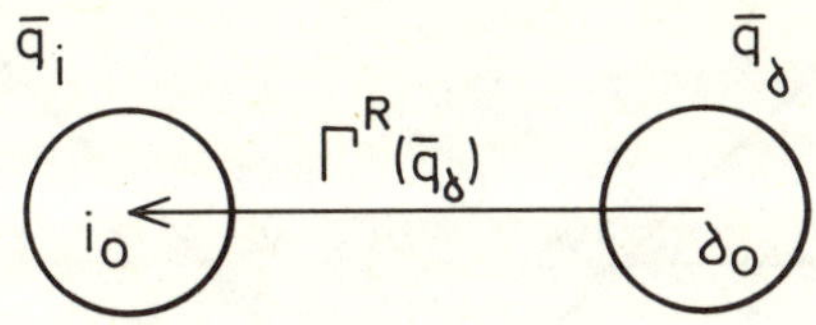

Figure 5.13

We call a component G_i isolated if $\Gamma^R(\bar{q}_i)$ is the empty set and

$$\bar{q}_i \notin \bigcup_{\substack{j \in \bar{p} \\ j \neq \underline{i}}} \Gamma^R(q_j)$$

i.e. no arc leaves or enters the set $\bar{q}_i$.

Recall that G_Σ was decomposed into p strongly connected components $G \equiv (\bar{q}_i, \Gamma_i)$. The figure below shows a graph with 2 strongly connected components.

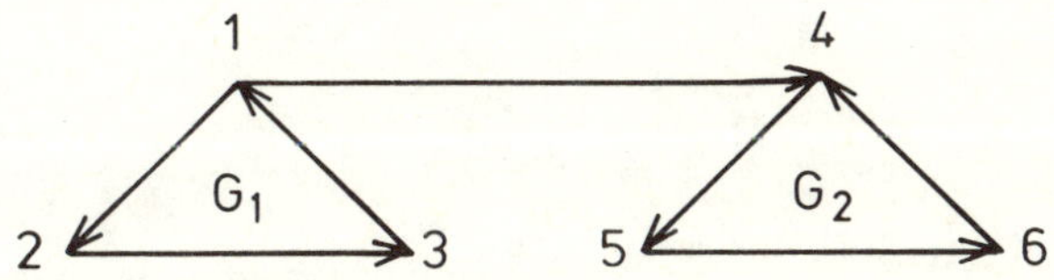

Figure 5.14

Note that we may not have the arc (3.5) since otherwise the whole graph would be strongly connected.

With this observation we can say that the reduced graph G^R contains no cycles i.e. if $\bar{q}_i \in \Gamma(\bar{q}_j)$ then $\bar{q}_j \notin \Gamma^R(\bar{q}_i)$. A cycle is a closed arc progression in which no arc is repeated.

Since the set $\{\bar{q}_1, \bar{q}_2, \ldots, \bar{q}_p\}$ has no cycles, we can order this set in such a way that

$$\Gamma^R(\bar{q}_1) \subset \{\bar{q}_1, \bar{q}_2, \ldots, \bar{q}_{i-1}\}$$

$$i=2,3,\ldots p$$

Figure 5.15 shows an example

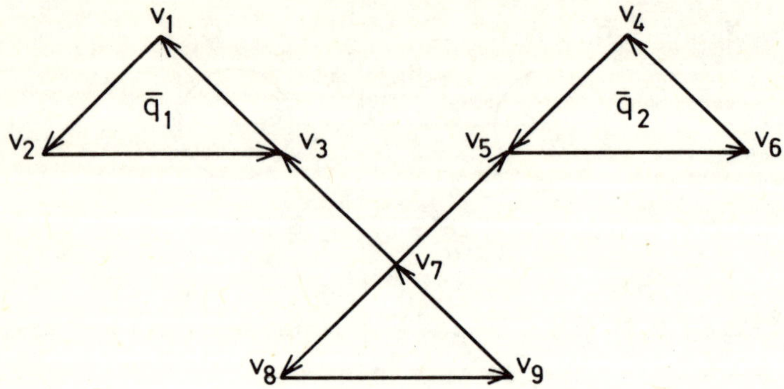

Figure 5.15

In this figure

$$\Gamma^R(\bar{q}_2) = \bar{q}_1 \;\; ; \;\; \Gamma^R(\bar{q}_3) = \{\bar{q}_1, \bar{q}_2\}$$

Associated with $\sum$, consider the kxk matrix $T_\sum$ defined by

$$t_{ij} = \begin{cases} 0 \text{ if } i \notin \Gamma(j) \\ 1 \text{ otherwise} \end{cases}$$

For the subgraph $(\bar{q}_1, \Gamma(j))$ the matrix is

$$\begin{matrix} & v_1 & v_2 & v_3 \\ v_1 \\ v_2 \\ v_3 \end{matrix} \begin{pmatrix} 0 & 0 & 1 \\ 1 & 0 & 0 \\ 0 & 1 & 0 \end{pmatrix} = T_{\sum 1}$$

If we label the arcs of the subgraph $(\bar{q}_1, \Gamma(j))$

$$\begin{matrix} & e_1 & e_2 & e_3 \\ v_1 \\ v_2 \\ v_3 \end{matrix} \begin{pmatrix} 0 & 1 & 0 \\ 0 & 0 & 1 \\ 1 & 0 & 0 \end{pmatrix} = A$$

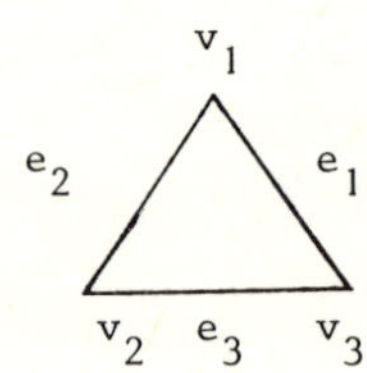

and use the rule $a_{ij} = 1$ when e_j leaves v_i, we obtain the incidence matrix A (transpose of $T_{\sum 1}$)

Let k_i, $i \in \bar{p} \equiv \{1,2,\ldots p\}$ be the number of elements in $\bar{q}_i$. It is always possible to relabel the channels of Σ in such a way that the resulting matrix T^* has a block triangular structure. For this purpose also use the fact that

$$\Gamma^R(\bar{q}_i) \subset \{\bar{q}_1, \bar{q}_2, \ldots \bar{q}_i\} \qquad i=2,3,\ldots p$$

For example we can write the following T^*_Σ for the last figure (5.15) as

	v_1	v_2	v_3	v_4	v_5	v_6	v_7	v_8	v_9
v_1	0	0	1	1					
v_2	1	0	0						
v_3	0	1	0						
v_4		↑		0	0	1	1		
v_5		T_{11}		1	0	0	1		
v_6				0	1	0			
v_7							0	0	1
v_8				T_{22}			1	0	0
v_9							0	1	0
							↑ T_{33}		

In general, the matrix T^*_Σ will have the form

$$T^* = \begin{pmatrix} T_{11} & \cdots & T_{1p} \\ & T_{22} \cdots & T_{2p} \\ & \ddots & T_{pp} \end{pmatrix}$$

Since the presence of an arc between v_i and v_j indicates that $C_i(\lambda I-A)^{-1}B_j = 0$, we can state that the overall matrix of Σ admits a similar block triangular structure; the ith diagonal block in the latter is the transfer matrix of the k_i-channel strongly connected system $\{C_j, A, B_j, \bar{q}_i\}$ described by the equations:

$$\left.\begin{aligned} \dot{x}_1(t) &= Ax_i(t) + \sum_{j \in \bar{q}_i} B_j u_j(t) \\ y_j(t) &= C_j x_i(t) \quad j \in \bar{q}_i \end{aligned}\right\} \tag{5.8.11}$$

Note that in general, this system is neither jointly controllable nor jointly observable. Nevertheless using realisation theory as outlined in the beginning of this chapter, we can extract a minimal realisation (state space of minimal dimension) and this minimal realisation corresponds to a completely controllable, completely observable system.

Let $\sum^i = \{\bar{C}_j, \bar{A}_i, \bar{B}_j, \bar{q}_i\}$ be such a realisation with the same overall transfer matrix as equation (5.8.11). We call $\sum^i$ the ith strongly connected sub-system of $\sum$.

We wish to relate the closed loop fystem of $\sum$ to the closed loop spectrum of the k-channel system determined by the $\sum^i$, when identical local controllers $\sum_i$ are applied to both systems. For this write $\Lambda(\sum_j, k)$ for the closed loop spectrum of $\sum$, and with identical controls applied to the $\sum^i$, let $\Lambda_i(\sum_j, q_i)$ denote its closed loop spectrum.

Since the transfer matrix of $\sum$ has a block triangular structure with the transfer matrix of $\sum^*$ being the ith diagonal block, the parameter matrices describing $\sum$ can be written as:

$$C_j = [\hat{C}_j, x \ldots x] \quad j \in \bar{q}_i$$

$$C_j = [0, \hat{C}_j, x \ldots x] \quad j \in \bar{q}_2$$

$$C_j = [0 \ldots 0, C_j] \quad j \in \bar{q}_p$$

$$A = \begin{pmatrix} \hat{A}_1 x \ldots x \\ 0 \; \hat{A}_2 x \ldots x \\ \vdots \quad \ddots \\ \vdots \;\; 0 \quad \ddots \; x \\ \vdots \qquad\quad \ddots \\ 0 - - - 0 A_p \end{pmatrix}$$

$$B_j = \begin{pmatrix} \hat{B}_j \\ 0 \\ \vdots \\ 0 \end{pmatrix} \quad j \in \bar{q}_1 \qquad B_j = \begin{pmatrix} x \\ \hat{B}_j \\ 0 \\ \vdots \\ 0 \end{pmatrix} \quad j \in \bar{q}_2 \ldots$$

$$B_j = \begin{pmatrix} x \\ x \\ \hat{B}_j \end{pmatrix} \quad j \in \bar{q}_p \ldots.$$

where the x^s denote matrices which are not of importance in the current analysis and $\sum^i \equiv \{C_j, A_i, B_j; \bar{q}_i\}$ is a system with the same overall transfer matrix as $\sum^i \equiv \{C_j, \bar{A}_i, \bar{B}_j; \bar{q}_i\}$.

Note that writing A as a block triangular matrix means that $(\lambda I - A)^{-1}$ is also block triangular with the ith diagonal block being given by $(\lambda I - \hat{A}_i)^{-1}$. The matrices

$$C_j = [0,0\ldots C_j, x\ldots x] \quad j \in \bar{q}_i$$
$$\qquad\qquad\uparrow i$$

$$B_j = \begin{bmatrix} x \\ x \\ B_j \\ 0 \\ 0 \end{bmatrix} \leftarrow i \quad j \in \bar{q}_i$$

must be such that

$$\hat{C}_j(\lambda I-\hat{A}_i)\hat{B}_j = \bar{C}_j(\lambda I-\bar{A}_i)\bar{B}_j$$

Since $\sum^i$ is jointly controllable and jointly observable, the parameter matrices of $\hat{\sum}^i$ can be written as:

$$\hat{C}_j = [0,\bar{C}_j,x] \qquad j \in \bar{q}_i$$

$$\hat{A}_i = \begin{bmatrix} A_{i1} & x & x \\ 0 & \bar{A}_i & x \\ 0 & 0 & A_{i2} \end{bmatrix} \qquad \hat{B}_j = \begin{bmatrix} x \\ \bar{B}_j \\ 0 \end{bmatrix}$$

such that

$$\hat{C}_j(\lambda I-\hat{A}_i)^{-1}\hat{B}_j = \bar{C}_j(\lambda I-\bar{A}_i)\bar{B}_j$$

Note that the set $\Lambda_o \equiv \bigcup_{i \in \bar{p}} \sigma(\ (A_{i1}) \bigcup \sigma(A_{i2}))$ is the complement of the union of the open loop spectra of the $\sum^i$ in the spectrum of $\sum$.

Union of open loop spectra of the $\sum^i \equiv \bigcup_{i \in \bar{p}} (\sigma(\bar{A}_i))$

For each $j \in \bar{k}$, let $\dot{Z}_j = H_jZ_j + L_jY_j, U_j = M_jZ_j + F_jY_j$ denote a local control system $\sum_j$ to be applied to the jth channel of $\sum$. Write $\hat{A}_i^*$ for the system matrix describing the overall closed loop system which results when controllers $\sum_j(j\epsilon\bar{p}_i)$ are applied to $\sum^i$. It can be shown (and we will show this below) that the system matrix associated with the closed loop system resulting from an application of $\sum_i$ to $\sum$ can be written as:

$$A^* = \begin{bmatrix} A_1 & x & \ldots & x \\ 0 & A_2^* & x\ldots & \\ \vdots & 0 & \ddots & x \\ 0 & \cdots & 0 & \hat{A}_p \end{bmatrix} \qquad \text{where}$$

$$A_i^* = \begin{pmatrix} A_{i1} & x & x \\ 0 & \bar{A}_i^* & x \\ 0 & 0 & A_{i2} \end{pmatrix}$$

Let us now show that A* can be written as a block triangular matrix by means of an example. Suppose k=2. Then we have 2 subgraphs $\bar{q}_1$ and $\bar{q}_2$ each strongly connected. The closed loop equation can be written as

$$\begin{aligned} \dot{x}(t) &= Ax(t) + BF\,y(t) + BM\,z(t) \\ &= (A+BFC)x(t) + BMz(t) \end{aligned}$$

where $B = [B_1 \;\; B_2]$; $F = \begin{pmatrix} F_1 & 0 \\ 0 & F_2 \end{pmatrix}$; $C = \begin{pmatrix} C_1 \\ C_2 \end{pmatrix}$

$$M = \begin{pmatrix} M_1 & 0 \\ 0 & M_2 \end{pmatrix} \; ; \quad z(t) = \begin{pmatrix} z_1(t) \\ z_2(t) \end{pmatrix}$$

Recall that A,B,C can be written as

$$B = \begin{pmatrix} \hat{B}_1 & x \\ 0 & B_2 \end{pmatrix} \; ; \quad C = \begin{pmatrix} \hat{C}_1 & x \\ 0 & C_2 \end{pmatrix} \; ; \quad A = \begin{pmatrix} \hat{A}_1 & x \\ 0 & A_2 \end{pmatrix}$$

Then

$$\begin{aligned} A^* &= \begin{pmatrix} \hat{A}_1 & x \\ 0 & B_2 \end{pmatrix} + \begin{pmatrix} \hat{B}_1 & x \\ 0 & B_2 \end{pmatrix} \begin{pmatrix} F_1 & 0 \\ 0 & F_2 \end{pmatrix} \begin{pmatrix} \hat{C}_1 & x \\ 0 & C_2 \end{pmatrix} \\ &= \begin{pmatrix} \hat{A}_1+\hat{B}_1F_1\hat{C}_1 & x \\ 0 & \hat{A}_2+\hat{B}_2F_2\hat{C}_2 \end{pmatrix} = \begin{pmatrix} \hat{A}_1^* & x \\ 0 & A_2^* \end{pmatrix} \end{aligned}$$

Now $\hat{A}_i^* = \begin{pmatrix} A_{i1} & x & x \\ 0 & \bar{A}_i & x \\ 0 & 0 & A_{i2} \end{pmatrix}$

Recall from equation 5.8.11 that subsystem i_1 is unobservable and subsystem i_2 is uncontrollable but the subsystem i (corresponding to A_i) is both controllable and observable. Thus output feedback only affects $\bar{A}_i$.

Hence we can write

$$\sigma(A^*) = \Lambda_o \bigcup_{i \in \bar{p}} \sigma(\bar{A}_i^*)$$

We are thus led to the following proposition which characterises the closed loop spectrum of $\sum$ in terms of the closed loop spectra of its strongly connected subsystems:

Proposition 5.8.5

$$\forall \text{ linear controls } \sum_j,\ j \in \bar{k},$$

$$\Lambda(\sum_j, \bar{k}) = \Lambda_o \bigcup_{i \in \bar{p}} \Lambda_i(\sum_j, q_i)$$

Proposition 5.8.5 shows that the closed loop spectrum can be freely assigned (respectively stabilised) just in case Λ_o is empty (respectively stable) and spectrum assignment (respectively stabilisation) is possible for each strongly connected subsystem of $\sum$. If these conditions are satisfied, then spectrum assignment (or stabilisation) can be achieved by applying the above methods to each one of the strongly connected subsystems of $\sum$.

In case Λ_o is empty, then the sum of the dimension of $\sum$'s strongly connected subsystems is equal to the dimension of $\sum$. In that case using n control variables we can achieve stability or assign a prescribed spectrum or in other words, the total dimension of the decentralised control system used never exceeds n.

The preceding observations can be summarised in the following way:

Theorem 5.8.8

Let $\sum \equiv \{C_i, A, B_i, \bar{k}\}$ be a jointly controllable, jointly observable system. Then, with decentralised control, the closed loop spectrum of $\sum$ can be:

i) freely assigned if and only if the sum of the dimensions of $\sum$'s strongly connected subsystems is equal to the dimension of $\sum$ and spectrum assignment is possible for each subsystem.

ii) stabilised if and only if the complement of the disjoint union of the spectra of $\sum$'s strongly connected subsystems in the spectrum of $\sum$ is stable and stabilisation is possible for each strongly connected system.

5.8.6 Comments

Theorem 5.8.8 and 5.8.7 have provided explicit conditions for both the spectrum assignment and stabilisation problems for k-channel linear systems. The analysis of Corfmat and Morse [10,11] is a deep one. However, as Sandell et al [12] point out, the Corfmat-Morse approach suffers from certain disadvantages as a practical design approach. They mention the fact that even if all the modes of a large scale system can be made controllable and observable from a single controller (or a few controllers if the system is not strongly connected) some of the modes may be very weakly controllable or observable. Thus impractically large gains may be required to assign all the poles from a single controller. Also, it is not clear that all the designers available degrees of freedom are used in the best possible way. Basically, the approach appears to require that all the disturbances in the system propogate to a single output where they may be observed and compensated for by control signals at an adjacent input. In many situations, it would be better to use controllers to suppress the propogation of disturbances through the system. In addition, concentration of all the complexity of the control structure at a single (or few) controllers may be undesirable.

In the next section we will outline an intuitively attractive approach which attempts to distribute the computational load over all the controllers and which thus utilises the available degrees of freedom of the designer in a better way.

5.9 DECENTRALISED CONTROL OF UNKNOWN SYSTEMS USING TUNING REGULATORS [13,14]

Davison considers the problem of decentralised control for a general servomechanism problem where the system to be regulated is unknown. The unknown system is subjected to disturbances and we seek a decentralised controller which is such that the system's outputs track given specified reference inputs despite the disturbances. Also, the control should be insensitive to small parameter variations which do not cause instability.

Davison assumes that the system is described by a finite linear time-invariant multivariable model and that the system is stable. He does not assume that the order of the system is known.

The solution procedure adopted is to impose the constraint that only one controller (at a given control station) may be adjusted at any time and that once adjusted , this controller will remain fixed and that the closed loop system must remain stable at all times during the tuning procedure.

Davison [14] has derived necessary and sufficient conditions for the existence of a solution to the problem. The solution can be obtained by performing a series of experiments on the plant and a decentralised controller can then be synthesised using the results of these experiments together with a series of one-dimensional parameter searches performed on the closed loop system.

For simplicity of exposition, we will next describe the approach for the special case where the desired set points are constant and any disturbances occuring in the system are constant.

Let us assume that an unknown plant consists of Γ control stations, the ith of which has output $u_i \in R^{m_i}$, output $y \in R^{r_i}$ and error $e_i \triangleq y_i - y_i^{ref}$ where y_i^{ref} is a constant set point for the system. It is desired to find a decentralised controller for the plant such that the closed loop system is asymptotically stable and such that $e_i(t) \rightarrow 0$ as $t \rightarrow \infty$ $(i=1,2,...)$ $\forall\, y_i^{ref} \in R^{r_i}$, $\forall\, w \in R^q$ where w is an unknown constant disturbance vector.

As we mentioned above, the only assumptions made regarding the plant are that:

a) The plant can be described by a finite dimensional linear time-invariant model.

b) The plant is open loop asymptotically stable.

In particular, we do not assume that we know the order or parameters of the plant. Nor do we assume that the plant is necessarily controllable and/or observable from any of its inputs or outputs.

When we implement the decentralised controller on the plant, we impose the following synthesis constraints:

1) A controller can only be implemented at one control station at a time

2) After a controller has been obtained and implemented at a control station, it is assumed that the controller will remain fixed

3) At all times during the controller synthesis, the resultant closed loop system must remain asymptotically stable.

5.9.1 Solution Procedure [13]

Let us assume, for ease of exposition that $m_i = r_i$, $i=1,2,\ldots\Gamma$. In this case, we wish to determine the tracking steady state gain parameters $T(i,j)$ in order to determine the existence of a solution to the problem and controller structure for the system. The following experiments need to be carried out in order to determine $T(i,j)$. We assume that no disturbances are applied to the plant during these experiments.

Step 1: Apply a constant input $u_i(t) = \bar{u}_i^1 = 0$ with $u_j(t) = 0$, $j=1,2,\ldots$, $j \neq i$ to the plant; then since the plant is stable, there exists a constant vector $\bar{y}_j^1 \in R^{r_j}$ such that

$$\lim_{t\to\infty} y_j(t) = \bar{y}_j^1 \quad j=1,2,\ldots\Gamma$$

Measure $\bar{y}_j^1$ $j=1,2,\ldots\Gamma$

Step 2: Apply a constant input $u_i(t) = \bar{u}_i^2$ with $u_j(t) = 0$ $j=1,2,\ldots\Gamma$ $j \neq i$ to the plant where $\bar{u}_i^2$ has been chosen such that $\bar{u}_i^2$ is linearly independent of $\bar{u}_i^1$; then, since the plant is stable, there exists a constant vector $\bar{y}_j^2 \in R^{r_j}$ such that

$$\lim_{t\to\infty} y_j(t) = \bar{y}_j^2, \quad j=1,2,\ldots\Gamma$$

Measure $\bar{y}_j^2$, $j=1,2,\ldots\Gamma$

Step 3: Apply a constant input $u_i(t) = \bar{u}_i^{m_i}$ with $u_j(t) = 0$, $j=1,2,\ldots\Gamma$ $j \neq i$ to the plant, where $\bar{u}_i^{m_i}$ has been chosen such that u^{m_i} is linearly independent of $u_i^1, u_i^2, \ldots \bar{u}_i^{m_i-1}$; then, since the plant is stable, there exists a constant vector $\bar{y}_j^{m_i} \in R^{r_j}$ such that

$$\lim_{t \to \infty} y_j(t) - \bar{y}_j^{m_i}, \quad j=1,2,\ldots\Gamma$$

Measure $\bar{y}_j^{m_i}$, $j=1,2,\ldots\Gamma$

From these off-line experiments, we can define the tracking steady state gain parameters T(i,j) of the system as:

$$T(i,j) = (\bar{y}_j^1, \bar{y}_j^2, \ldots \bar{y}_j^{m_i})\ (\bar{u}_i^1, \bar{u}_i^2, \ldots \bar{u}_i^{m_i-1}),$$

$$i=1,2,\ldots\Gamma$$

$$j=1,2,\ldots\Gamma$$

Then, Davison *[13]* proves the following theorem:

Theorem 5.9.1

There exists a solution to the decentralised servomechanism problem described above if and only if there exists a list of distinct integers $(s_1, s_2, \ldots s_\Gamma)$ (not necessarily unique) where $s_i \in [1,2,\ldots\Gamma]$ such that the following Γ conditions all hold:

1. $\text{rank } T(s_1, s_1) = r_{s_1}$

2. $\text{rank} \begin{pmatrix} T(s_2,s_2) & T(s_2,s_1) \\ T(s_1,s_2) & T(s_1,s_1) \end{pmatrix} = r_{s_1} + r_{s_2}$

(Γ) $\text{rank} \begin{pmatrix} T(s_\Gamma,s_\Gamma), & T(s_\Gamma,s_{\Gamma-1})\ldots, & T(s_\Gamma,s_1) \\ \\ T(s_\Gamma,s_\Gamma), & T(s_\Gamma,s_{\Gamma-1})\ldots, & T(s_1,s_1) \end{pmatrix}$

$$= \sum_{i=1}^{\Sigma} r_{s_i}$$

If this theorem holds, then we can synthesise the controller as we show next.

5.9.2 Controller synthesis

Step 1: Apply the following controller to control agent number s_1:

$$u_{s_1} = \times - \varepsilon_{s_1} T^{-1}(s_1,s_1) \int_0^t e_{s_1}(\tau)d\tau \qquad (5.9.1)$$

where ε_{s_1} is a positive scalar. ε_{s_1} is then adjusted such that the maximum speed of response occurs i.e. the controller is applied to the plant with ε_{s_1} very small (for example 10^{-10}) and a step response of the closed loop system is then obtained. A one dimensional search is next carried out on $\varepsilon_s > 0$ in order to maximise the speed of response (without excessive oscillations occuring). We will denote by $\varepsilon^*_{s_1}$ the "optimum" choice of ε_{s_1}.

Step 2: With the controller (5.9.1) connected to the plant (and with ε_{s_1} set at $\varepsilon^*_{s_1}$), we repeat the experiments of section 5.9.1 in order to calculate $T(s_2,s_2)$. Let us denote the resultant matrix by $T_{\varepsilon^*_{s_1}}(s_2,s_2)$
We next apply the following controller to control agent number s_2:

$$u_{s_2} = - \varepsilon_{s_2} T^{-1}_{\varepsilon^*_{s_1}}(s_2,s_2) \int_0^t \varepsilon_{s_2}(\tau)d\tau \tag{5.9.2}$$

whereas in step 1, ε_{s_2} is a positive scalar parameter which needs to be adjusted by on-line tuning in order to achieve a maximum speed response. Let us denote by $\varepsilon^*_{s_2}$ this "optimum" choice of ε_{s_2}.

Step Γ: With the controllers of control agent number $s_1,s_2,\ldots s_{\Gamma-1}$ connected (with $\varepsilon_{s_i} = \varepsilon^*_{s_i}$ $i=1,2,\ldots -1$), we repeat the experiments of section 5.9.1 in order to calculate $T(s_\Gamma,s_\Gamma)$; let us denote the resultant matrix obtained by $T_{\varepsilon^*_{s_1}\ldots\varepsilon^*_{s_{\Gamma-1}}}(s_\Gamma,s_\Gamma)$. We next apply the following controller to control agent no. s_Γ:

$$u_{s_\Gamma} = -\varepsilon_{s_\Gamma} T^{-1}_{\varepsilon^*_{s_1}\ldots\varepsilon^*_{s_{-1}}}(s_\Gamma,s_\Gamma) \int_0^t \varepsilon_s(\tau)d\tau \tag{5.9.3}$$

where ε_{s_Γ} is a positive scalar which is adjusted on-line in order to achieve maximum speed of response; we denote the optimum choice of ε_{s_Γ} by $\varepsilon^*_{s_\Gamma}$.
This concludes the controller synthesis; thus the final decentralised controller is given by:

$$u_{s_1} = -\varepsilon^*_{s_1} T^{-1}(s_1,s_1) \int_0^t \varepsilon_{s_1}(\tau)d\tau \tag{5.9.4}$$

$$u_{s_i} = -\varepsilon^*_{s_i} T^{-1}_{\varepsilon^*_{s_1}\ldots\varepsilon_{s_{i-1}}}(s_i,s_i) \int_0^t \varepsilon_{s_i}(\tau)d\tau \qquad i=2,3,\ldots\Gamma \tag{5.9.5}$$

We note that in order to solve the problem, the controller synthesis must proceed in a certain preassigned order i.e. the controller for control agent number s_1 must first be applied, then the controller for control agent number s_2 etc. up to the controller agent number s_Γ. Davison *[13]* points out that failure to do so may make it impossible to solve the problem.

This is a practical design technique and Davison has used it to solve problems of power systems control *[13]* and the basic technique has also been used to tackle problems of control of heat exchangers *[15]*.

5.10 COMMENTS

In this chapter we have outlined the implications of the pioneering work of Wang and Davison and of Corfmat and Morse. There have been a number of other works in this area but the work of Wang and Davison and those of Corfmat and Morse are by far the most significant.

The notion of fixed modes arises from similar considerations as the difficulties that we encountered in Chapter 3 on examining stochastic problems with non-classical information patterns. Thus, it is the existence of the non-classical information pattern which ensures that the standard pole placement results for centralised controllers are no longer applicable to the decentralised case. It is possible, however, in special cases, to get solutions. The case of strongly connected systems is one such case. However, such cases are rare and we do need robust design tools in order to tackle practical decentralised control problems. We saw some such techniques for the stochastic case in Chapter 4 and we will examine the deterministic techniques in Chapter 6.

5.11 ACKNOWLEDGEMENTS

I have drawn heavily on the works of Wang and Davison and of Corfmat and Morse in writing this chapter. I have also used the notes of my Ph.D. student, Mr. Vinicius Armentano, whose help is gratefully acknowledged.

5.12 REFERENCES

Specific References

1. Wonham W.M. "on pole assignment in multi-input controllable linear systems" IEEE Trans AC 12, 660-665, 1967.

2. Brasch F.M. and Pearson J.B. "Pole placement using dynamic compensators" IEEE Trans AC 15, 34-43, 1970.

3. Luenberger D.G. "observers for multivariable systems" IEEE Trans AC 11, 190-197, 1966.

4. Wang S. "An example in decentralised control" IEEE Trans AC 23, 5, p. 938, 1979.

5. Wang S. and Davison E.J. "on the stabilisation of decentralised control systems" IEEE Trans AC 18, 5, pp. 473-478, 1973.

6. Davison E.J. "Decentralised stabilisation and regulation in linear multivariable systems" in "Directions in decentralised control" Ho Y. and Mitter S. (editors), Pergamon Press, 1976.

7. Bardiaux, Docteur Ingeneiur thesis, Nantes, France, 1976.

8. Singh M.G., Elloy J.P., Mezencev R., and Munro N., "Applied Industrial Control", Pergamon Press, Oxford, 1980.

9. Moog C., Private communication.

10. Corfmat J. and Morse A., "Decentralised control of linear multivariable systems", Automatica, Vol. 12, pp. 479-495, 1976.

11. Corfmat J. and Morse A., "Control of linear systems through specified input channels", SIAM J. Control and Optimisation Vol. 14, 1, pp. 163-175, 1976.

12. Sandell N., Varaiya P., Athans M., and Safanov M. "Survey of decentralised control methods for large scale systems", IEEE Trans AC-23, 2, pp. 108-128, 1978.

13. Davison E.J. and Tripathi N. "Decentralised tuning regulators: an application to solve the load and frequency control problem for a large power system", Proc. 18th IEEE CDC Conference, December 1979.

14. Davison E.J. "Decentralised robust control of unknown systems using tuning regulators", IEEE Trans AC-23, 2, pp. 276-289, 1978.

15. Davison E.J., Taylor P. and Wright J. "On the application of tuning regulators to control a commercial heat exchanger", Report No. 7902, Control Science and Engineering, University of Toronto, 1979.

General References on Decentralised Stabilisation

1. Aoki M. "On feedback stabilisability of decentralised dynamic systems", Automatica, 8, 2, 1972.

2. McFadden D. "On the controllability of decentralised microeconomic systems, the assignment problem" in "Mathematical Systems theory and applications", New York, Springer Verlag, 1967.

3. Aoki M. and Li M.T. "Controllability and stabilisability of decentralised dynamic systems" in Proc. 1973 JACC, Columbus, 1973.

4. Kobayaski H., Hanafusu H. and Yoshikawa T. "Controllability of continuous time decentralised control systems", IEEE Trans AC, 1978.

5. Davison E.J. "The robust decentralised control of a general servomechanism problem", IEEE Trans AC-21, pp. 14-24, 1976.

6. Corfmat R. and Morse A. "Stabilisation with decentralised feedback control", IEEE Trans AC-18, pp. 679-682, December 1973.

CHAPTER 6

DESIGN TECHNIQUES

6.1 INTRODUCTION

In the previous chapter, we studied mainly techniques for the analysis of decentralised systems. In this final chapter, we provide an overview of the design techniques that are available.

We recall that the fundamental result in the analysis of decentralised controllers was the result of Wang and Davison on fixed modes [1] i.e. it is possible to stabilise linear time invariant systems having multiple control stations by decentralised feedback only if the system has no unstable fixed modes. Furthermore, we saw how this condition could be checked numerically. In this chapter, we go on to the development of practical design techniques. We note that once we have a system which has no unstable fixed modes, we would seek to obtain good performance out of the system. Thus, we would seek to achieve a compromise between the degree of stability and optimality. In this chapter, the techniques that we outline all rely on notions of optimality.

The first approach that we will describe is due to Geromel and Bernussou [2]. However, before we outline this approach, we group together some preliminary concepts.

6.2 PRELIMINARY RESULTS [3]

Theorem 6.2.1:

If A(t) is continuous for $t \geqslant o$, there exists a unique solution to the vector differential equation

$$\dot{x}(t) = A(t)x(t) \quad ; \quad x(o)=c \tag{6.2.1}$$

This solution exists for $t \geqslant o$ and may be written in the form

$$x(t) = x(t)\ c \tag{6.2.2}$$

where x(t) is the unique matrix satisfying the matrix differential equation

$$\dot{X}(t) = A(t)X(t) \quad ; \quad X(o)=I \tag{6.2.3}$$

or

$$X(t) = I + \int_o^t A(s)X(s)ds \tag{6.2.4}$$

6.2.1 The matrix exponential

In the scaler case, the equation $\frac{du}{dt} = a\,u$, $u(o) = C$ has the solution $u = e^{at}C$. For $A(t)$ = constant, it would be convenient to find an analogous solution to the matrix differential equation

$$\dot{X}(t) = AX(t) \quad ; \qquad X(o) = C$$

which has the form:

$$X = e^{AT}C$$

It can be shown that this is so since $\frac{d}{dt} e^{At} = Ae^{At} = e^{At} A$.

6.2.2 Solution of inhomogenous equations with constant coefficients

Consider the system

$$\dot{x}(t) = Ax(t) + f(t);\ x(o) = C \tag{6.2.5}$$

Let us examine the derivative

$$\frac{d}{dt}(e^{-At}x(t)) = -Ae^{-At}x(t) + e^{-At}\dot{x}(t) \tag{6.2.6}$$

$$= e^{-At}(\dot{x}(t) - Ax(t)$$

$$= e^{-At}f(t) \tag{6.2.7}$$

Hence

$$e^{-At}\underline{x}(t) = \underline{C} \int_o^t e^{-As} f(s)ds$$

or

$$\underline{x}(t) = e^{At}\underline{C} + \int_o^t e^{A(t-s)} f(s)ds \tag{6.2.8}$$

6.2.3 Perturbation Theory

Given the matrix $e^{(A+\varepsilon B)t}$, we wish to evaluate it as a power series in ε i.e.

$$e^{(A+\varepsilon\ B)t} = e^{At} + \sum_{n=1}^{\infty} \varepsilon^n Q_n(A(t),B(t)) \tag{6.2.9}$$

Assume that $e^{(A+\varepsilon B)t}$ is the solution of the differential equation

$$\frac{dX(t)}{dt} = (A+\varepsilon B)X(t) \text{ or}$$

$$\dot{X}(t) = AX(t) + \varepsilon\ BX(t) \text{ with } X(o) = I \tag{6.2.10}$$

It follows from equation (6.2.8) that X(t) satisfies the linear integral equation

$$X(t) = e^{At} + \varepsilon \int_{o}^{t} e^{A(t-s)} BX(s)ds$$

or $$e^{(A+\varepsilon B)t} = e^{At} + \varepsilon \int_{o}^{t} e^{A(t-s)} Be^{As} ds + \ldots. \tag{6.2.11}$$

which is the approximation to first order in ε of $e^{(A+\varepsilon B)t}$.

Theorem 6.2.2

The solution of

$$\dot{X}(t) = AX(t) + X(t)\,B \qquad X(o) = C$$

is given by $X(t) = e^{At}\, Ce^{Bt}$

This is easy to check since

$$\dot{X}(t) = Ae^{At}\, Ce^{Bt} + e^{At}\, CBe^{Bt} = Ae^{At}Ce^{Bt} + e^{At}Ce^{Bt}B$$
$$= AX(t) + X(t)B$$

6.2.4 Theorem 6.2.3

If the expression

$$X = -\int_{o}^{\infty} e^{At}\, Ce^{Bt} dt \qquad \text{exists for all } C,$$

it represents the unique solution of

$$AX + XB = C \tag{6.2.12}$$

Proof

Consider the equation

$$\dot{Z}(t) = AZ(t) + Z(t)B \ ; \quad Z(o) = C$$

Let us integrate both sides of this equation between o and ∞ under the assumption that $\lim_{t\to\infty} Z(t) = o$. The result is:

$$-C = A\int_{o}^{\infty} Z(s)ds + \left(\int_{o}^{\infty} Z(s)ds\right)B \tag{6.2.13}$$

We can see that

$$-\int_{o}^{\infty} Z(s)ds = -\int_{o}^{\infty} e^{At}\, Ce^{Bt} dt$$

satisfies equation (6.2.12) (see also theorem 6.2.2).

The uniqueness of the solution follows from the linearity of the equation (6.2.12). Considering this matrix equation as a set of n^2 linear equations arising from the elements x_{ij} of X, we see that the existence of a solution for all C implies that the determinant of the coefficients of x_{ij} is non zero. Thus there is a unique solution.

Theorem 6.2.4

A necessary and sufficient condition for the equation

$$AX + XB = C$$

to have solution for all C is that $\lambda_i + \mu_j \neq o$ where λ_i and μ_j are eigenvalues of A and B respectively.

The proof of this theorem consists basically in isolating the matrix of the coefficients and writing

$$D\bar{X} = C$$

where $\bar{X}$ is a vector made of the elements of the matrix X and recognising that the eigenvalues of D are $\lambda_i + \mu_j$.

Corollary

A necessary and sufficient condition for the equation

$$AX + XA^T = C$$

to have a unique solution for all C is that $\lambda_i + \mu_j \neq o$ where λ_i, μ_j are the eigenvalues of A.

Theorem 6.2.5 (Lyapunov's theorem)

Let M be determined by the relation

$$A^TM + MA = I$$

Then a necessary and sufficient condition for the real matrix A to have all eigenvalues with negative real parts is that M be positive definite.

With these preliminary notions, we can now go on to stating the optimisation problem.

6.3 THE OPTIMISATION PROBLEM

Consider the linear time invariant system

$$\dot{x}(t) = Ax(t) + Bu(t) \qquad (6.3.1)$$

$$y = Cx(t) \qquad (6.3.2)$$

where $x \in R^n$, $u \in R^r$ and $y \in R^m$ are respectively the state, control and output vectors.

Consider the cost function

$$J = \int_0^\infty [x^T(t)\ Qx(t) + u^T(t)\ Ru(t)]dt \tag{6.3.3}$$

Suppose now that we introduce the constraints that the control $\underline{u}(t)$ must be generated by linear output feedback with time-invariant feedback gains i.e.

$$u(t) = -Fy(t) \tag{6.3.4}$$

or
$$u(t) = -FCx(t) \tag{6.3.5}$$

under this constraint, we can rewrite system (6.3.1) and (6.3.2) as:

$$\dot{x}(t) = [A-BFC]x(t) \tag{6.3.6}$$

thus $\underline{x}(t)$ is given by

$$x(t) = \phi(t,o)x(o) \tag{6.3.7}$$

where $\phi(t,o)$ is the transition matrix for the system (6.3.6) i.e.

$$\phi(t,o) = \exp(A-BFC) \tag{6.3.8}$$

Substituting (6.3.7) in the cost function (6.3.3) we obtain

$$J = x(o)^T \left[\int_0^\infty \phi^T(t,o)(Q+C^TF^TRFC)\phi(t,o)dt\right]x(o) \tag{6.3.9}$$

By using the property of the traces of matrices $b^Ta = tr(ab^T)$ we can rewrite (6.3.9) as:

$$J = tr\ \{\int_0^\infty \phi(t,o)(Q+C^TF^TRFC)\phi(t,o)dt\ (x(o)x(o)^T)\} \tag{6.3.10}$$

Because of the structural constraint (6.3.4) imposed on the control, the minimisation problem depends on the initial conditions. In order to get around this dependence, it is possible to consider the initial state as a random variable and subsequently to consider the average value of the performance index.

Let us assume that x(o) is a uniformly distributed random variable where the uniform distribution is over the n dimensional unit sphere. Thus we let

$$V_o = E\ \{x(o)x(o)^T\} = \frac{1}{n}\ I \tag{6.3.11}$$

where E is the expected value operator. Then, the cost function in (6.3.10) can be rewritten as

$$J = \frac{1}{n}\ tr[P] \tag{6.3.12}$$

where
$$P = \int_0^\infty \phi^T(t,o)(Q+C^TF^TRFC)\phi(t,o)dt \tag{6.3.13}$$

We recall from theorem 6.2.3 that P is the solution of

$$o = P[A-BFC] + [A-BFC]^T P+Q+C^TF^TRFC \tag{6.3.14}$$

Before we consider analytical results associated with minimizing the cost function (6.3.12), it is interesting to observe one immediate and simple link between the non-standard optimal case (like the one considered here which has arisen due to the constrained controller structure) and the standard optimal case.

Let us assume that C=I and u = -Fx and let us consider the standard state regulator problem:

$$\left.\begin{aligned} &\min J = \int_o^\infty (x^TQx + u^T Ru)\,dt \\ &\text{s.t. } \dot{x}(t) = Ax(t) + Bu(t) \end{aligned}\right\} \tag{6.3.15}$$

We know *[4]* that for this standard regulator problem, the optimal control is given by u*(t) = -Fx*(t) and that the minimal cost is given by

$$J = x(o)^T P^*x(o) \tag{6.3.16}$$

Now, let $\tilde{F}$ be the matrix which minimises the cost function (6.3.12) i.e.

$$\tilde{u}(t) = -\tilde{F}x(t)$$

with the minimum cost being given by

$$J = \frac{1}{n} \text{tr } [\tilde{P}] \tag{6.3.17}$$

Now what is the relationship between (6.3.16) and (6.3.17) and between F* and $\tilde{F}$?

We will next show that if there are no structure constraints, then the matrix $\tilde{F}$ which minimises (6.3.12) equals F*.

Proof (by contradiction)

Let us assume that $\tilde{F}$ and F* are different. Then, by definition

$$x^T(o)\ \tilde{P}x(o) \geqslant x^T(o)\ P^*x(o)$$

i.e. $\tilde{P} - P^*$ is positive semi definite. Hence $tr(\tilde{P} - P^*) > o$ and thus $tr(\tilde{P}) > tr(P^*)$. But this contradicts the optimality of $\tilde{F}$ with respect to the cost function (6.3.12). Hence $\tilde{F} = F^*$ and $\tilde{P} = P^*$. When structure constraints are put on the system, then, in general $\tilde{F} \neq F^*$. Thus Tr $\tilde{P}$ - Tr$[\tilde{P}]$ - tr$[P^*]$ gives some measure of how "close" a constrained optimal regulator is to the standard one.

Let us now go on with our development. We rewrite the performance index (6.3.17) (after dropping $\frac{1}{n}$ since $\frac{1}{n}$ being a constant, it does not affect the minimisation).

$$\hat{J}(F) = trP = tr\left[\int_o^\infty \exp([A-BFC]^T t)(Q+C^TF^TRFC)\exp([A-BFC]t)\,dt\right. \tag{6.3.18}$$

It should be noted that $\hat{J}(F)$ is a real function of rxm variables, the f_{ij}.

It can be shown using the perturbation results in section 6.2 that

$$\frac{\partial \hat{J}(F)}{\partial F} = 2(RFCLC^T - B^T PLCT^T) \qquad (6.3.19)$$

where
$$P = \int_0^\infty \exp([A-BFC)^T t)(Q+C^T F^T RFC)\exp([A-BFC]t)dt \qquad (6.3.20)$$

and P is the solution of

$$0 = P[A-BFC] + [A-BFC]^T P + Q + C^T F^T RFC \qquad (6.3.21)$$

Also
$$L = \int_0^\infty \exp([A-BFC]t)\exp([A-BFC]^T t)dt \qquad (6.3.22)$$

where L is also a solution of

$$0 = L[A-BFC]^T + [A-BFC]L + I \qquad (6.3.23)$$

We note that equation (6.3.21) and (6.3.23) are Lyapunov equations and the solutions P and L will be positive definite if and only if $|A-BFC|$ is a stable matrix. We are of course assuming here that an appropriate matrix F exists which makes the eigenvalues of $[A-BFC]$ have negative real parts.

The equation (6.3.19), (6.3.21) and (6.3.23) are necessary in order to compute the gradient

$$\frac{\partial \hat{J}(F)}{\partial F}$$

6.3.1 Solving the optimisation problem

One way of solving the optimisation problem is to guess an initial F_o such that $(A-BF_oC)$ is stable. We can then calculate P and L from equations (6.3.21) and (6.3.23) and the gradient using equation (6.3.19). F_o can be modified by successive iteration until the gradient $\frac{\partial J(F)}{\partial F} \simeq o$ giving a local minimum.

To see the link between this and the standard regulator, let us set the gradient equal to zero in (6.3.19). Then, for the optimal F we obtain

$$\dot{F} = R^{-1}B^T PLC^T (CLC^T)^{-1} \qquad (6.3.24)$$

If C has rank n or C^{-1} exists, then from (6.3.24),

$$\dot{F} = R^{-1}B^T PC^{-1} \qquad (6.3.25)$$

and (6.3.21) becomes

$$0 = P^T A + A^T P + Q - PBR^{-1}B^T P \qquad (6.3.26)$$

We recognise that for the special case where C^{-1} exists, the necessary conditions for the constrained problem are identical to the necessary and sufficient conditions for the standard linear regulator.

6.3.2 The Algorithm of Levine and Athans [5]

This algorithm does not use the gradient. It uses the following iterative procedure:

Let $$F_{n-1} = R^{-1}B^TP_{n-1}L_{n-1}C^T[CL_{n-1}C^T]^{-1} \tag{6.3.27}$$

where K_n is the solution of

$$0 = P_n[A-BF_{n-1}C] + [A-BF_{n-1}C]^TK_n + Q + C^TF^T_{n-1}RF_{n-1}C \tag{6.3.28}$$

and L_{n-1} is the solution of

$$0 = L_{n-1}[A-BF_{n-1}C]^T + [A-BF_{n-1}C]L_{n-1} + I \tag{6.3.29}$$

We start the procedure by guessing an F_o which stabilises (A-BFC). Then P_1 is the unique positive definite solution of (6.3.28). This value of P_1 is computed and then substituted into (6.3.27). F_1 is still unknown but it now depends only on L_1. Substituting (6.3.27) into (6.3.28) results in our having only one unknown L_1. Equation (6.3.28) is then solved for a positive definite L_1. This value of L_1 is substituted into (6.3.27) to give F_1. This completes the iteration. The value of F_1 obtained initialises the next iteration.

The drawback with this algorithm is that we have to solve a non-linear equation at each iteration. Also, once F_1 is computed in the first iteration, it is possible for $A-BF_1C$ to be an unstable matrix so the algorithm could fail. From this point of view, the next algorithm that we describe i.e. the algorithm of Geromel and Bernussou [2] is better since we have to solve 2 linear equations (Lyapunov) at each iteration and these are easier to calculate. Also, stability can be guaranteed at each iteration.

6.3.3 The algorithm of Geromel and Bernussou [2]

Consider the large scale system described by the state equations

$$S_i \left\{ \begin{aligned} \dot{x}_i &= \sum_{j=1}^{N} A_{ij}x_j + B_iu_i \\ x_1(o) &= x_{io} \qquad i=1,2,\dots N \end{aligned} \right\} \tag{6.3.30}$$

where $x_i \in R^{n_i}$, $u_i \in R^{r_i}$. This system can be rewritten in the form

$$(S) \left\{ \begin{aligned} \dot{x} &= Ax + Bu \\ x(o) &= x_o \end{aligned} \right. \tag{6.3.31}$$

where $A = \{A_{ij},\ i=1,\dots N;\ \ j=1,\dots N\} \in R^{nxn}$; B = block diag $(B_1, B_2, \dots B_N) \in R^{nxr}$;

$$n = \sum_{i=1}^{N} n_i \;, \quad r = \sum_{i=1}^{N} r_i$$

we seek a linear static decentralised feedback controller of the type

$$u_i = - k_i \, x_i \qquad i=1,2,\ldots N \tag{6.3.32}$$

Let $\Omega(S)$ be the set of all block diagonal gain matrices i.e.

$$\Omega(S = \{K \in R^{rxn} \,|K = \text{block diag } (k_1,k_2,\ldots k_N); \;\; k_i \in R^{r_i x n_i}\}$$

Then the problem is to find a decentralised control of the form given by equation (6.3.32) so as to minimise a quadratic cost function i.e.

$$\min_{K \in \Omega(s)} \int_o^\infty (x^T Q x + u^T R u) \tag{6.3.33}$$

$$\text{s.t.} \qquad \dot{x} = Ax + Bu \tag{6.3.34}$$

$$u = -Kx \tag{6.3.35}$$

where it is assumed that

$$Q = CC^T \geq 0 \text{ and } R > 0$$

Let $\phi(t) = \exp|A-BK|$ be the transition matrix associated with $\dot{x} = (A-BK)x$. Then the cost function J can be rewritten as

$$J(L) = x(o)^T \int_o^\infty \phi^T(t)(Q + K^T RK)\phi(t)dt \; x(o)$$

Using the same procedure as before we can write a new cost function by considering x(o) to be a random variable which is uniformly distributed over an n-dimensional sphere of unit radius i.e.

$$\min_{K \in \Omega(S)} \quad \hat{J}(K) = \text{tr } P$$

$$P = \int_o^\infty \exp([A-BK]^T t)(Q + K^T RK)\exp([A-BK]^T t)dt$$

The gradient matrix $\frac{\partial \hat{J}(K)}{\partial K}$ is given by the system of equations:

$$\left.\begin{aligned} &\frac{\partial \hat{J}(K)}{\partial K} = 2\;(KK-B^T P)L \\ &(A-BK^T P + P(A-BK) + Q + K^T RK = 0 \\ &(A-BK)L + L(A-BK)^T + I = 0 \end{aligned}\right\} \tag{6.3.36}$$

for those $K \in \Omega(S)$ such that (A-BK) is a matrix with eigenvalues in the left half plane.

Let us denote by $D = \text{block diag } (D_1,D_2,\ldots D_N) \in R^{nxr}$ a matrix which comprises of those entries of $\frac{\partial \hat{J}(K)}{\partial K}$ which correspond to the unconstrained entries of K and the other entries are zero. This is equivalent to projecting the gradient matrix on the set $\Omega(S)$, since $D \in \Omega(S)$. Then Geromel and Bernussou propose the following algorithm:

Assume that the initial matrix $K \in \Lambda(S)$ where $\Lambda(S) = \{K | K \in \Omega(S)$ and $(A-BK)$ stable $\}$

Step 1: Using (6.3.36), determine the gradient matrix $\frac{\partial \hat{J}(K)}{\partial K}$ and the feasible direction matrix $D = \text{block diag } (D_1, D_2, \ldots D_N)$.

Step 2: Test for convergence by checking if $|\{d_{pq}\}_i| < \varepsilon \;\forall\; p=1,\ldots r_i$, $q=1,\ldots n_i$; $i=1,\ldots N$; where $\{d_{pq}\}$ is the $(p,q)^{th}$ entry of the matrix D_i and ε is a small pre-chosen constant. If the convergence test is satisfied, stop. Otherwise go to step 3.

Step 3: Set $K = K-aD$ where the step size $a \geqslant o$ must be chosen such that $\hat{J}(K-aD) < \hat{J}(K)$. Go to step 1.

The properties of this algorithm are summarised in the following lemma:

Lemma

If $\hat{J}(K)$ is a function of class C^1 for every $K \in \Lambda(S)$, then:

1) at each iteration there exists a step size $a > o$ such that

$$\hat{J}(K-aD) < \hat{J}(K)$$

2) provided that (A,C) is completely observable or $Q > o$, then for every a which is such that $J(K-aD) < J(K) < \infty$ one has $(K-aD) \in \Lambda(S)$

Remarks

1) The step size "a" can be determined using any conventional uni-dimensional search technique. The authors *[2]* used an adaptive step size routine. Thus if ℓ denotes the iteration index, then the step size obeys the following rule:

$$a^{\ell+\ell} = \pi a^{\ell} \text{ if } \hat{J}(K^{\ell}-a^{\ell}D^{\ell}) < \hat{J}(K)$$

and $$\ell(K^{\ell}-a^{\ell}D^{\ell}) > o$$

$$a^{\ell} = \nu a^{\ell} \text{ otherwise}$$

where $\pi > 1$, $o < \nu < 1$ and one starts from an arbitrary initial value of a.

2) We note that it is not feasible to compute $\frac{\partial \hat{J}(K)}{\partial K_i}$ $i=1,\ldots N$ i.e. the partial variations of $\hat{J}(K)$ with respect to K_i directly. We recall from equation (6.3.36) that $\frac{\partial \hat{J}(K)}{\partial K}$ is a full matrix, although we only need the block diagonal bits of it. It would be attractive if we could calculate only the block diagonal elements. Unfortunately, this is not feasible.

3) The method needs an initial stable gain matrix K_o. The authors use a complicated procedure based on Lyapunov theory. Unfortunately, this only gives a sufficient condition which is relatively conservative. In the next

section we describe a different approach to finding this initial stable gain matrix.

6.4 INITIALISING THE ALGORITHM OF GEROMEL-BERNUSSOU [9]

We consider only those systems which have no unstable fixed modes under decentralised state feedback. The particular approach that we will consider was previously used in [20] for output feedback. However, the approach as applied to our decentralised control problem is much more powerful since unlike the static output feedback where no convenient necessary and sufficient conditions exist, here we are assured that provided there are no unstable fixed modes, we can find a stable decentralised gain matrix. The other condition that is required is that the pairs (A_{ii}, B_i) are controllable.

The first step in this method is to obtain the gradient of an arbitrary eigenvalue λ of a matrix with respect to the variation of one of its elements a_{ij}.

Let v and w^T be the right and left eigenvectors, respectively, associated with the same eigenvalue λ, that is

$$(A-\lambda I)\ v = o \tag{6.4.1}$$

$$w^T(A-\lambda I) = o \tag{6.4.2}$$

By differentiating (6.4.1) we obtain

$$\frac{dA}{da_{ij}} - \frac{d\lambda}{da_{ij}}\ v \qquad + (A-\lambda I)\ \frac{dv}{d\lambda} = o \tag{6.4.3}$$

Pre-multiplying by w^T and using (6.4.2), we have

$$\frac{d\lambda}{da_{ij}} = \frac{w^T \dfrac{dA}{da_{ij}} v}{w^T v} \tag{6.4.4}$$

Equation (6.4.4) must be applied with some caution if λ is a multiple eigenvalue for the associated eigenvectors may no longer be unique. Lancaster [21] has proved that if λ is a simple eigenvalue (multiplicity one), then the eigenvectors v and w^T may be defined to be continuous functions of a_{ij}.

In order to compute the derivative expression for our problem, consider the matrix C, where

$$C = \begin{bmatrix} I_1 & 0 & \cdots & 0 \\ 0 & I_2 & & \vdots \\ \vdots & & \ddots & \vdots \\ 0 & \cdots & \cdots & I_N \end{bmatrix}$$

and I_i is the $n_i x n_i$ identity matrix.

With this definition for the matrix C, we have

$$BK = BKC$$

But

$$BKC = \sum_{i,j} b^i k_{ij} c_j$$

where b^i is the ith column of B and c_j is the jth row of C.

Since the eigenvalue λ depends upon the various parameters of K, we have from (6.4.4)

$$\frac{\partial\lambda}{\partial k_{ij}} = \frac{w^T b^i c_j v}{w^T v} \quad (6.4.5)$$

Taking into account the form of c_j, we obtain

$$\frac{\partial\lambda}{\partial k_{ij}} = \begin{cases} v_j \dfrac{w^T b^i}{w^T v} & \\ 0 & \text{if } k_{ij} = 0 \end{cases} \quad (6.4.6)$$

where v_j is the jth component of v.

The gradient of the real part of λ with respect to the variable feedback gains can be expressed as a matrix G, where

$$g_{ij} = R_e \left\{\frac{\partial\lambda}{\partial k_{ij}}\right\} \quad (6.4.7)$$

The expression (6.4.7) is applied to the dominant eigenvalue λ_d. This leads to the following algorithm:

Step 0 : Select an arbitrary $K \in \bar{K}$

Step 1 : compute the dominant eigenvalue λ_d of A+BK. If it is negative, stop. Otherwise, compute the right and left eigenvectors v_d and w_d^T associated with λ_d.

Step 2 : compute the gradient matrix G.

Step 3 : do a steepest descent search in the direction of -G and update $K \leftarrow K - \alpha G$, where α is the step size. Go to step 1.

Remarks:

a) If the dominant eigenvalue with positive or null real part turns out to be a local minimum, the algorithm must be started again from a different matrix.

b) The unidimensional search may be performed by using quadratic interpolation.

c) If at any iteration the dominant eigenvalue has multiplicity one, perturb slightly K *[22]* in such a way that it becomes simple.

Next, we give two examples to illustrate the procedure.

6.4.1 Examples

This first example is taken from *[2]* and consists of three linear interconnected systems. The absence of fixed modes with respect to $\bar{K}$ was verified by using the procedure given in Chapter 5.

$$A = \begin{pmatrix} 0 & 1 & 0.5 & 1 & 0.6 & 0 \\ -2 & -3 & 1 & 0 & 0 & 1 \\ 0.5 & 1 & 0 & 2 & 1 & 0.5 \\ 0 & 0.5 & 1 & 3 & 1 & -0.5 \\ 1 & 0 & 1 & 0 & 0 & 1 \\ 0 & 0.5 & 0.5 & 0 & -3 & -4 \end{pmatrix} \quad B = \begin{pmatrix} 1 & 0 & 0 & 0 \\ 1 & 0 & 0 & 0 \\ 0 & 3 & 0 & 0 \\ 0 & 0 & 4 & 0 \\ 0 & 0 & 0 & 2 \\ 0 & 0 & 0 & 3 \end{pmatrix}$$

Taking K_o as

$$K_o = \begin{pmatrix} -2 & -1 & 0 & 0 & 0 & 0 \\ 0 & 0 & -3 & -4 & 0 & 0 \\ 0 & 0 & 3 & 4 & 0 & 0 \\ 0 & 0 & 0 & 0 & -2 & -1 \end{pmatrix}$$

we obtain $\lambda_d = 5.84$ as the dominant eigenvalue of $A-BK_o$.

After three iterations we obtain

$$K = \begin{pmatrix} -0.41 & -1.20 & 0 & 0 & 0 & 0 \\ 0 & 0 & -2.82 & -4.13 & 0 & 0 \\ 0 & 0 & 3.27 & 3.81 & 0 & 0 \\ 0 & 0 & 0 & 0 & 0.14 & 0.66 \end{pmatrix}$$

with Re $\lambda_d = -0.11$.

The algorithm was also applied to the large system representing the part of the energy cycle of a container ship that we considered previously in Chapter 5.

For this system $A \in R^{20x20}$, $B \in R^{20x5}$. Here B is not block diagonal. The global system is divided into three subsystems with the following dimensions

subsystem 1 : $A_{11}(10x10)$; $B_{11}(10x2)$

subsystem 2 : $A_{22}(4x4)$; $B_{22}(4x2)$

subsystem 3 : $A_{33}(6x6)$; $B_{33}(6x1)$

The matrices A and B are given in tables 5.1 and 5.2 in Chapter 5.

The stabilising block diagonal $K = (K_1, K_2, K_3)$ where $K(20x5)$, $K_1(2x10)$, $K_2(2x4)$, $K_3(1x6)$ is given in table 6.1.

The cost function value for this K is

$$J(K) = 5x10^9$$

By using the Geromel-Bernussou algorithm *[2]* on starting from this gain matrix, we obtained an improved gain matrix, after 101 iterations which gave

$$J(\hat{K}) = 2.0393x10^9$$

In this section we have outlined a new approach to the decentralised controller initialisation problem. The resulting stable controller can be used as the initial controller to be improved by, for example, the method in *[2]*.

The method can be improved by applying Fletcher-Powell method, say, instead of a pure gradient. We would like to point out that this method as well as the method in *[2]* will offer difficulties for very large systems (for example when A has dimension 100) since both are based on optimisation in the parameter space. However, for many problems of practical interest, it represents a viable design technique.

In the next section we suggest an alternative approach which uses a hierarchical structure.

6.5 A HIERARCHICAL STRUCTURE FOR COMPUTING DECENTRALISED CONTROL *[6]*

The problem of the control of N large-scale linear interconnected dynamical subsystems can be written as

$$\min J = \tfrac{1}{2} \int_0^{t_f} (||x||^2_{Q*} + ||u(t)||^2_R) dt$$

subject to

$$\dot{x} = Ax + Bu + Cz$$

$$z = Lx$$

MATRIX K_1

-0.285E+01	-0.337E+03	-0.288E+03	-0.159E+03	-0.609E+02
-0.146E+02	0.137E+02	0.134E+01	-0.768E+01	0.160E+02
0.422E+01	-0.287E+03	-0.245E+03	-0.135E+03	-0.519E+02
-0.140E+02	-0.130E+02	0.345E+01	0.105E+02	0.133E+02

MATRIX K_2

-0.198E+03	-0.295E+03	0.746E+03	0.514E+02
-0.717E+02	0.614E+02	0.340E+04	0.299E+03

MATRIX K_3

-0.107E+03	-0.243E+03	-0.169E+04	0.161E+03	0.346E+03	0.176E+04

TABLE 6.1

where $||.||_S^2 = .\ T_S\ .$, Q^*, R, A, B, C are all appropriate block diagonal matrices while L is a full matrix. It is assumed that there are N blocks in A, B, C corresponding to the N interconnected subsystems with L providing the interconnections and that the blocks in Q^*, R correspond to those in A, B, C.

Now, as shown in [23], it is possible to find a control u of the following form which solves the above problem:

$$u(t) = -G_b x - Tx \tag{6.5.1}$$

If we substitute this control u into the cost function, we obtain the minimum cost J_{opt}:

$$\begin{aligned} J_{opt} &= \tfrac{1}{2}\int_0^{t_f} (||x||_{Q^*}^2 + ||G_b x + Tx||_R^2)\,dt \\ &= \tfrac{1}{2}\int_0^{t_f} \{||x||_{Q^*}^2 + ||x||_{W^*}^2\}\,dt \end{aligned} \tag{6.5.2}$$

where

$$W^* = G_b^T R G_b + G_b^T R T + T^T R G_b + T^T R T \tag{6.5.3}$$

In the above expressions, since T is a full matrix, the control implementation requires feedback of all states, and this is undesirable from a practical point of view for systems which are geographically scattered. To avoid this information transfer between the subsystem controllers, let us constrain T to be the diagonal matrix T_d. Then the problem reduces to choosing a diagonal matrix T_d in order to minimise

$$J = \tfrac{1}{2}\int_0^{t_f} (||x||_{Q^*}^2 + ||x||_W^2)\,dt \tag{6.5.4}$$

where $\quad W = G_b^T R G_b + G_b^T R T_d + T_d^T R G_b + T_d^T R T_d$

J in (6.5.4) is to be minimised subject to the constraints

$$\dot{x} = (A - BG_b)x + Cz - BT_d x \tag{6.5.5}$$

$$z = Lx$$

where B is of dimension nxn.

Let us rewrite W in the form

$$W = (G_b^d + T_d + G_b^o)^T R (G_b^d + T_d + G_b^o) = (F + G_b^o)^T R (F + G_b^o)$$

where

G_b^d matrix composed of the diagonal elements of G_b.

G_b^o matrix composed of the off-diagonal elements of G_b.

Let us define

A	matrix composed of the diagonal elements of $(A-BG_b)$.
A^o	matrix composed of the off-diagonal elements of $(A-BG_b)$.
Q, B_d	matrices composed of the diagonal elements of Q^* and B.
Q^o, B^o	matrices composed of the off-diagonal elements of Q^* and B

Then we can rewrite the cost function in (6.5.4) as

$$J = \tfrac{1}{2}\int_o^{t_f} \{||x||_Q^2 + ||x||_{Q^o}^2 + x^T F^T RFx + x^T F^T RG_b^o x$$

$$+ x^T G_b^o RFx + x^T G_b^o RG_b^o x\}dt$$

$$= \tfrac{1}{2}\int_o^{t_f} \{||x||_Q^2 + x^T RFx + g(x,F,G_b^o)\}dt$$

where
$$g(x,F,G_b^o) = ||x||_{Q^o}^2 + x^T F^T RG_b^o x$$

$$+ x^T G_b^{oT} RFx + x^T G_b^{oT} RG_b^o x$$

The dynamic equation (6.5.5) can also be rewritten in a more compact form as

$$\dot{x} = Ax - B_d T_d x + y(x,z,T_d)$$

where
$$y(x,z,T_d) = A^o x + Cz - B^o T_d x$$

Next we will provide a three-level computational structure for solving this problem.

6.5.1 The three-level control computation structure

As in the prediction methods for nonlinear problems developed in *[7]*, we will "fix" certain variables at the second level and use these at the first level to decompose the problem into a number of independent subproblems.

Let us fix

$$x^* = x$$

and
$$T_d^* = T_d \qquad (6.5.6)$$

Then by introducing (6.5.6) and (6.5.7) as additional equality constraints, our optimisation problem can be rewritten as follows:

$$\min J = \tfrac{1}{2}\int_o^{t_f} \{||x||_Q^2 + x^{*T} F^T RFx^* + g(x^*,F^*,G_b^o)\}dt$$

where $g(x^*,F^*,G_b^o) = ||x^*||^2_{Q^o} + x^{*T}F^{*T}RG_b^o x^* + x^{*T}G_b^{oT}RF^*x^* + x^{*T}G_b^{oT}RG_b^o x^*$

subject to

$$\dot{x} = Ax - B_dT_dx^* + y(x^*,z,T_d^*)$$

where $y(x^*,z,T_d^*) = A^{\,o}x^* + Cz - B^oT_d^*x^*$

and the constraints

$$z = Lx$$

$$T_d^* = T_d$$

$$x^* = x$$

In order to solve this problem, let us write the Hamiltonian as

$$H = \tfrac{1}{2}||x||^2_Q + \tfrac{1}{2}x^{*T}F^TRFx^* + \tfrac{1}{2}g(x^*,F^*,G_b^o)$$

$$+ \lambda^T(Ax - B_dT_dx^* + y(x^*,z,T_d^*)) + \pi^T[Lx - z]$$

$$+ \beta^T(x - x^*) + \sum_{i=1}^{n} v_i(T_{d_i} - T^*_{d_i})$$

where π, β, v_i are Lagrange multipliers and λ is the costate variable. From the necessary conditions for optimality we obtain

$$\frac{\partial H}{\partial \pi} = o \text{ or } z = Lx \tag{6.5.7}$$

$$\frac{\partial H}{\partial z} = o \text{ or } \pi = C^T\lambda \tag{6.5.8}$$

or $k = L^T\pi$, which is used merely to simplify the calculations.

$$\frac{\partial H}{\partial \beta} = o \text{ or } x^* = x \tag{6.5.9}$$

$$\frac{\partial H}{\partial v_i} = o \text{ or } T^*_{d_i} = T_{d_i} \tag{6.5.10}$$

$$\frac{\partial H}{\partial x^*} = o \text{ or}$$

$$\beta = F^TRFx^* + Q^o(x^* + F^{*T}RG_b^ox^* + G_b^{oT}RF^*x^* + G_b^{oT}RG_b^ox^* - T_d^TB_d^T\lambda + (A^{oT} - T_d^*B^{oT})\lambda$$

or

$$\beta = (F^TRF + Q^o + F^{*T}RG_b^oRF^* + G_b^oRG_b^o)x^* + (A^{\,oT} - T_d^*B^{oT} - T_d^TB_d^T)\lambda \tag{6.5.11}$$

$$\frac{\partial H}{\partial T_d^*} = o$$

or

$$v = \text{diag}\left[(RG_b^o x^* - B^{oT}\lambda)x^{*T}\right] \tag{6.5.12}$$

Now if we assume that x^*, T_d^*, β, v have been provided by the second level then the Hamiltonian can be decomposed such that each subproblem has only one variable.

Thus for $\partial H | \partial T_{d_i} = o$.

$$x_i^{*2}(G_{b_i}^d + T_{d_i})R_i - B_{d_i}\lambda_i x_i^* + v_i = o$$

or

$$T_{d_i} = -G_{b_i}^d - \frac{1}{R_i x_i^{*2}}\left[- B_{d_i}\lambda_i x_i^* + v_i\right] \tag{6.5.13}$$

$$\frac{\partial H}{\partial \lambda_i} = \dot{x}_i$$

or

$$\dot{x}_i = A_i x_i - B_{d_i}\left\{- G_{b_i}^d - \frac{1}{R_i x_i^{*2}}\left[v_i - B_{d_i}\lambda_i x_i^*\right]\right\}_d x_i^* + y_i(x^*,z,T_d^*) \tag{6.5.14}$$

$$\frac{\partial H}{\partial x_i} = - \dot{\lambda}_i$$

or

$$- \dot{\lambda}_i = Q_i x_i + A_i \lambda_i + k_i + \beta_i \tag{6.5.15}$$

Also

$$\dot{x}_i = A_i x_i + B_{d_i} G_{b_i}^d x_i^* + \frac{B_{d_i}}{R_i x_i^*}(v_i - B_{d_i}\lambda_i x_i^*) + y(x^*,z,T_d^*)$$

or

$$\dot{x}_i = A_i x_i + B_{d_i} G_{b_i}^d x_i^* + \frac{B_{d_i}}{R_i x_i^*} v_i - \frac{B_{d_i}^2}{R_i}\lambda_i + y(x^*,z,T_d^*) \tag{6.5.16}$$

Let $\lambda_i = P_i x_i + \eta_i$. Then

$$\dot{P}_i x_i + P_i \dot{x}_i + \dot{\eta}_i = - Q_i x_i - A_i P_i x_i - A_i \eta_i - k_i - \beta_i$$

from which after minor manipulations we obtain

$$\dot{P}_i = - 2A_i P_i + \frac{B_{d_i}^2}{R_i} P_i^2 - Q_i \tag{6.5.17}$$

with $P_i(t_f) = o$ and

$$\eta_i = -\left(A_i + \frac{P_i B_{d_i}^2}{R_i}\right)\eta_i - P_i\left(B_{d_i} G_{b_i}^* x_i^* + \frac{B_{d_i} v_i}{R_i x_i^*} + y(x^*,z,T_d^*)\right) - k_i - \beta_i \tag{6.5.18}$$

This suggests the following three-level algorithm.

Step o: Guess the initial trajectories z^l and k^l at level 3 for the iteration index $l = 1$.

Step 1: Guess the initial trajectories $x^{*\alpha}$, $T^{*\alpha}_d$, β^α, v^α at level 2, and set the iteration index $x = 1$.

Step 2: Using $x^{*\alpha}$, $T^{*\alpha}_d$, β^α, v^α obtained from step 1, calculate P_i, η_i from (6.5.17) and (6.5.18) and thus x from (6.5.16) and λ(from $\lambda = {}^iPx^i + \eta$). Calculate also T_d.

Step 3: Substitute the x, λ, obtained at level 1, and π, z from level 3 into the right sides of (6.5.11) to (6.5.12) to obtain $x^{*\alpha+1}$, $T^{*\alpha+1}_d$, $\beta^{\alpha+1}$ and $v^{\alpha+1}$. If the integral of the norm of the difference of $x^{*\alpha+1}$ and $x^{*\alpha+1}$ as well as that of $T^{*\alpha+1}_d$ and $T^{*\alpha+1}_d$, $\beta^{\alpha+1}$ and β^α and v^α is not sufficiently small, go to step 2. Otherwise go to level 3 to calculate new k^{l+1} and z^{l+1} from (6.5.8) and (6.5.7). If the norm of the difference between k^{l+1} and k^l and between z^{l+1} and z^l is small, record T_d as the decentralised gain matrix; otherwise go to step 1 using k^{l+1}, z^{l+1} as the new guess.

Remarks:

1) This is a prediction type of algorithm, and its convergence can be proved using a technique similar to that used by Hassan and Singh *[7]* for nonlinear systems.

2) The algorithm is very simple since at the lowest level mere substitution into single-variable formulae is required.

3) At the optimum, we obtain T_d and thus the decentralised gain matrix.

4) It should be noted that the entire calculation is done off-line, and only the decentralised gains are used on-line to compute and implement the optimal decentralised control.

5) Although the decentralised gains are initial-state dependent, computational experience indicates that the control is relatively insensitive to small variations in initial conditions. This arises from the fact that on examining (6.5.13) for T_{d_i} we find that T_{d_i} depends on the ratio of the states and costates. Intuitively, since the dynamics dictate the relationships between the various states and costates and these do not change significantly for small variations in initial conditions, the optimal T_{d_i} will only be changed slightly if the initial conditions change so that not changing T_{d_i} would still give us a good suboptimal control, as we see in our example in Section 6.5.3.

6) The disadvantage of this technique is that T_d is initial-state dependent although it is not sensitive to small variations in initial conditions as stated above.

Next let us put a bound on the suboptimality of the controller.

6.5.2 Suboptimality Bounds

Since our control is suboptimal, it would be useful to have an expression which bounds the possible performance degradation. For the optimal system we have

$$\dot{x}_i = (A - BG_b)\dot{x} + CL\dot{x} - BT\dot{x}$$

or $$\dot{\hat{x}}(t) = \phi(t,o)xo$$

For the suboptimal case let $T_d = T - T''_d$, then $x_{sub}(t)$ can be written as

$$x_{sub}(t) = \phi(t,o)x(o) + \int_o^{t_f} \phi(t,\tau)BT''_d x_{sub}(\tau)d\tau$$

$$= x(t) + \left[\int_o^{t_f} \phi(t, \;)BT''_d \;'(t, \;)d\right] x(o)$$

$$= \hat{x}(t) + x(t)$$

$$= |1 + \lambda t,o)|x(t) = \Gamma(t,o)\hat{x}(t)$$

$$\phi(t,\tau) = (A - BG_b + CL - BT_d)\phi'(t,\tau)$$

$$\lambda(t,o) = \left[\int^{t} \phi(t,\tau)BT''_d\phi'(t,\tau)d\tau\right]\phi(o,t)$$

Thus we have

$$J_{sub} = \tfrac{1}{2}\int_o^{t_f} ||x_{sub}||^2_Q + x^T_{sub}(G_b + T - T''_d)^T R$$

$$(G_b + T - T''_d)x_{sub}dt$$

$$= \tfrac{1}{2}\int_o^{t_f} \{||x_{sub}||^2_Q + x^T_{sub}(G_b + T)^T R(G_b + T)x_{sub}$$

$$+ x^T_{sub}T''^T_d RT''_d x_{sub} - 2x^T_{sub}T''^T_d R$$

$$\cdot\ (G_b + T)x_{sub}\}dt$$

$$J_{sub} = \tfrac{1}{2}\int_o^{t_f} ||x_{sub}||^2_{Q+w*} + x^T_{sub}T''^T_d RT''_d x_{sub}$$

$$- 2x^T_{sub}T''^T_d R(G_b + T)x_{sub}dt$$

$$= \tfrac{1}{2}\int_o^{t_f} ||\hat{x}||^2_{Q+w*} + ||\hat{x}||^2_{Q+w*} + 2x^T(Q + w*)\hat{x}$$

$$+ ||x_{sub}||_{T_d''TRT_d''} - 2||x_{sub}||^2_{T_d''^T R(G_b + T)}dt$$

$$J_{sub} = \frac{1}{2}\int_o^{t_f} ||\hat{x}||^2_{Q+w^*}dt + \frac{1}{2}\int_o^{t_f}\{||\hat{x}||^2_{(\Gamma - I)T(Q+w^*)(\Gamma - I)}$$

$$+ 2||\hat{x}||^2_{(\Gamma - I)^T(Q+w^*)} + ||\hat{x}||^2_{\Gamma^T T_d{''}^T R T_d{''}\Gamma}$$

$$- 2||\hat{x}||^2_{\Gamma^T T_d{''}^T R(G_b + T)\Gamma\}} \quad dt$$

$$= J_{opt} + \frac{1}{2}\int_o^{t_f} ||\hat{x}||^2_{F(t)}$$

where

$$F(t) = (\Gamma(t,o) - I)^T(Q + w^*)(\Gamma - I)$$

$$+ [\Gamma(t,o) - I)^T(Q + w^*)$$

$$+ \Gamma^T(t,o)T_d''^T R T_d''^T \Gamma(t,o)$$

$$- 2\Gamma^T(t,o)T_d''^T R(G_b + T)\Gamma$$

Now since

$$\int_o^{t_f}||\hat{x}||^2_F dt \leq \mu_M(F)\int_o^{t_f}||\hat{x}||^2 dt$$

and

$$\int_o^{t_f}||\hat{x}||^2 dt \leq \frac{1}{\mu_m(Q + w^*)}\int_o^{t_f}||\hat{x}||^2_{Q+w^*} dt$$

$$J_{sub} \leq 1\ (1 + \varepsilon)J_{opt}$$

$$o < \varepsilon \leq \frac{\mu_m(F)}{\mu_m(Q + w^*)}$$

where

$\mu_M(F)$ max eigenvalue of F,

$\mu_m(Q + w^*)$ minimum eigenvalue of $Q + w^*$.

Next we apply the algorithm to a river pollution control model to provide decentralised control.

6.5.3 Example: River Pollution Control

The problem of control of pollution in a river system has previously been treated in [8]. The problem is essentially to control optimally the treated sewage discharges from sewage stations on a river system in order to reach a compromise between the cost of treatment and the cost incurred by having too polluted a river. Singh et al *[8]* have shown that this problem can be treated as a "linear quadratic" problem. If this problem is solved as is done in *[8]* then the resulting optimal control requires feedback of all the states. The implementation of the feedback between the subsystems is very costly in this case because the subsystems, which are "reaches" of the river, are often many kilometres distant from each other, and the cost of telemetry links is the main cost of implementation of the control. If we implement the decentralised controller developed here, we could clearly have substantial savings, and the controller would moreover be more robust in the sense of not being dependent on the functioning of links between the local controllers. In the treatment in this section we will use the two-reach no-delay model of *[8]*.

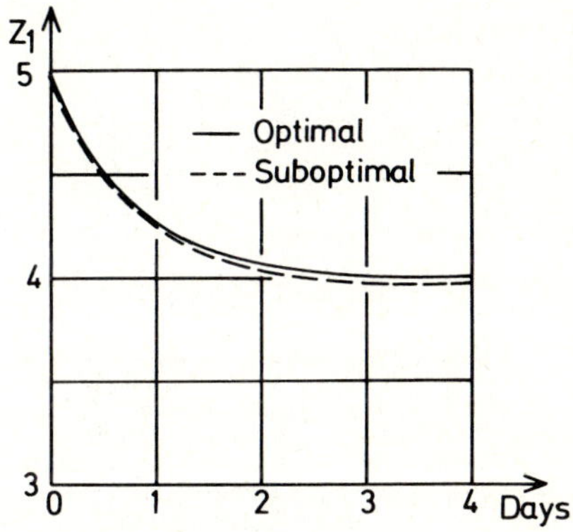

Figure 6.1

The original problem here is to minimise

$$J = \int_0^8 \{2(x_1 - 4.06)^2 + (x_2 - 8)^2 + 2(x_3 - 5.94)^2 + (x_4 - 6)^2 + u_i^2 + u_2^2\}dt$$

subject to

$$\frac{d}{dt}\begin{pmatrix} x_1 \\ x_2 \\ x_3 \\ x_4 \end{pmatrix} = \begin{pmatrix} -1.32 & 0 & 0 & 0 \\ -0.32 & -1.2 & 0 & 0 \\ 0.9 & 0 & -1.32 & 0 \\ 0 & 0.9 & -0.32 & -1.2 \end{pmatrix}\begin{pmatrix} x_1 \\ x_2 \\ x_3 \\ x_4 \end{pmatrix} + \begin{pmatrix} 0.1 & 0 \\ 0 & 0 \\ 0 & 0.1 \\ 0 & 0 \end{pmatrix}\begin{pmatrix} u_1 \\ u_2 \end{pmatrix} + \begin{pmatrix} 5.35 \\ 10.9 \\ 4.19 \\ 1.9 \end{pmatrix}$$

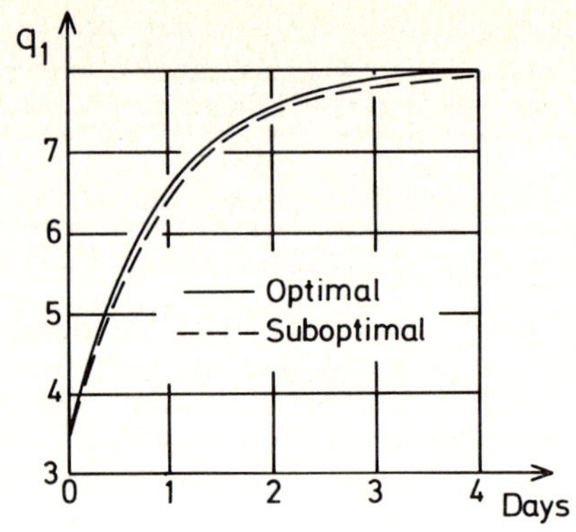

Figure 6.2

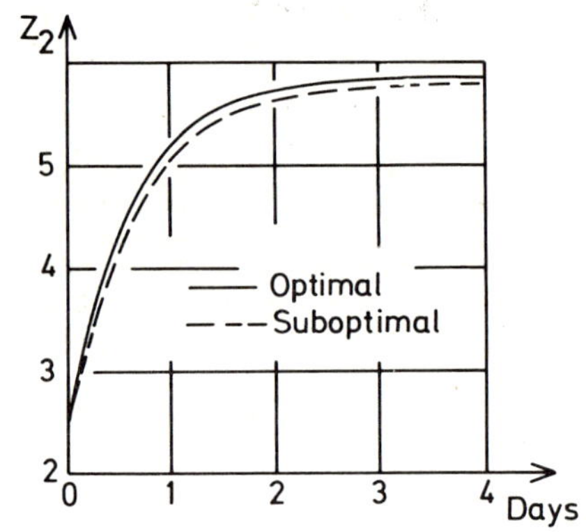

Figure 6.3

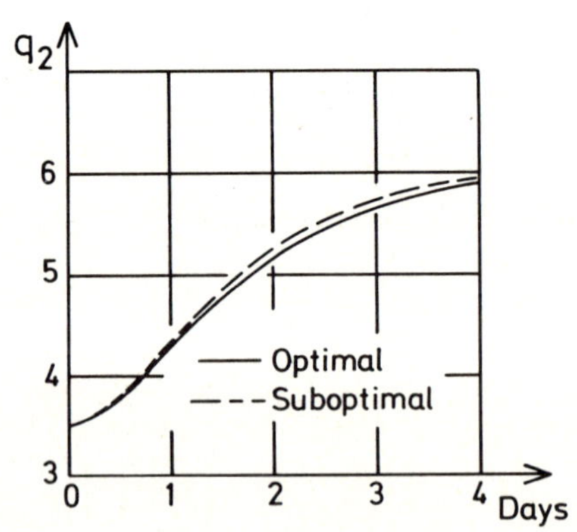

Figure 6.4

For this case, the optimal and the decentralised control gains were calculated. The diagonal gain matrix T_d was calculated for the initial conditions $[10 \ \ 7 \ \ 5 \ \ 7]$, and convergence to 10^{-4} was achieved in four iterations of the third level, which took 18.65s to execute on the IBM 370-165 computer. To test the sensitivity of the decentralised controller to variations in initial conditions, the initial conditions were perturbed by 50% to $[5 \ \ 3.5 \ \ 2.5 \ \ 3.5]$. For this case, the optimal centralised control gave a cost of 6.79, while the decentralised controller $(G_b + T_d)$ calculated using the first set of initial conditions gave a cost of 7.98 showing that the decentralised control is, indeed, relatively insensitive to small variations in the initial conditions. For comparison, the completely decentralised control obtained by ignoring all interactions gave a cost of 8.45. Figures 6.1 to 6.4 give the trajectories of the optimal and suboptimal states. As we can see, these are very close to each other, thus showing that the suboptimal controller is a good one.

Next let us see how we can modify the algorithm to ensure that the system has a pre-specified degree of stability.

6.6 HIERARCHICAL STRUCTURE FOR DECENTRALISED CONTROL WITH A PRE-SPECIFIED DEGREE OF STABILITY *[14]*

Consider an interconnected dynamical system comprising N subsystems described by

$$\dot{x}_i = A_i x_i + B_i u_i + \sum_{j=1}^{N} A_{ij} x_j + d_i$$

To incorporate the structural perturbations Siljak *[9]* the above equation can be rewritten as

$$\dot{x}_i = A_i x_i + B_i u_i + \sum_{j=1}^{N} e_{ij} A_{ij} x_j + d_i \qquad i=1,2,\ldots N \qquad (6.6.1)$$

where e_{ij} are the elements of the interconnection matrix E which is continuous in time with $0 \leq e_{ij}(t) \leq 1$. This equation defines an interconnected dynamical system where the e_{ij}'s enable us to incorporate any structural perturbations which may occur during the operation of the system. We would like to design a controller for this system by minimising a quadratic performance index and ensuring at the same time that the controlled system exhibits a pre-specified degree of stability α (in the sense of Anderson and Moore *[10]* chapter IV) i.e.

$$\min J = \sum_{i=1}^{N} \tfrac{1}{2} \int_0^T e^{2\alpha T}\{||x_i - x_i^d||^2_{Q_i} + ||u_i||^2_R\}dt \qquad (6.6.2)$$

where $||.||^2_S = .T_S.$

T is more than 4 time constants of the system such that the system arrives at its steady state.

In order to convert this problem into the standard 'Linear Quadratic' form, let us define

$$\hat{x}(t) = e^{\alpha t}x(t); \quad \hat{u}(t) = e^{\alpha t}u(t) \tag{6.6.3}$$

Then we can rewrite our problem as

$$\text{Min } J = \sum_{i=1}^{N} \tfrac{1}{2} \int_{o}^{t} (||\hat{x}_i - x_i^d||^2_{Q^*i} + ||\hat{u}_i||^2_{R_i})dt \tag{6.6.4}$$

subject to

$$\dot{\hat{x}}_i = \hat{A}_i\hat{x}_i + B_i\hat{u}_i + \sum_{j=1}^{N} e_{ij}A_{ij}\hat{x}_j + \hat{d}_i(t) \tag{6.6.5}$$

where $\hat{A}_i = (A_i + \alpha I_i)$ (I is the identity matrix) and

$$\hat{d}_i(t) = d_i e^{\alpha t} \tag{6.6.6}$$

For this system, we can write optimal control $\hat{u}_i(t)$ as

$$\hat{u}_i(t) = \sum_{j=1}^{N} [e_{ij}T_{ij}\hat{x}_j] - \hat{\xi}_i \tag{6.6.7}$$

where $\quad G_{bi} = - R_i^{-1}B_i^T P_i$

where P_i is the solution of the local (i.e. decomposed) Riccati equation for the ith subsystem.

Define $\tau = [e_{ij}T_{ij}]$. Then in the global form

$$\hat{u}(t) = G_b\hat{x} - \tau\hat{x} - \xi \tag{6.6.8}$$

and the minimal value of the cost function can be written as

$$J_{opt} = \tfrac{1}{2} \int_{o}^{t} \Big[||\hat{x}_o - \hat{x}^d||^2_{Q^*} + ||\hat{x}_o||^2_{W^*} + 2\hat{x}_o^T G_b^T \hat{\xi} + 2\hat{x}_o^T \tau^T R\hat{\xi} + \hat{\xi}^T R\hat{\xi} \Big] dt \tag{6.6.9}$$

where $\hat{x}_o$ is the optimal state resulting from the application of the globally optimal control given by equation (6.6.8).

Let us write

$$W^* = G_b^T R G_b + G_b^T R\tau + \tau^T R G_b + \tau^T R\tau \tag{6.6.10}$$

In order to compute a completely decentralised control let us replace in equation (6.6.10) the matrix τ by T_d which is constrained to be a diagonal matrix. We would therefore like to choose the matrix T_d in order to minimise the cost function

$$J = \tfrac{1}{2} \int_{o}^{t} (||\hat{x} - \hat{x}^d||^2_{Q^*} + ||\hat{x}||^2_{W} + 2\hat{x}^T G_b R\hat{\xi} + 2xT_d^T R\hat{\xi} + \hat{\xi}^T R\hat{\xi})dt \tag{6.6.10a}$$

subject to

$$\dot{\hat{x}} = (\hat{A} - BG_b)\hat{x} + C\hat{z} + BT_d\hat{x} + \hat{D} \tag{6.6.11}$$

$$\hat{z} = L\hat{x} \tag{6.6.12}$$

where $W = G_b^T R G_b + G_b^T R T_d + T_d^T R G_b + T_d^T R T_d$

B is a matrix of dimension nxn (if in practice B is of lower dimension than nxn we can introduce additional fictitious controls such that the corresponding elements in B which multiply these controls are zero)

$$\hat{D}(t) = \hat{d}(t) - B\hat{\xi}$$

Let us rewrite W as

$$W = (G_b^d + T_d + G_b^o)R(G_b^d + T_d + G_b^o) = (F + G_b^o)^T R(F + G_b^o) \tag{6.6.13}$$

where G_b^d is the matrix consisting of the diagonal elements of G_b, G_b^o is the matrix consisting of the off-diagonal elements of G_b, A and A^o are the matrices consisting of the diagonal and off-diagonal elements of $(A - BG_b)$ respectively, Q, B_d are the matrices consisting respectively of the diagonal elements of Q* and B and Q^o, B^o are the matrices consisting of the off-diagonal elements of Q* and B respectively.

Then we can write our optimisation problem as

$$\text{Min } J = \tfrac{1}{2}\int_o^t \{||\hat{x} - \hat{x}^d||_Q^2 + \hat{x}^T F^T R F\hat{x} + 2\hat{x}^T G_b^T R$$

$$+ 2\hat{x}^T T_d^T R\hat{\xi} + \hat{\xi}^T R\hat{\xi} + g(\hat{x}, F, G_b^o)\}dt \tag{6.6.14}$$

Subject to

$$\dot{\hat{x}} = A\hat{x} - B_d T_d\hat{x} + y(\hat{x}, \hat{z}, T_d) \tag{6.6.15}$$

$$\hat{z} = L\hat{x} \tag{6.6.16}$$

where $g(\hat{x}, F, G_b^o) = ||\hat{x} - \hat{x}^d||_{Q^o}^2 + \hat{x}^T F^T R G_b^o \hat{x}$

$$+ \hat{x}^T G_b^{oT} R F\hat{x} + \hat{x}^T G_b^{oT} R G_b^o \hat{x} \tag{6.6.17}$$

$$y(\hat{x}, \hat{z}, T_d) = A^o\hat{x} + C\hat{z} - B^o T_d\hat{x} + \hat{D} \tag{6.6.18}$$

6.6.1 The hierarchical computational structure to compute decentralised control

We will use a prediction principle type algorithm (cf. Hassan and Singh, [17]) to solve this problem. The prediction principle approach to hierarchical optimisation involves adding certain additional linear constraints in order to decompose the optimisation problem into a number of independent subproblems for certain fixed trajectories supplied by the second level and to improve these latter trajectories using a fixed point type of algorithm.

Let us introduce the additional linear constraints

$$T_d^* = T_d$$
$$\hat{x}^* = \hat{x}$$

Let us substitute these values into our problem so that it becomes

$$\text{Min } J = \tfrac{1}{2}\int_o^t (||\hat{x} - \hat{x}^d||_Q^2 + \hat{x}^{*T}F^T\hat{R}Fx + 2\hat{x}^T G_b^T$$

$$R\xi + 2\hat{x}^{*T}T_d^T R\hat{\xi} + \hat{\xi}^T R\hat{\xi}^T + g(x^*,F^*,G_b^o))dt \tag{6.6.19}$$

subject to

$$\dot{\hat{x}} = \hat{x} - B_d T_d x^* + y(x^*,z,T_d) \tag{6.6.20}$$

$$\hat{z} = L\hat{x} \tag{6.6.21}$$

$$T_d^* = T_d \tag{6.6.22}$$

$$\hat{x}^* = \hat{x} \tag{6.6.23}$$

where $$g(x^*,F^*,G_b^o) = ||\hat{x}^* - \hat{x}^d||_{Q^o}^2 + \hat{x}^{*T}F^{*T}RG_b^o\hat{x}^*$$

$$+ \hat{x}^{*T}G_b^o RF^*\hat{x}^* + \hat{x}^{*T}G_b^o RG_b^o\hat{x}^* \tag{6.6.24}$$

$$y(\hat{x}^*,\hat{z},T_d^*) = A^o x^* + C\hat{z} - B^o T_d^* x^* + \hat{D}(t) \tag{6.6.25}$$

In order to solve this modified problem let us write the overall Hamiltonian as

$$H = \tfrac{1}{2}||\hat{x} - \hat{x}^d||_Q^2 + \tfrac{1}{2}\hat{x}^*F^T RF\hat{x}^* + \hat{x}^T G_b^T R\hat{\xi} + \hat{x}^{*T}T_d^T R\hat{\xi}$$

$$+ \tfrac{1}{2}\hat{\xi}^T R\hat{\xi} + \tfrac{1}{2}g(\hat{x}^*,F^*,G_b^o) + .\lambda^T[\hat{x} - B_d T_d x^*$$

$$+ y(x^*,z,T_d^*)] + \pi^T[L\hat{x} - \hat{z}] + \beta^T(\hat{x} - \hat{x}^*)$$

$$+ \sum_{i=1}^n \nu_i(T_{d_i} - T_{d_i}^*) \tag{6.6.26}$$

where π, β, ν_i are Lagrange multipliers used to satisfy the equality constraints $\hat{z} = L\hat{x}$, $\hat{x} = \hat{x}^*$ and $T_{d_i} = T_{d_i}^*$. Note that because we had constrained T_d to be a diagonal matrix, we can attach the scalar Lagrange multipliers $T_{d_i} = T_{d_i}^*$ which means that we do not have to improve matrix functions at the second level.

The necessary conditions for optimality for this problem can be written as

$$\frac{\partial H}{\partial \pi} = 0 \text{ or } \hat{z} = L\hat{x} \tag{6.6.27}$$

$$\frac{\partial H}{\partial \hat{z}} = 0 \text{ or } \pi = C^T\lambda \tag{6.6.28}$$

$$\frac{\partial H}{\partial \beta} = 0 \text{ or } \hat{x}^* = \hat{x} \tag{6.6.29}$$

$$\frac{\partial H}{\partial \nu_i} = 0 \text{ or } T^*_{d_i} = T_{d_i} \tag{6.6.30}$$

$$\frac{\partial H}{\partial x^*} = 0 \text{ or}$$

$$\beta = F^TRF\hat{x} + T_d^TR\xi + Q^o(\hat{x}^* - \hat{x}^d) + F^{*T}RG_b^o\hat{x}^* + G_b^{oT}RFx^* + G_b^{oT}RG_b^ox^* - T_d^TB_d^T\lambda + (A^{oT} - T_d^*B_o^T)\lambda$$

or

$$\beta(t) = \left[F^TRF + Q^o + F^{*T}RG_b^o + G_b^oRF^* + G_B^{oT}RG_b^o\right]\hat{x}^* - Q^o\hat{x}^d + T_d^TR\hat{\xi} + (A^{oT} - T_d^*B^{oT} - T_d^TB_d^T)\lambda \tag{6.6.31}$$

$$\frac{\partial H}{\partial T_d^*} = 0$$

or $$\nu = \text{diag}\left[\tfrac{1}{2}RG_b^o\hat{x}^*\hat{x}^{*T} + \tfrac{1}{2}RG_b^o\hat{x}^*\hat{x}^{*T} - B^{oT}\lambda\hat{x}^{*T}\right]$$

or $$\nu = \text{diag}\left[RG_b^o\hat{x}^* - B^{oT}\lambda)\hat{x}^{*T}\right] \tag{6.6.32}$$

Now if we assume that $\hat{x}^*$, T_d^*, β, ν, z and $\hat{k} = L^T\pi$ are given then we can decompose the Hamiltonian into subsystems where each subsystem has only one variable.

Let $R\xi = v$, $G_b^TR\xi = r$. Then $(\partial H/\partial T_{d_i}) = 0$ gives

$$T_{d_i} = -G_{b_i}^d - \frac{1}{R_i\hat{x}_i^{*2}}\left[\hat{x}_i^*v_i - B_{d_i}\lambda_i\hat{x}_i^* + \nu_i\right]$$

$$-\frac{\partial}{\partial \lambda_i} = \dot{\hat{x}}_i$$

or $$\dot{\hat{x}}_i = A_i\hat{x}_i - B_{d_i}\left\{-G_{b_i}^d - \frac{1}{R_i\hat{x}_i^{*2}}\left[\hat{x}_i^*v_i + \nu_i - B_{d_i}\lambda_i\hat{x}_i^*\right]\right\}\hat{x}_i^* + y_i(\hat{x}^*, \hat{z}, T_d^*) \tag{6.3.33}$$

$$\frac{\partial H}{\partial \hat{x}_i} = -\dot{\lambda}_i$$

or $$-\dot{\lambda}_i = Q(\hat{x}_i - \hat{x}_i^d) + r_i + A_i\lambda_i + k_i + \beta_i$$

or $$\dot{\lambda}_i = -Q(\hat{x}_i - \hat{x}_i^d) - r_i - A_i\lambda_i - k_i - \beta_i \tag{6.6.34}$$

and $$\dot{\hat{x}}_i = A_i\hat{x}_i + B_{d_i}G^d_{bi}\hat{x}^*_i + \frac{B_{d_i}}{R_i}v_i + \frac{B_{d_i}}{R_i\hat{x}^*_i}v_i - \frac{B^2_{d_i}}{R_i}\lambda_i + y_i(\hat{x}^*,\hat{z},T^*_d) \tag{6.6.35}$$

Now to solve the two point boundary value problem given by equations (6.6.3 (6.6.34), (6.6.35) with known terminal conditions, let us put

$$\lambda_i = P_i x_i + \eta_i$$

Then it is easy to see that P_i and η_i are given by

$$\dot{P}_i = -2A_iP_i + \frac{B^2_{d_i}}{R_i}P_i^2 - Q_i \tag{6.6.36}$$

$$\dot{\eta}_i = \left(-A_i + \frac{P_iB^2_{d_i}}{R_ix^*_i}\right)\eta_i - P_i(B_{d_i}G^d_{b_i}\hat{x}^*_i) + B_{d_i}\frac{\nu_i}{R_i} + \frac{B_{d_i}\nu_i}{R_i\hat{x}^*_i} + y_i(\hat{x}^*,\hat{z},T^*_d) + Q_i\hat{x}_i^d - r_i - k_i - \beta_i \tag{6.6.37}$$

Thus we envisage a three level hierarchical algorithm for solving this problem. This algorithm is given by

Step 1: Guess the initial trajectories $\hat{z}^\rho$ and k^ρ at level 3 for the iteration index $\rho = 1$.

Step 2: Guess the initial trajectories $x^{\gamma *}$, $T_d^{*\gamma}$, β^γ, at the second level and set the iteration index $\gamma = 1$.

Step 3: Using $x^{*\gamma}$, $T_d^{*\gamma}$, β^γ, ν^γ obtained from Step 2, calculate P_i, η_i from equations (6.6.36) and (6.6.37) and thus $\hat{x}$ from equation (6.6.35) and from equation (6.6.34). Calculate also T_d.

Step 4: Substitute the x, λ obtained at level one into the R.H.S. of equations (6.6.29), (6.6.32) to obtain $\hat{x}^{*\gamma+1}$, $T_{d_i}^{*\gamma+1}$, $\beta^{\gamma+1}$ and $\nu^{\gamma+1}$. If the norm of the difference between $x^{*\gamma+1}$, and $x^{*\gamma}$ as well as between $(T_{d_i}^{*\gamma+1}, T_d^{*\gamma})$, $(\beta^{\gamma+1},\beta^\gamma)$, $(\nu^{\gamma+1}, \nu^\gamma)$ is not sufficiently small, go to Step 2. Otherwise go to level 3 to calculate $k^{\rho+1}$, $\hat{z}^{\rho+1}$ from equations (6.6.28) and (6.6.27). If the norm of the difference between $k^{\rho+1}$, k^ρ) and between $(\hat{z}^{\rho+1}, z^\rho)$ is sufficiently small, record T_d as the decentralised gain matrix otherwise go to Step 2 using k^{1+1}, z^{1+1} as the new guess.

Remarks

1) This is a prediction type algorithm whose convergence can be proved using a technique similar to the one used by Hassan and Singh [7].

2) The algorithm is very simple since at the lowest level mere substitution into single variable formulae is all that is required.

3) At the optimum, we obtain T_d and thus the decentralised gain matrix.

4) The entire calculation is done off line and only the decentralised gains are implemented on line to produce the decentralised control.

5) Although the decentralised gains are initial state dependent, computational experience indicates that the control is relatively insensitive to small variations.

6) In practice we will use a set of 'average' initial conditions to compute the decentralised gains off-line and then implement the gains to control the system even when the initial conditions change.

7) The controller has a prespecified degree of stability α and this stability aspect will be studied in the next sub-section.

Note that when the interconnections matrix e_{ij} is considered the closed loop system will be

$$\dot{x}_i = A_i x_i - B_i G_{bi} x_i - B_i e_{ii} T_{d_i} x_i + \sum_{j=1}^{N} e_{ij} A_{ij} x_j + d_i - B_i \xi_i \qquad (6.6.38)$$

where T_{d_i} is the part of T_d corresponding to subsystem i.

Next let us study the stability properties of this decentralised control system.

6.6.2 Stability of decentralised control system

The stability of interconnected dynamical systems with completely decentralised controllers have been previously studied by Siljak et al. and they derived a sufficient condition the satisfaction of which guarantees the stability of the system. Also, for systems which do not satisfy this condition they proposed a second level controller which can improve the stability as well as the suboptimality index of the system. Following their procedure we shall derive, in this section, a sufficient condition for the stability of the interconnected dynamical systems with the implementation of these completely decentralised controllers in which T_d is diagonal and we shall show that the addition of the diagonal gain matrix T_d will have the effect of stabilising the system in addition to improving the suboptimality of the system.

Theorem 6.6.1

Part I: The system represented by equation (6.6.38) will be exponentially stable of degree α if the following condition is satisfied.

$$\sum_{i=1}^{N}\sum_{j=1}^{N} e_{ij}f_{ij} + \sum_{i=1}^{N}\sum_{j=1}^{N} b_{ij} \leq \frac{1}{2}\frac{\min\limits_i \lambda_m(w_i)}{\max\limits_i \lambda_m(P_i)} + \frac{\min\limits_i a_i \sum\limits_{i=1}^{N} e_{ii}}{\max\limits_i\{\lambda_M(P_i)\}} \tag{6.6.39}$$

where $W_i = Q_i + P_iB_iR_i^{-1}B_i^TP_i$.

P_i is the solution of the decomposed Riccati equation from which we obtained G_{bi}

$$\lambda_m(\pi_i) = \text{min eigenvalue of } \pi_i$$

$$\lambda_M(\pi_i) = \text{max eigenvalue of } \pi_i$$

$$f_{ij} = ||A_{ij}||$$

$$a_i = ||\Gamma_i||$$

$$\Gamma_i = P_iB_iT_{d_i}$$

$$b_{ij} = ||H_{ij}||$$

$$H = (-\hat{A} + BG_b + B\tau)$$

Part II: The implementation of the decentralised controllers containing the diagonal gain matrix T_d will have the effect of stabilising the global system if it was unstable under the operation of the decentralised gain matrix G_b only.

Proof: Consider first the minimisation of equation (6.6.4) subject to equation (6.6.5) under the assumption that the structural connection between the subsystems is ideal. An ideal structure implies that there are no structural perturbations. In this case we obtain the closed loop system

$$\dot{\hat{x}} = (\hat{A} - BG_b - B\tau)\hat{x} - \hat{B}\xi(t) + d(t)$$

or $$\dot{\hat{x}} = (\hat{A} - BG_b - B\tau)\hat{x} + \hat{D}(t)$$

If this system is stable and if the final time T is sufficiently large such that $\dot{\hat{x}} \to 0$ and $\hat{x} \to \hat{x}^d$ when $t \to T$ then at $t = T$ we have

$$\hat{D}(T) = -(\hat{A} - BG_B - B\tau)\hat{x}^d(T)$$

or $$\hat{D}(t) = -(\hat{A} - BG_b - B\tau)\hat{x}^d(t)$$

Now we will study the stability of the system when only the local feedback gain G_{bi} is applied and the interactions are ignored and where the gain T_d is used as well so that the interactions are taken into account.

Case 1: Stability of the system controlled using the gain B_{bi} only

Consider the Lyapunov function

$$V = \sum_{i=1}^{N} \hat{x}_i^T P_i \hat{x}_i \tag{6.6.40}$$

where P_i is the solution of the decomposed Riccati equation from which we obtained G_{bi}. Since P_i is positive definite, let us find the condition under which $\dot{V}$ is at least negative semidefinite. Let us assume that the final time $\underline{T}$ is sufficiently large and the interconnection matrix E has elements e_{ij} which are zero or one. Then

$$\begin{aligned}\frac{dV}{dt} &= \sum_{i=1}^{N} \{[\hat{x}_i^T(\hat{A}_i^T - P_iB_iR_i^{-1}B_i^T) + \sum_{j=1}^{N} \bar{e}_{ij}\hat{x}_j^T\hat{A}_{ij}^T + \hat{D}_i^T] \\ &\quad \chi\, P_i\hat{x}_i + \hat{x}_i^T P_i[(\hat{A}_i - B_iR_i^{-1}B_i^T)\hat{x}_i + \sum_{j=1}^{N} e_{ij}A_{ij}\hat{x}_j\hat{D}_i]\} \\ &= \sum_{i=1}^{N} \hat{x}_i^T(-Q_i - P_iB_iR_i^{-1}B_i^T)\hat{x}_i + \sum_{i=1}^{N} 2\hat{x}_i^T P_i \sum_{j=1}^{N} \bar{e}_{ij}A_{ij}\hat{x}_j \\ &\quad + 2\sum_{i=1}^{N} \hat{x}_i^T P_i \hat{D}_i \\ &= -\sum_{i=1}^{N} ||\hat{x}_i||^2_{W_i} + 2\sum_{i=1}^{N} \hat{x}_i^T P_i \sum_{j=1}^{N} \bar{e}_{ij}A_{ij}\hat{x}_j \\ &\quad 2\sum_{i=1}^{N} \hat{x}_i^T P_i \sum_{j=1}^{N} H_{ij}\hat{x}_j^d(t)\end{aligned} \tag{6.6.41}$$

where $H = (-A + BG_b + B\tau)$; $W_i = Q_i + P_iB_iR_i^{-1}B_i^TP_i$

Now since

$$\begin{aligned}\sum_{i=1}^{N} \hat{x}_i^T P_i \sum_{j=1}^{N} \bar{e}_{ij}A_{ij}\hat{x}_j &\leq \sum_{i=1}^{N} \lambda_M(P_i) \sum_{j=1}^{N} \bar{e}_{ij}f_{ij}||\hat{x}_j||^2 \\ &\leq \max \lambda_M(P_i) \sum_{i=1}^{N}\sum_{i=1}^{N} \bar{e}_{ij}f_{ij}||\hat{x}_j||^2\end{aligned} \tag{6.6.42}$$

where $\lambda_M(P_i)$ max eigenvalue of P_i

$$f_{ij} = \lambda_M^{1/2}(A_{ij}^T A_{ij})$$

$$\sum_{i=1}^{N} ||x_i||^2_{W_i} \geq \min_i \lambda_m(W_i) \sum_{i=1}^{N} ||\hat{x}_i||^2 \tag{6.6.43}$$

where $\lambda_m(w_i)$ = min eigenvalue of W_i

$$\sum_{i=1}^{N} \hat{x}_i^T P_i \sum_{j=1}^{N} H_{ij} x_j^d(t) \leq \max \lambda_M(P_i) \sum_{i=1}^{N} \sum_{i=1}^{N} b_{ij} ||\hat{x}_j||^2$$

where $b_{ij} = \lambda_M^2(H_{ij}^T H_{ij})$

Then for $dV/dt \leq 0$ i.e. for the system to be asymptotically stable we have

$$2 \max_i \lambda_M(P_i) \sum_{i=1}^{N} \sum_{i=1}^{N} \bar{e}_{ij} f_{ij} + 2 \max_i \lambda_M(P_i) \sum_{i=1}^{N} \sum_{j=1}^{N} b_{ij} \leq \min_i \lambda_M(W_i)$$

or
$$\sum_{i=1}^{N} \sum_{j=1}^{N} \bar{e}_{ij} f_{ij} + \sum_{i=1}^{N} \sum_{j=1}^{N} b_{ij} \leq \frac{1}{2} \frac{\min_i \lambda_m(W_i)}{\max_i \lambda_M(P_i)} \tag{6.6.44}$$

Since $|e_{ij}| \leq |\bar{e}_{ij}|$ element by element, the above equation holds for any interconnection matrix and the system will be asymptotically stable with degree α.

Case 2: Stability of the system when the gain T_d is applied as well. Considering the same Lyapunov function as above we have

$$\frac{dV}{dt} = -\sum_{i=1}^{N} ||\hat{x}_i||_{W_i}^2 + 2 \sum_{i=1}^{N} \hat{x}_i^T P_i \sum_{j=1}^{N} \bar{e}_{ij} A_{ij} x_j + 2 \sum_{i=1}^{N} \hat{x}_i^T P_i \times \sum_{j=1}^{N} H_{ij} \hat{x}_j^d(t) - 2 \sum_{i=1}^{N} \hat{x}_i^T P_i B_i \hat{e}_{ii} T_{d_i} \hat{x}_i \tag{6.6.45}$$

and since

$$\sum_{i=1}^{N} ||\hat{x}_i||_{\Gamma_i}^2 \bar{e}_{ii} \geq \min_i a_i \sum_{i=1}^{N} ||x_i||^2 \bar{e}_{ii} \tag{6.6.46}$$

where $\Gamma_i = P_i B_i T_{d_i}$ and $a_i = \lambda_M^{1/2}(\Gamma_i^T \Gamma_i)$ then for dV/dt to be negative semi-definite

$$\sum_{i=1}^{N} \sum_{j=1}^{N} \bar{e}_{ij} f_{ij} + \sum_{i=1}^{N} \sum_{j=1}^{N} b_{ij} \leq \frac{1}{2} \frac{\min_i \lambda_m(W_i)}{\max_i \lambda_M(P_i)} + \frac{\min_i a_i \sum_{i=1}^{N} \bar{e}_{ii}}{\max_i \lambda_M(P_i)} \tag{6.6.47}$$

here again since $|e_{ij}| \leq |\bar{e}_{ij}|$ element by element, the above expression holds for any interconnection matrix E. In addition if it is satisfied, the original system will be exponentially stable with degree α. This proves the first part of the theorem.

Comparing the expression in equations (6.6.4) and (6.6.4), we see that T_d has the effect of stabilising the system since the expression (6.6.4) is less restrictive. This proves the second part of the theorem.

6.6.3 Example

In order to illustrate this approach and its stability property we shall consider the following simple problem

$$\min J = \tfrac{1}{2}\int_0^{10} ||x||_Q^2 + ||u||_R^2 dt \qquad (6.6.48)$$

s.t. $\dot{x} = Ax + Bu$

where

$$A = \begin{bmatrix} 0 & 1 & 1 \\ -2 & -3 & -0.5 \\ 1 & 2 & 3 \end{bmatrix} \quad B = \begin{bmatrix} 1 & 0 \\ 1 & 0 \\ 0 & 1 \end{bmatrix} \quad Q = I_3,\ R = I_2,\ \alpha = 0.0$$

where I_i is the identity matrix of order i.

This problem has been solved for three different cases:

i) by using the globally optimal closed loop gain matrix

ii) by decomposing the problem into two subproblems, the first one with

$$A = \begin{bmatrix} 0 & 1 \\ -2 & -3 \end{bmatrix} \quad B_1 = \begin{bmatrix} 1 \\ 1 \end{bmatrix}$$

while the second with

$$A_2 = [3] \qquad B_2 = [1]$$

and solving these two problems as if they were completely decoupled to obtain the block diagonal gain matrix G_b. The matrix G_b, was used in the feedback loop and the suboptimal closed loop trajectories were obtained.

iii) by decomposing the problem into two subproblems and calculating the diagonal gain matrix T_d which has been used in addition to the matrix G_b in the feedback loop. The suboptimal trajectories for this case have also been calculated.

Figures 6.5 to 6.7 show the optimal and suboptimal responses for the above three cases. From these, one can conclude the following:

1) The system is globally stable in the three cases.

2) Although for the second case the system is also stable, one can see from the trajectories that the system in this case has the tendency to oscillate which illustrates Part II of Theorem 6.6.1.

3) The suboptimal trajectories with the closed loop gain matrix $G_b + T_d$ are very close to the optimal ones.

An alternative design approach to decentralised control relies on constructing simplified models of the environment of each subsystem. We next consider this approach.

6.7 DECENTRALISED CONTROL USING INTERACTION MODELS [15]

The optimisation problem of a system, comprising N linear-interconnected dynamical subsystems, can be written as

$$\min J = \tfrac{1}{2} \sum_{i=1}^{N} \int_{o}^{\infty} (||x_i - x_i^d||^2_{Q_i} + ||u_i||^2_{R_i})dt \qquad (6.7.1)$$

where $||.||^2_S = .^T S .$, Q_i are assumed to be at least positive semidefinite matrices while R_i are assumed to be positive definite matrices.

J in equation (6.7.1) is to be minimised subject to the dynamical constraints of the subsystem, i.e.

$$\dot{x}_i = A_i x_i + B_i u_i + C_i z_i + d_i \qquad i=1,2,\ldots,N \qquad (6.7.2)$$

where z_i is the interaction vector which is assumed to be a linear combination of the states of the other subsystem, i.e.

$$z = Lx \qquad (6.7.3)$$

where L is a complete matrix.

Here, the control vector u_i for the ith subsystem is of dimension r_i, the interaction vector z_i is of dimension m_i and B_i, and C_i are matrices of ranks r_i and m_i, respectively.

The optimal controller for each subsystem is given by

$$u_i = -R_i^{-1}B_i^T P_i x_i - R_i^{-1}B_i^T \xi_i \qquad (6.7.4)$$

where P_i is a solution of a Riccati-type equation and ξ_i is a solution of a linear differential equation. ξ_i is a function of the states of all the other subsystems and it is obtained by the hierarchical calculation described by Singh et al [12], i.e.

$$\xi_i = \sum_{i=1}^{N} T_{ij} x_j(t) + \theta_i \qquad j=1,2,\ldots.N \qquad (6.7.5)$$

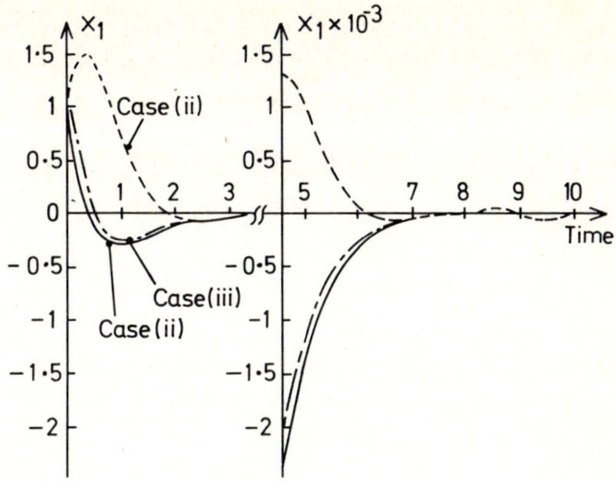

Figure 6.5 : Evolution of the state variable x_1 with respect to time t.

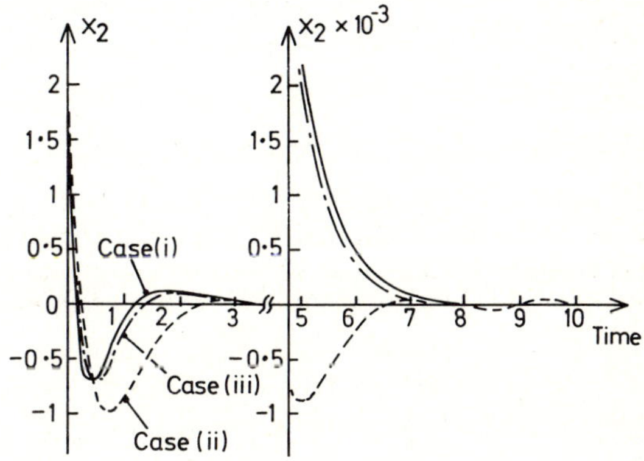

Figure 6.6 : Evolution of the state variable x_2 with respect to time t.

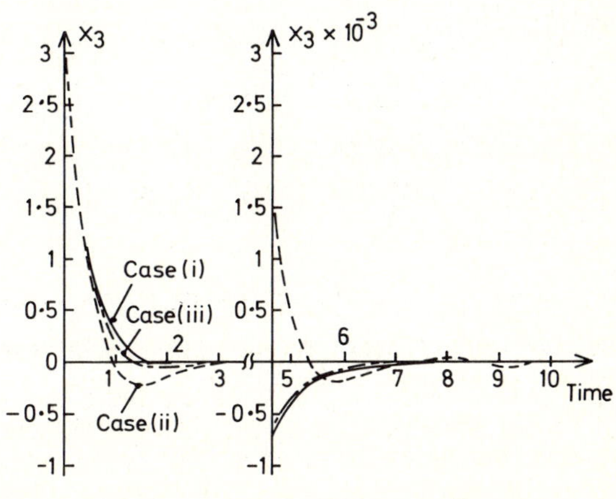

Figure 6.7 : Evolution of the state variable x_3 with respect to time t.

Now, if the interactions between the subsystems coming in through equation (6.7.3) are ignored, then equation (6.7.4) can be reduced to a completely decentralised-control law given by

$$u_i(t) = R_i^{-1}B_i^TP_ix_i(t) - R_i^{-1}B_i^T\theta_i = G_ix_i(t) - \nu_i \quad (6.7.6)$$

the application of which will give

$$\dot{x}_i' = (A_i - B_iG_ix_i' + C_iz_i' + d_i - B_i\nu_i \quad (6.7.7)$$

In order to take the interactions into account more fully in the offline state, let us consider the control u_i in equation (6.7.2) to comprise two components, i.e.

$$u_i = u_{1i} + u_{2i}$$

where u_{1i} is the control u_i given in equation (6.7.6) and u_{2i} is an additional control for the ith subsystem that we shall compute as described below.

Now, one possibility is to consider the interactions as unknown disturbances and to minimise a cost function of the form

$$J = \tfrac{1}{2}\sum_{i=1}^{N}\int_0^\infty [||x_i' - x_i^d||^2_{Q_i} + \psi_i(u_{i2},\dot{u}_{i2})]dt \quad (6.7.8)$$

where x_i' in this case are the suboptimal trajectories resulting from the application of the decentralised-control law given by equation (6.7.6) and where ψ_i is an appropriate function of the control and its derivative.*

The cost function J in equation (6.7.8) is to be minimised subject to the dynamics

$$\dot{x}_i' = A_i^*x_i' + B_iu_{2i} + C_iz_i' + \mathcal{D}_i$$

Now, let us choose the vector γ_i which minimises the Euclidean norm of the vector $B_iu_{2i} - C_i\gamma_i$; then we have

$$\gamma_i = (C_i^TC_i)^{-1}C_i^TB_iu_i \quad (6.7.9)$$

Then, substituting for B_iu_{2i} by $C_i\gamma_i$, we have

$$\dot{\hat{x}}_i = A_i^*\hat{x}_i + C_i\{[C_i^TC_i]^{-1}C_i^TB_iu_{2i} + \hat{z}_i\} + \mathcal{D}_i \quad (6.7.9)$$

where $\hat{x}_i$ is the approximate solution of the states resulting from these substitutions

---

* In conventional regulator theory (with no disturbances) we usually choose ψ to be of the form u^TRu. When disturbances are present, and cannot be measured, it is possible to choose a function $\psi(u,\dot{u},\ddot{u},\ldots)$ so that the corresponding control ensures that x approaches x^d in the steady state and it is not necessary to measure the disturbances [13]

$$\left.\begin{aligned} A_i^* &= A_i - B_i K_i^{-1} B_i^T P_i \\ &\text{and} \\ \mathcal{D}_i &= d_i - B_i R_i^{-1} B_i^T \theta_i \end{aligned}\right\} \qquad (6.7.10)$$

Now, if the overall system matrix of the system is A_G, then we could approximate $\hat{z}_i$ by the linear differential equation

$$\dot{\hat{z}}_i = A_{z_i} \hat{z}_i + d_{z_i}$$

where A_{z_i} is that part of A_G which corresponds to the elements of $\hat{z}_i$. Let us write

$$y_i = (C_i^T C_i)^{-1} C_i^T B_i u_{2i} + \hat{z}_i \qquad (6.7.11)$$

Then

$$\dot{y}_i = \mathcal{B}_i \dot{u}_{2i} + \dot{\hat{z}}_i$$

where $\mathcal{B}_i = (C_i^T C_i)^{-1} C_i^T B_i$

substituting for $\dot{\hat{z}}_i$

$$\begin{aligned} \dot{y}_i &= \mathcal{B}_i \dot{u}_{2i} + A_{z_i} \hat{z}_i + d_{z_i} \\ &= \mathcal{B}_i \dot{u}_{2i} + A_{z_i} y_i - A_{z_i} \mathcal{B}_i u_{2i} + d_{z_i} \qquad (6.7.12) \\ &= A_{z_i} y_i + \mathcal{B}_i \dot{u}_{2i} - A_{z_i} \mathcal{B}_i u_{2i} + d_{z_i} \end{aligned}$$

Let us write

$$v_i = \mathcal{B}_i \dot{u}_{2i} - A_{z_i} \mathcal{B}_i u_{2i} \qquad (6.7.13)$$

and let us define our function ψ in the cost function of equation (6.7.9) as

$$\psi_i(u_{2i}, \dot{u}_{2i}) = ||\mathcal{B}_i \dot{u}_{2i} - A_{z_i} \mathcal{B}_i u_{2i}||^2_{R_i}$$

Then the optimal decentralised-control problem can be rewritten as

$$\min J = \tfrac{1}{2} \sum_{i=1}^{N} \int_0^\infty (||\tilde{x}_i - \tilde{x}_i^d||^2_{Q_i} + ||v_i||^2_{R_i}) dt \qquad (6.7.14)$$

subject to

$$\dot{\tilde{x}}_i = \left(\begin{array}{c|c} A_i^* & C_i \\ \hline 0 & A_{z_i} \end{array}\right) \tilde{x}_i + \left(\begin{array}{c} 0 \\ \hline I \end{array}\right) u_i + \left(\begin{array}{c} \mathcal{D}_i \\ \hline d_{z_i} \end{array}\right)$$

where

$$\tilde{x}_i = \begin{pmatrix} x_i \\ \hline y_i \end{pmatrix}, \quad \tilde{Q}_i = \begin{pmatrix} Q_i & 0 \\ \hline 0 & 0 \end{pmatrix}, \quad \tilde{x}_i^d = \begin{pmatrix} x_i^d \\ 0 \end{pmatrix}$$

Let us define

$$\tilde{A}_i = \begin{pmatrix} A_i^* & C_i \\ \hline 0 & A_{z_i} \end{pmatrix}, \quad \tilde{B}_i = \begin{pmatrix} 0 \\ \hline I \end{pmatrix}, \quad \tilde{D}_i = \begin{pmatrix} \mathcal{D}_i \\ \hline d_{z_i} \end{pmatrix}$$

The optimal solution v_i of this problem can be written as

$$v_i = -\tilde{R}_i^{-1}\tilde{B}_i^T P_i x_i - \tilde{R}_i^{-1}\tilde{B}_i^T \eta_i$$

or (6.7.16)

$$v_i = \tilde{G}_{i1} x_1 - \tilde{G}_{i2} y_i - \tilde{R}_i^{-1}\tilde{B}_i^T \eta_i$$

where $\tilde{P}_i$ is the solution of the Riccati equation

$$\dot{\tilde{P}}_i = -\tilde{P}_i A_i - \tilde{A}_i^T \tilde{P}_i + \tilde{P}_i \tilde{B}_i \tilde{R}_i^{-1} \tilde{B}_i^T \tilde{P}_i - \tilde{Q}_i$$

and η_i is the solution of the linear-vector differential equation

$$\dot{\eta}_i = (\tilde{A}_i^T - \tilde{P}_i \tilde{B}_i \tilde{R}_i^{-1} B_i^T)\eta_i - \tilde{P}_i \tilde{D}_i + \tilde{Q}_i \tilde{x}_d$$

Substituting v_i in equation (6.7.16) into equation (6.7.12), and using equation (6.7.13), we have

$$\dot{x}_i^* = A_i^* x_i^* + C_i y_i^* + \mathcal{D}_i \tag{6.7.17}$$

$$\dot{y}_i^* = A_{Z_i} y_i^* - \tilde{G}_{i1} x_i^* - \tilde{G}_{i2} y_i^* - \tilde{R}_i^{-1}\tilde{B}_i^T \eta_i + d_{z_i}$$

i.e.

$$\dot{y}_i^* = (A_{2i} - \tilde{G}_{i2}) y_i^* - \tilde{G}_{i1} x_i^* - \tilde{R}_i^{-1} B_i^T \eta_i + d_{z_i} \tag{6.7.18}$$

where x_i^* and y_i^* are the trajectories resulting from substituting v_i in equation (6.7.16).

From equation (6.7.13) we have

$$\mathcal{B}_i u_{2i} = v_i + A_{z_i} \mathcal{B}_i u_{2i} \tag{6.7.19}$$

Again, by making an approximation similar to that made in equation (6.7.9), i.e. by calculating the two vectors α_i, β_i which minimise the Euclidean norms of the two vectors

$$\mathcal{B}_i\alpha_i - A_{z_i}\, \mathcal{B}_i u_{2i}$$

$$\mathcal{B}_i\beta_i - v_i$$

and substituting them in equation (6.7.19) we have

$$\dot{u}'_{2i} = (\mathcal{B}_i^T \mathcal{B}_i)^{-1} \mathcal{B}i^T A_{z_i}\, {}_i u'_{2i} + (\mathcal{B}_i^T\mathcal{B}_i)^{-1} \mathcal{B}_i^T$$

$$(- \tilde{G}_{i1}x_i^* - \tilde{G}_{i2}y_i^* - \tilde{R}_i^{-1}\tilde{B}_i^T\eta_i)$$

or (6.7.20)

$$\dot{u}'_{2i} = A_{u_i} u'_{2i} - H_i x_i^* - F_i y_i^* - S_i$$

where u'_{2i} is an approximate solution for u_{2i} resulting from this substitution.

$$A_{u_i} = (\mathcal{B}_i^T\mathcal{B}_i)^{-1}\mathcal{B}_i^T A_{z_i} \mathcal{B}_i$$

$$H_i = (\mathcal{B}_i^T\mathcal{B}_i)^{-1}\mathcal{B}_i^T\tilde{G}_i$$

$$F_i = (\mathcal{B}_i^T\mathcal{B}_i)^{-1}\mathcal{B}_i^T G_{i2}$$

$$S_i = (\mathcal{B}_i^T\mathcal{B}_i)^{-1}\mathcal{B}_i^T R_i^{-1}\mathcal{B}_i^T\eta_i$$

Then, using equation (6.7.19) we can compute u'_{2i} as shown in the analogue diagram in Fig. 6.8.

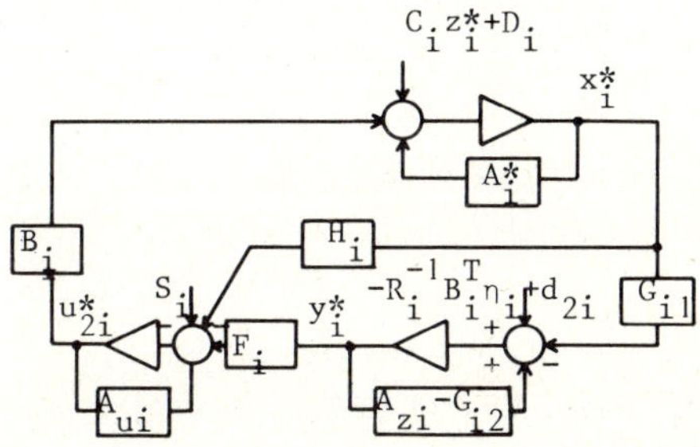

Figure 6.8

Remarks

a) The computation of the control is very simple. G_{i1} and G_{i2} are computed by solving the problem offline within a hierarchical structure and u'_{2i} is computed online for each subsystem in an analogue fashion with the various blocks in the diagram having been precalculated offline using the above equation.

b) The control is independent of initial conditions.

c) The controller can accommodate unknown nonzero constant disturbances and still reach the desired steady-state value x_i^d.

d) The controller is completely decentralised.

Next, we illustrate the new controller on the 3-reach 22nd-order river-pollution control problem previously considered by Singh et al [8].

6.7.1 Application of controller to the river problem

The river problem is a good example for the application of decentralised-control techniques, because here, as discussed by Singh et al [8] the control centres (sewage-treatment stations) are often many kilometres apart. Thus, the cost of providing telemetry links for the passage of state information between the sewage stations is by far the biggest cost in the control implementation. The particular problem considered here was previously considered by Singh et al [8] and is of a 3-reach distributed delay model of the river Cam near Cambridge.

A reach of a river is a section of the river of a convenient length with a single polluter, and if we consider the dynamics of the two most important variables in river pollution control, i.e. biochemical oxygen demand (b.o.d.) and dissolved oxygen (d.o.), it is possible to write the following delay differential equations to model the dynamic behaviour of the system:

$$\dot{z}_i = 1.32z_1 + 0.1m_1 + 0.9z_0 + 5.35 + 0.1u_1 \qquad (6.7.21)$$

$$\dot{q}_i = -0.32z_1 - 1.29q_1 + 0.9q_0 + 1.9 \qquad (6.7.22)$$

$$\dot{z}_2 = 0.9 \sum_{j=1}^{s} a_j z_1(t - \tau_j) - 1.32z_2 + 0.1m_2 + 4.19 + 0.1u_2 \qquad (6.7.23)$$

$$\dot{q}_2 = 0.9 \sum_{j=1}^{s} a_j q_1(t - \tau_j) - 0.32z_2 - 1.29q_2 + 1.9 \qquad (6.7.24)$$

$$\dot{z}_3 = 0.9 \sum_{j=1}^{s} a_j z_2(t - \tau_j) - 1.32z_3 + 0.1m_3 + 4.19 + 0.1u_3 \qquad (6.7.25)$$

$$\dot{q}_3 = 0.9 \sum_{j=1}^{s} a_j q_2(t - \tau_j) - 0.32z_3 - 1.29q_3 + 1.9 \qquad (6.7.26)$$

where q_1, q_2, q_3 are the concentrations of dissolved oxygen in the three reaches z_1, z_2, z_3 are the concentrations of b.o.d. in the three reaches and m_i, u_i (i=1,3) are the controls. Suitable values of s, τ_1, a for a river are

$$s = 3,\ \tau_1 = 0,\ \tau_2 = \tfrac{1}{2}\text{ day},\ \tau_3 = 1\text{ day},\ z_o = 0, q_o = 10,$$

$$a_1 = 0.15,\ a_2 = 0.7,\ a_3 = 0.15.$$

The system, as described in equation (6.7.21 - 6.7.26) is nominally of infinite dimension in the state space. It is possible to obtain a good finite-dimensional approximation by expanding the delayed terms in a Taylor series and taking the first two terms e.g. equation (6.7.23) which can be rewritten as

$$\dot{z}_2 = 0.9[0.15z_1(t) + 0.7z(t - 0.5) + 0.15z_1(t - 1)]$$
$$- 1.32z_2 + 0.1m_2 + 4.19 + 0.1u_2$$

Here, we have two delays, so it is necessary to introduce four additional states. Let these be given by z_4, z_5, z_6, z_7. Let $z_4(t) = z_1(t - 0.5)$; then $z_4(s) = z_1(s)e^{-0.5s}$ where s is the Laplace variable.

Now

$$\dot{z}_1(s)e^{-0.5s} = z_1(s)[1 + 0.5s + 0.25/2s^2 + \ldots]^{-1}$$

Taking only the first three terms

$$z_1(t) = z_4(t) + 0.5\dot{z}_4(t) + 0.125\ddot{z}_4(t) \qquad (6.7.27)$$

Let

$$\dot{z}_4 = z_5$$

Then

$$\dot{z}_5 = 8z_1 - 8z_4 - 4z_5 \qquad (6.7.28)$$

Similarly, for the other delay we let

$$z_1(t - 1) = z_6$$

Then

$$\dot{z}_6 = z_7 \qquad (6.7.29)$$

$$\dot{z}_7 = 2z_1 - 2z_6 - 2z_7 \qquad (6.7.30)$$

Thus, we can rewrite equation (6.7.23) as

$$\dot{z}_2 = 0.135z_1 + 0.63z_4 + 0.135z_6 - 1.32z_2 + 0.1m_2$$
$$+ 4.19 + 0.1u_2$$

Similarly, we can introduce four additional variables, each for the delays in equations (6.7.24 to 6.7.26). This makes the overall system of order 22.

6.7.2 Solution of river-pollution problem using new method

For the above system, a cost function with weight R = 1000

$$Q = \left[\begin{array}{l} \qquad\qquad \overset{6}{\longleftrightarrow} \\ 0.002 \quad 0 \;\ldots\ldots\; 0 \;\ldots\ldots\ldots\ldots \\ 6 \updownarrow \ldots\ldots\ldots\ldots\ldots 0.001 \;\ldots\ldots 0 \ldots\ldots \\ 0 \qquad\qquad\qquad\quad \vdots \end{array}\right]$$

was chosen. For the model reduction, A_z was chosen to be

$$A_z = \begin{bmatrix} 1.30232 & -0.00053 \\ -0.32 & -1.2 \end{bmatrix}$$

The initial conditions used were

$$x(0)^T = [10.0 \quad 7.0 \quad 5.94 \quad 10.0 \quad 10.0 \quad 10.0 \quad 10.0 \quad 6.0 \quad 7.0 \quad 7.0 \quad 7.0 \quad 7.0 \quad 5.2370 \quad 5.94 \quad 5.94 \quad 5.94 \quad 5.94 \quad 4.69]$$

$$y(0)^T = [10.0 \quad 7.0 \quad 5.94 \quad 6.0]$$

$$u(0)^T = [0 \quad 0]$$

Figures 6.9 to 6.14 show the optimal and decentralised suboptimal trajectories for b.o.d. and d.o. for the three reaches. We see that the optimal and suboptimal trajectories are very close, showing that the control is, indeed, near optimal.

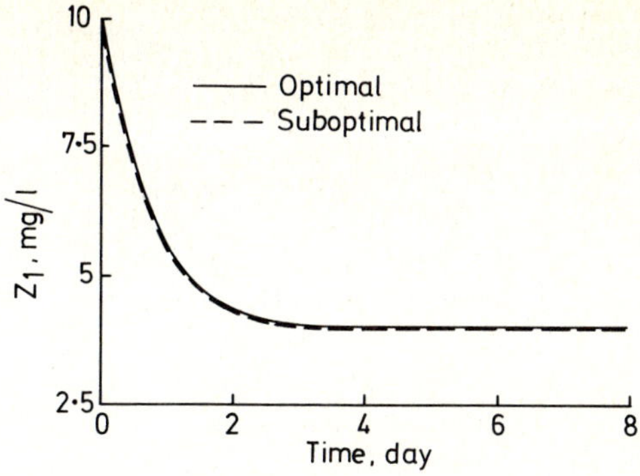

Figure 6.9 : Optimal and suboptimal b.o.d. for reach 1

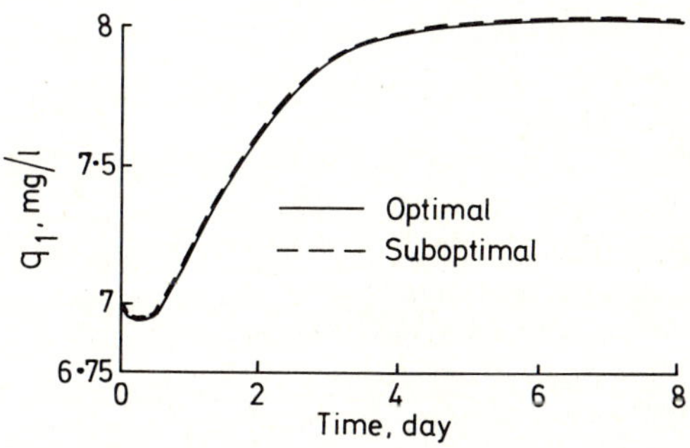

Figure 6.10 : Optimal and suboptimal d.o. for reach 1

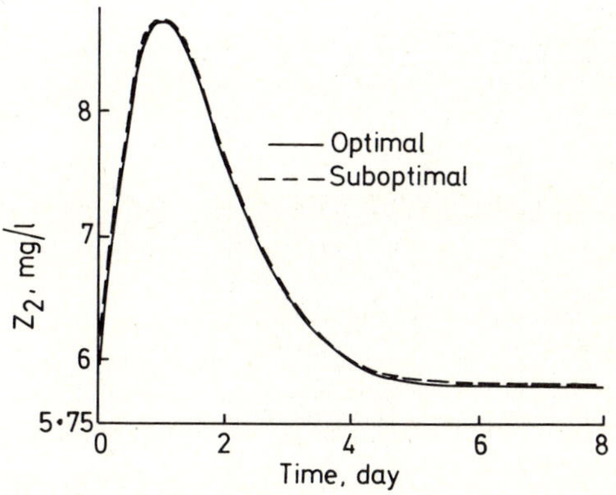

Figure 6.11 : Optimal and suboptimal b.o.d. for reach 2

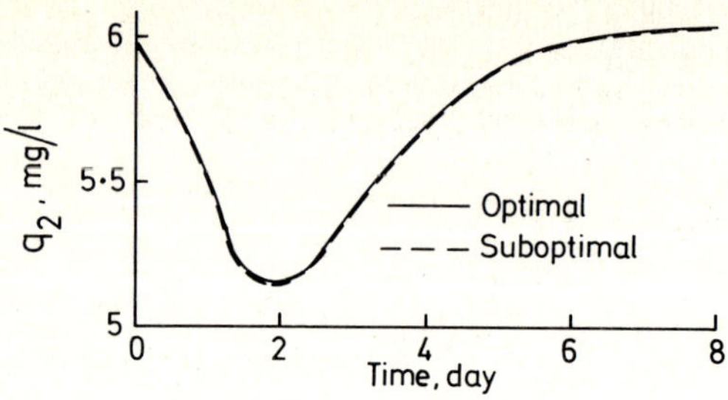

Figure 6.12 : Optimal and suboptimal d.o. for reach 2

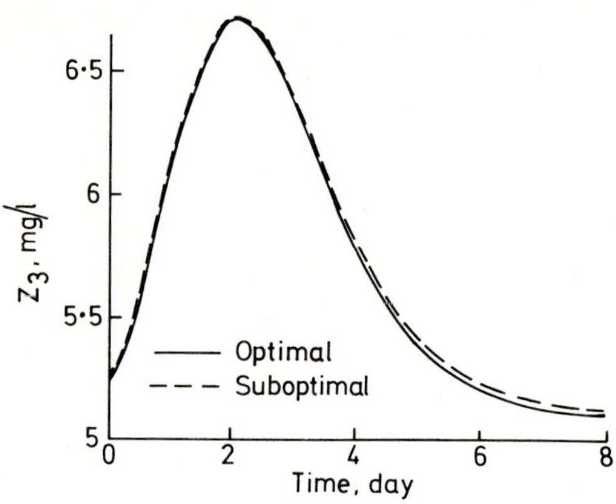

Figure 6.13 : Optimal and suboptimal b.o.d. for reach 3

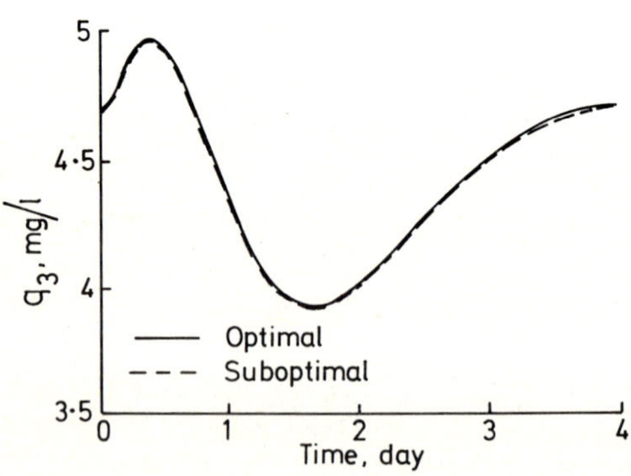

Figure 6.14 : Optimal and suboptimal d.o. for reach 3

6.8 APPLICATION TO THE SHIP EXAMPLE

In order to get more of a feel for the approach, let us apply it to our ship example. We recall that the global ship system is divided into three sub-systems as follows

$$\frac{d}{dt}\begin{pmatrix} x_1 \\ x_2 \\ x_3 \end{pmatrix} = \begin{pmatrix} A_1 & & \\ & A_2 & \\ & & A_3 \end{pmatrix}\begin{pmatrix} x_1 \\ x_2 \\ x_3 \end{pmatrix} + \begin{pmatrix} B_1 & & \\ & B_2 & \\ & & B_3 \end{pmatrix}\begin{pmatrix} u_1 \\ u_2 \\ u_3 \end{pmatrix}$$

with the following dimensions of x and u (i.e. state and control) respectively.

For subsystem 1 : $\begin{cases} \dim(x) = 10 \\ \dim(u) = 2 \end{cases}$

For subsystem 2 : $\begin{cases} \dim(x) = 4 \\ \dim(u) = 2 \end{cases}$

For subsystem 3 : $\begin{cases} \dim(x) = 6 \\ \dim(u) = 1 \end{cases}$

The following steps are considered in order to write the algorithm for the design of the decentralised controller.

Step 1 : Construct the following matrices

$$\tilde{A}_i = \begin{pmatrix} A_i & C_i \\ 0 & A_{z_i} \end{pmatrix}, \quad \tilde{B}_i = \begin{pmatrix} B_i & 0 \\ 0 & I_i \end{pmatrix}, \quad \tilde{R}_i = \begin{pmatrix} R_{i1} & 0 \\ 0 & R_{i2} \end{pmatrix}$$

$$\tilde{Q}_i = \begin{pmatrix} Q_i & 0 \\ 0 & 0 \end{pmatrix} \qquad i=1,2,3$$

where C_i : is the interconnection matrix between the subsystems

R_{i1} : is chosen from the global solution of the ship problem

R_{i2} : is chosen arbitrarily

Q_i : is chosen from the global solution

$\left.\begin{matrix} A_{z2} \\ A_{z3} \\ A_{z1} \end{matrix}\right\}$ is chosen as described previously except that we do ensure that of these models is stable

Step 2 : Solve the following Riccati matrix equation

$$-\tilde{P}_i\tilde{A}_i - \tilde{A}_i^T\tilde{P}_i + \tilde{P}_i\tilde{B}_i\tilde{R}_i^{-1}\tilde{B}_i^T\tilde{P}_i - \tilde{Q}_i = 0$$

Step 3 : Evaluate the gain matrix from the following equation

$$G = -\tilde{R}_i^{-1}\tilde{B}_i^T\tilde{P}_i = \begin{pmatrix} -G_{1i1} & -G_{2i1} \\ -G_{1i2} & -G_{2i2} \end{pmatrix}$$

Step 3 : Compute the following equations

1) $\mathcal{B}_i = (C_i^T C_i)^{-1} C_i^T B_i$

2) $A_{ui} = (\mathcal{B}_i^T \mathcal{B}_i)^{-1} \mathcal{B}_i^T A_{zi} \mathcal{B}_i$

3) $H_i = (\mathcal{B}_i^T \mathcal{B}_i)^{-1} \mathcal{B}_i^T G_{1i2}$

4) $F_i = (\mathcal{B}_i^T \mathcal{B}_i)^{-1} \mathcal{B}_i^T G_{2i2}$

5) $A_{u\pi} = (A_i - B_i\tilde{G}_{1i1})$

6) $AZGI = (A_{zi} - \tilde{G}_{2i2})$

7) $GIHI = (\tilde{G}_{1i1} + H_i)$

8) $GIFI = -(\tilde{G}_{2i1} + F_i)$

Step 5 : Solve the system of differential equations given below

$$\underline{\dot{x}} = (A - BG)\underline{x} \tag{6.8.1}$$

$$\underline{\dot{x}}_i = (AZGI)\underline{x}_i - B_i\bar{G}_{2i1}\underline{y}_i + CT_i\underline{z}_i + B_i\underline{u}_{2i} \tag{6.8.2}$$

$$\underline{\dot{y}}_i = (AZGI)\underline{y}_i - \bar{G}_{1i2}\underline{x}_i \tag{6.8.3}$$

$$\underline{\dot{u}}_{2i} = A_{ui}\underline{u}_{2i} - H_i\underline{x}_i - F_i\underline{y}_i \tag{6.8.4}$$

$$\underline{\dot{u}}_i = (GIHI)\underline{x}_i + (GIFI)\underline{y}_i + A_{ui}\underline{u}_{2i} \tag{6.8.5}$$

where $\underline{z}_i$: are the states which interconnect each subsystem with the other subsystems and are computed from the global system i.e. equation (6.8.1)

CT_i : is the interconnection relation matrix

The solution of the above system can be done by using the Runge-Kutta-Metson's method. The system can be written in the following form

$$\dot{x} = (MATI)\underline{x}$$

where $x = (x\ \dot{x}_i,\ y_i,\ u_{2i},\ u_i)$

and

$$MATI = \begin{pmatrix} A - BG & 0 & 0 & 0 & 0 \\ GTI & AuII & -BIGI & BI & 0 \\ 0 & -GII2 & AZGI & 0 & 0 \\ 0 & -HI & -FI & AUI & 0 \\ 0 & -GIHI & -GIFI & AUI & 0 \end{pmatrix}$$

<u>Generalised inverse of matrix of $(B_i^T B_i)$</u>

When the value of A_{zi} is a scalar and the control u_i is a vector then the product of $(B_i^T B_i)$ is a singular matrix and therefore the inverse of this matrix cannot be found. This situation occurs when the generalised inverse of the matrix of $(B_i^T B_i)$ is computed for the second subsystem, i.e. when i-2. We can get around this by using the following two theorems which are standard:

<u>Theorem 6.8.1</u>

If A = BC where A,B,C are respectively mxn, mxu, and uxn, and all three matrices are of rank u, then the generalised inverse matrix can be defined by

$$A^+ = C^T(CC^T)^{-1}\ (B^TB)^{-1}B^T$$

<u>Theorem 2</u>

If A is mxn and of rank u and the partition of matrix A is possible and is of the form

$$A = \begin{pmatrix} A_{11} & A_{12} \\ A_{21} & A_{22} \end{pmatrix}$$

where A_{11} is a nonsingular matrix of rank u then

$$A = \begin{pmatrix} I \\ P \end{pmatrix} [A_{11}, A_{12}] = \begin{pmatrix} I \\ P \end{pmatrix} A_{11}\ [IQ] = \begin{pmatrix} A_{11} \\ A_{21} \end{pmatrix} [IQ]$$

where $P = A_{21}A_{11}^{-1}$ and $Q = A_{11}^{-1}A_{12}$

Therefore

$$A^+ = C^T(CC^T)^{-1}\ (B^TB)^{-1}B^T$$

where

$$A = \begin{pmatrix} A_{11} & A_{12} \\ A_{21} & A_{22} \end{pmatrix}, \quad C = (IQ \quad , \quad B = \begin{pmatrix} A_{11} \\ A_{21} \end{pmatrix}$$

$$Q = A_{11}^{-1}\ A_{12}$$

and by replacing the corresponding values the generalised inverse matrix of (B^TB) for the second subsystem is given by

$$(B_2^TB)^{-1} = \begin{pmatrix} 0.33069 & -0.08 \\ -0.08 & 0.01932 \end{pmatrix}$$

6.8.1 Practical considerations on the choice of the interaction model

There are two important considerations which must be borne in mind when we apply the approach to the ship model. Firstly, the basic model has a full B matrix and not a block diagonal one as demanded by the theory in section 6.7. The second one is: how do we choose the interaction model. In a sense, both these aspects are related. As far as the interaction model is concerned, it is of the form

$$\dot{z}_i = A_{z_i} z_i \qquad i=1,2,3.$$

One way of choosing the interaction model is to use the sparsity that exists in the global A matrix in order to eliminate columns where the off-diagonal elements are zero. This is fine as far as A_{z2} and A_{z3} are A_{z3} are concerned. However, this choice of A_{z1} does not give a good response. Nevertheless, the important thing to remember is that the model is somewhat arbitrary - in a sense it defines the response that we seek from our original system. Thus we have some room for manoeuvre. In this example, we chose, after some trial and error:

$$A_{z_1} = \begin{pmatrix} -60 & -0.3 & -0.4 \\ -0.45 & -60 & -0.25 \\ -0.35 & -0.7 & -60 \end{pmatrix}$$

and $\quad Q = \text{diag}\ (0.1\text{x}10^6)$

As for the full B matrix, we only use the block diagonal parts of B in the decentralised controller design. The off-diagonal parts of B (as well as A) form a part of the real interaction vector z.

Figures 6.15 to 6.17 show the responses of the 20 states under decentralised control.

We see that the controller provides the system with a good response in that all the states approach their steady state values within 300 seconds.

Next, we outline another approach which uses an interaction model. In this case, we use a crude interaction model which is improved on-line using a model following type of technique.

6.9 THE MODEL FOLLOWER [16]

We reconsider once again the following optimisation problem:

$$\min J = \sum_{i=1}^{N} \frac{1}{2} \int_0^\infty \{||x_i||^2_{Q_i} + ||u_i||^2_{R_i}\}dt \tag{6.9.1}$$

subject to

$$\dot{x}_i = A_i x_i + B_i u_i + C_i z_i \tag{6.9.2}$$

$$z_i = \sum_{j=1}^{N} L_{ij} x_j \tag{6.9.3}$$

where x_i is the n_i-dimensional state vector of the ith subsystem, u_i is the m_i-dimensional control vector and z_i is the q_i-dimensional interconnection vector. We will assume that the pair (A_i, B_i) is completely controllable.

Let us suppose initially that z_i can be represented by the following dynamical model*:

$$\dot{z}_i = A_{z_i} z_i \tag{6.9.4}$$

where A_{z_i} can be chosen to be the block of the global A matrix corresponding to the vector z_i.

To clarify this idea of the choice of A_{z_i}, let us consider the integrated forms of equations (6.9.2) and (6.9.3)

$$\dot{x} = Ax + Bu + Cz$$

$$z = Lx$$

---

*z_i depends on u_i, x_i, x_j and u_j, however. We shall initially consider the model given by equation (6.9.4) as an approximate one.

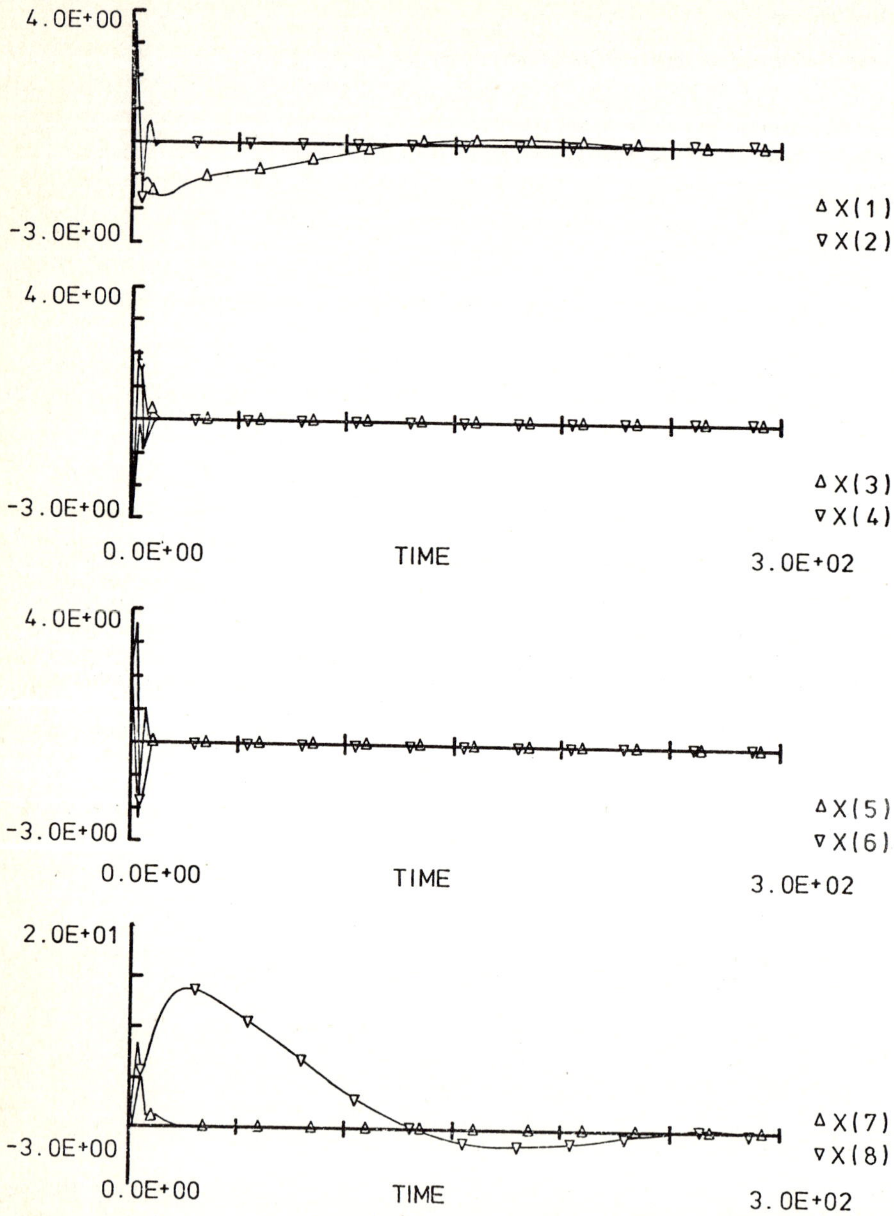
4.0E+00
-3.0E+00
Δ X(1)
∇ X(2)
4.0E+00
-3.0E+00
Δ X(3)
∇ X(4)
0.0E+00
TIME
3.0E+02
4.0E+00
-3.0E+00
Δ X(5)
∇ X(6)
0.0E+00
TIME
3.0E+02
2.0E+01
-3.0E+00
Δ X(7)
∇ X(8)
0.0E+00
TIME
3.0E+02

Figure 6.15

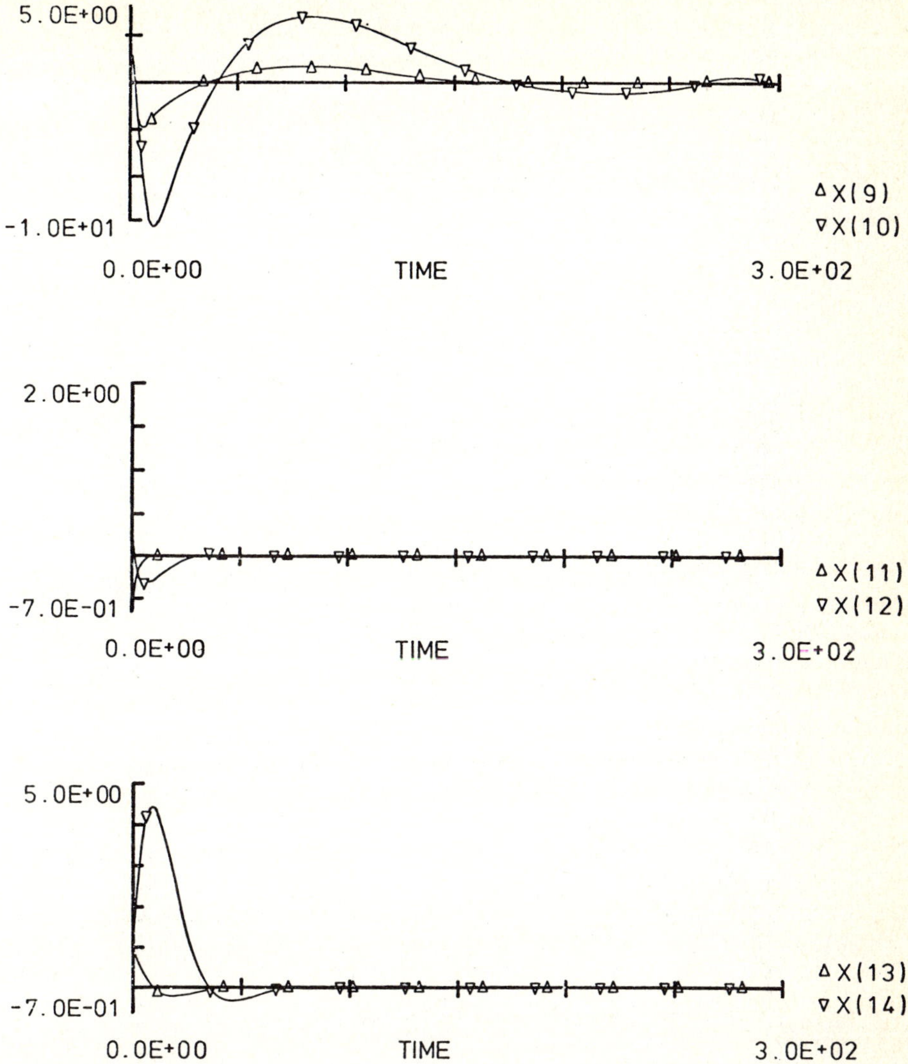
5.0E+00
-1.0E+01
0.0E+00
TIME
3.0E+02
X(9)
X(10)
2.0E+00
-7.0E-01
0.0E+00
TIME
3.0E+02
X(11)
X(12)
5.0E+00
-7.0E-01
0.0E+00
TIME
3.0E+02
X(13)
X(14)

Figure 6.16

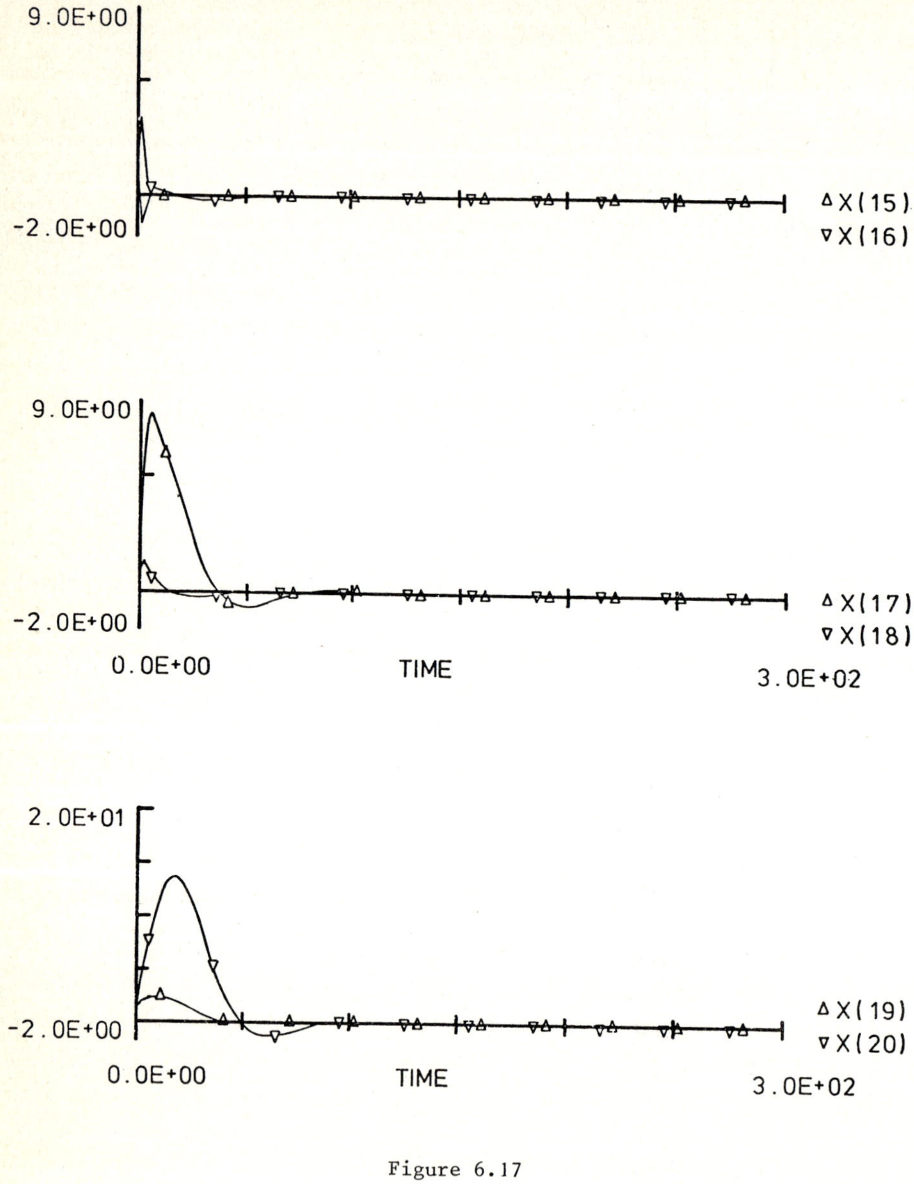
9.0E+00
-2.0E+00
ΔX(15)
∇X(16)
9.0E+00
-2.0E+00
ΔX(17)
∇X(18)
0.0E+00
TIME
3.0E+02
2.0E+01
-2.0E+00
ΔX(19)
∇X(20)
0.0E+00
TIME
3.0E+02

Figure 6.17

Let us assume, as an example, that the global A matrix $A_g = A + CL$ is given by

$$\left[\begin{array}{cc|ccc} a_{11} & a_{12} & 0 & 0 & a_{15} \\ a_{21} & a_{22} & a_{23} & 0 & 0 \\ \hline a_{23} & 0 & a_{33} & a_{34} & a_{35} \\ a_{41} & 0 & a_{43} & a_{44} & a_{45} \\ 0 & 0 & a_{53} & a_{54} & a_{55} \end{array}\right]$$

If this system is split into two subsystems along the dotted lines then

$$z_1 = \begin{bmatrix} x_5 \\ x_3 \end{bmatrix} \qquad A_{z_i} = \begin{bmatrix} a_{55} & a_{53} \\ a_{35} & a_{33} \end{bmatrix}$$

and $\qquad z_2 = x_1 \qquad A_{z_2} = a_{11}$

Thus, the rule used in the choice of the crude model is to take account of the sparsity of A_g by eliminating columns where the off-diagonal parts are zero. However, sometimes, this procedure could lead to an unstable model. One way of getting round this is to recognise that in fact the interaction model is a crude one which basically should be chosen so as to give a good response with the decentralised control. Since the order of A_{z_i} is governed by the choice of the decomposition, we can choose A_{z_i} to be a diagonal matrix of appropriate order with appropriate negative elements of the diagonal. We will examine this question in more detail when we consider our ship example.

Then, we modify our optimisation problem to the approximate form

$$\min J = \frac{1}{2} \sum_{i=1}^{N} \int_0^{\infty} (||y_i||_Q^2 + ||u_i||_{R_i}^2) dt \tag{6.9.5}$$

subject to

$$\dot{y}_i = \tilde{A}_i y_i + \tilde{B}_i u_i \tag{6.9.6}$$

where

$$y_i = \begin{bmatrix} x_i \\ \cdots \\ z_i \end{bmatrix} ; \tilde{A}_i = \begin{bmatrix} A_i & \vdots & C_i \\ \cdots & \vdots & \cdots \\ 0 & \vdots & A_{z_i} \end{bmatrix} ; \tilde{B}_i = \begin{bmatrix} B_i \\ \cdots \\ 0 \end{bmatrix}$$

$$\tilde{Q}_i = \begin{bmatrix} Q_i & \vdots & 0 \\ \cdots & \vdots & \cdots \\ 0 & \vdots & 0 \end{bmatrix}$$

The idea is to minimise equation (6.9.5) subject to equation (6.9.6). However, in that case, the resulting gain matrix will depend on the chosen matrix A_{z_i} and on the trajectories of z_i, since the control u_i will be a function of both x_i and z_i. Now, since in most of the realistic applications z_i is not available for measurement, in this paper, we will not try to improve the model for z_i. Rather, we shall try to modify the trajectories of z_i.

Now let us assme that the pair $(\tilde{A}_i, \tilde{B}_i)$ is controllable. Then, the solution of equation (6.9.5) subject to equation (6.9.6) is given by

$$u_i^* = -R_i^{-1}\tilde{B}_i' P_i y_i$$

where P_i is the solution of a matrix Riccati equation. We can rewrite the optimal control as

$$u_i^* = G_{i1}x_i - G_{i2}z_i \tag{6.9.7}$$

From the above equation, we see that the gain matrix is a function of the approximate system model, and the optimal control u_i^* is a function of the subsystem state x_i and the interconnection input vector z_i.

Now, although we can obtain z_i from our interconnection model, this model is a relatively crude one. It is therefore necessary to improve the model online. To do this, let us consider a subsystem model whose input $\hat{z}_i$ is provided by the crude model, i.e.

$$\dot{\hat{x}}_i = A_i\hat{x}_i + B_iu_i + C_i\hat{z}_i \tag{6.9.8}$$

where $\hat{x}_i$ is the state vector of the subsystem model. Then, if we substitute for u_i in equations (6.9.2) and (6.9.8) after having replaced z_i by $\hat{z}_i$ in equation (6.9.7), we have

$$\dot{x}_i' = (A_i - B_iG_{i1})x_i' + C_iz_i - B_iG_{i2}\hat{z}_i \tag{6.9.9}$$

or
$$\dot{x}_i' = \hat{A}_ix_i' + C_iz_i - B_iG_{i2}\hat{z}_i \tag{6.9.10}$$

where
$$\hat{A}_i = A_i - B_iG_{i1} \tag{6.9.11}$$

x_i' is the suboptimal state of subsystem i resulting from using $\hat{z}_i$ instead of z_i and

$$\dot{\hat{x}}_i' = \hat{A}_i\hat{x}_i' + C_i z_i - B_i G_{i2}\hat{z}_i \quad (6.9.12)$$

Let

$$\tilde{x}_i = x_i' - \hat{x}_i' \quad (6.9.13)$$

$$\tilde{z}_i = z_i - \hat{z}_i \quad (6.9.14)$$

Thus, subtracting equation (6.9.12) from equation (6.9.10) we obtain

$$\dot{\tilde{x}}_i = \hat{A}_i\tilde{x}_i + C_i\tilde{z}_i \quad (6.9.15)$$

and our aim is to minimise the error vectors $\tilde{x}_i$ and $\tilde{z}_i$. For this problem, we can construct another optimisation problem, which is:

$$\min \tilde{J}_i = \frac{1}{2}\int_0^\infty (||\tilde{x}_i||^2_{H_i} + ||\tilde{z}_i||^2_{S_i})dt \quad (6.9.16)$$

subject to the constraint equation (6.9.15), where H_i and S_i are positive-semidefinite and positive-definite matrices, respectively.

The solution of the problem is

$$\tilde{z}_i^* = -S_i^{-1}C_i^T K_i\tilde{x}_i \quad (6.9.17)$$

where K_i is a solution of an appropriate Riccati-type equation.

Using equations (6.9.17) and (6.9.14), we obtain

$$z_i = \hat{z}_i - S_i^{-1}C_i^T K_i\tilde{x}_i \quad (6.9.18)$$

and we will use this to generate the control. Thus the dynamical equation for the subsystem will be

$$\dot{x}_i = -\hat{A}_i x_i + C_i z_i - B_i G_{i2}[\hat{z}_i - S_i^{-1}C_i^T K_i\ x_i - \hat{x}_i)] \quad (6.9.19)$$

$$\dot{\hat{x}}_i = \hat{A}_i\hat{x}_i + (C_i - B_i G_{i2})\hat{z}_i \quad (6.9.20)$$

$$\dot{\hat{z}}_i = A_{z_i}\hat{z}_i \quad (6.9.21)$$

and we could represent these equations by Fig. 6.18.

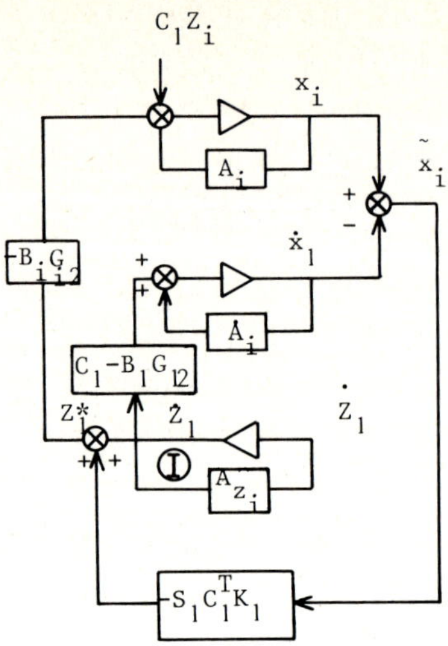

Figure 6.18

To enable the controller to adapt to structural perturbations, we define the interconnection variable with the elements of the z_i vector i.e. e_{ij}, where

$$0 \leq e_{ij} \leq 1, \quad i=1,2,\ldots,N, \qquad j=1,2,\ldots,q_i$$

Thus, we let

$$Z_i = \begin{pmatrix} e_{i1} & z_{i1} \\ \\ e_{iq_i} & z_{iq_i} \end{pmatrix}$$

or
$$Z_i = \sum_{j=1}^{N} E_{ij}x_j \tag{6.9.22}$$

where E_{ij} is the interconnection matrix with elements 0 or e_{ij}.

We will define a scalar ε_i with each subsystem

$$\varepsilon_i = \frac{\sum_{j=1}^{q_i} e_{ij}}{q_i} \tag{6.9.23}$$

and we will multiply the gain matrix B_iG_{i2} by this variable. Thus, the equation becomes

$$\dot{x}_i = \hat{A}_i x_i + C_i z_i - \varepsilon_i B_i G_{i2}[z_i - S_i C_i^T K_i (x_i - \hat{x}_i)] \tag{6.9.24}$$

Remarks

a) The decentralised controller that we have designed above has gains which are independent of the initial conditions.

b) The suboptimality in the controller design arises from the fact that

(i) the controller does not take into account all the states of the overall system

(ii) the decentralised gain matrix G_i depends on the approximate interconnection model

(iii) although we improve the outputs of the interconnection model, there always remains a difference between z^* and z.

c) The controller is easy to compute, as we see from Fig. 6.18. We only need to work subsystem by subsystem, and the global solution is not required. The blocks in Fig. 6.18 are trivially easy to calculate.

d) From equation (6.9.24), we see that, on average, the controller takes into consideration the possibility of structural perturbations.

Now, let us study the stability properties of our controller.

6.9.1 Stability of global system using decentralised controllers

To study the stability of the above controller, consider the following theorem.

Theorem 6.9.1

The interconnected system when controlled by the decentralised controllers developed above will be asymptotically stable if the following sufficient condition is satisfied:

$$\sum_{i=1}^{N} \gamma_i + \max_i \lambda_M(P_i) \sum_{i=1}^{N} \sum_{i=1}^{N} \beta_{ij} \leqslant \frac{1}{2} \min_i \lambda_m(W_i) \tag{6.9.25}$$

where $\gamma_i = ||A_{i_3}^T P_i||$

$P_{ij} = ||C_i E_{ij}||$

$$P_i = \begin{pmatrix} P_i & 0 \\ 0 & K_i \end{pmatrix}; \quad w_i = \begin{pmatrix} \tilde{Q}_i & 0 \\ 0 & H_i \end{pmatrix}$$

$$+ \begin{pmatrix} P_i \tilde{B}_i R_i^{-1} \tilde{B}_i^T P_i & 0 \\ 0 & K_i C_i S_i^{-1} C_i^T K_i \end{pmatrix}$$

$\lambda_M(P_i)$ = maximum eigenvalue of the matrix P_i

$\lambda_m(W_i)$ = minimum eigenvalue of the matrix w_i

Proof

From equations (6.9.24, 6.9.20 and 6.9.21), we have

$$\dot{X}_i = \dot{A}_i X_i + C_i z_i \tag{6.9.26}$$

where

$$_i = \begin{pmatrix} A_i - B_i G_{i1} + \varepsilon_i B_i G_{i2} S_i^{-1} C_i^T K_i & -\varepsilon_i B_i G_{i2} & -\varepsilon_i B_i G_{i2} S_i^{-1} C_i^T K_i \\ 0 & A_{z_i} & 0 \\ 0 & C_i - B_i G_{i2} & A_i - B_i G_{i1} \end{pmatrix}$$

$$C_i = \begin{pmatrix} C_i \\ 0 \\ 0 \end{pmatrix} \quad X_i = \begin{pmatrix} x_i \\ \hat{z}_i \\ \hat{x}_i \end{pmatrix}$$

$$z_i = \sum_{j=1}^{N} E_{ij} x_j = \sum_{j=1}^{N} E_{ij} X_j \tag{6.9.27}$$

where $E_{ij} = [E_{ij} \quad 0 \quad 0]$

Therefore

$$\dot{X}_i = A_i X_i + C_i \sum_{j=1}^{N} E_{ij} X_j \tag{6.9.28}$$

Let us now consider the following Lyapunov function:

$$v = \sum_{i=1}^{N} X_i^T P_i X_i \tag{6.9.29}$$

where

$$P_i = \left[\begin{array}{c|c} P_i & 0 \\ \hline 0 & K_i \end{array}\right]$$

where P_iK_i are the solutions of the Riccati equations for the optimisation problem: equation (6.9.5) subject to equation (6.9.6) and equation (6.9.16) subject to equation (6.9.15) respectively.

Since v is positive definite, we will seek the condition for which $\dot{v}$ will be negative semidefinite. Thus, we have

$$\dot{v} = \sum_{i=1}^{N} \dot{X}_i^T P_i X_i^T + X_i^T P_i \dot{X}_i$$

On using the augmented equation (6.9.28) of the dynamical system, we obtain

$$\dot{v} = \sum_{i=1}^{N} \left\{ X_i^T A_i^T P_i X_i + \sum_{j=1}^{N} X_j^T E_{ij}^T C_i P_i X_i + X_i^T P_i A_i X_i + X_i^T P_i C_i \sum_{j=1}^{N} E_{ij} X_j \right\} \tag{6.9.30}$$

Let us put $\varepsilon_i = 1 - \xi_i$ and rewrite A_i as $A_i = A_{i1} - A_{i2} + A_{i3}$, where

$$A_{i1} = \left[\begin{array}{c|c|c} A_i & C_i & 0 \\ \hline 0 & A_{z_i} & 0 \\ \hline 0 & 0 & A_i - B_i G_{i1} \end{array}\right]$$

$$A_{i2} = \left[\begin{array}{c|c|c} B_i G_{i1} & B_i G_{i2} & 0 \\ \hline 0 & 0 & 0 \\ \hline 0 & 0 & C_i S^{-1} C_i^T K_i \end{array}\right]$$

$$A_{i3} = \left[\begin{array}{c|c|c} G_iB_iG_{i2}S_i^{-1}C_i^TK_i & \xi_iB_iG_{i2} - C_i & \varepsilon_iB_iG_{i2}S_i^{-1}C_i^TK_i \\ \hline 0 & 0 & 0 \\ \hline 0 & C_i - B_iG_{i2} & C_iS_i^{-1}C_i^TK_i \end{array}\right]$$

Then, for a globally stable system, and substituting for A_i by $A_{i1} - A_{i2} + A_{i3}$, equation (6.9.30) becomes

$$\sum_{i=1}^{N} \left\{ X_i^TA_iP_iX_i - X_i^TA_{i2}^TP_iX_i + X_i^TA_{i3}^TP_iX_i + X_i^TP_iA_{i1}X_i - X_i^TP_iA_{i2}X_i + X_i^TP_iA_{i3}X_i + 2X_i^TP_iC_i \sum_{i=1}^{N} E_{ij}X_j \right\} \leq 0 \qquad (6.9.31)$$

On using the solution P_i, K_i of the Riccati equations we have

$$\sum_{j=1}^{N} \left\{ -X_i^T\bar{Q}_iX_i - X_i^TP_iA_{i2}X_i + 2X_i^TA_{i3}^TP_iX_i + 2X_i^TP_iC_i \sum_{j=1}^{N} E_{ij}X_j \right\} \leq 0$$

Let $\quad w_i = \bar{Q}_i + \quad i\ i2$

$$= \left[\begin{array}{c|c} \tilde{Q}_i & 0 \\ \hline 0 & H_i \end{array}\right] + \left[\begin{array}{c|c} P_i\tilde{B}_iR_i^{-1}\tilde{B}_i^TP_i & 0 \\ \hline 0 & K_iC_iS_i^{-1}C_i^TK_i \end{array}\right]$$

Then

$$\sum_{i=1}^{N} 2 \left\{ X_i^TA_{i3}^TP_iX_i + X_i^TP_iC_i \sum_{j=1}^{N} E_{ij}X_j \right\} \leq \sum_{i=1}^{N} X_i^Tw_iX_i \qquad (6.9.32)$$

$$\sum_{i=1}^{N} X_i^TP_iC_i \sum_{j=1}^{N} E_{ij}X_j \leq \sum_{i=1}^{N} \lambda_M(P_i) \sum_{j=1}^{N} \beta_{ij}||X_j||^2$$

Since

$$\sum_{i=1}^{N} X_i^T P_i C_i \sum_{j=1}^{N} E_{ij} X_j \leq \sum_{i=1}^{N} \lambda_M(P_i) \sum_{j=1}^{N} \beta_{ij} ||X_j||^2$$

$$\leq \max_i \lambda_M(_i) \sum_{i=1}^{N} \sum_{j=1}^{N} \beta_{ij} ||X_j||^2$$

where $\beta_{ij} = ||C_i E_{ij}||$

and writing

$$\gamma_i = ||A_{i3}^T P_i||$$

we have the final result

$$\sum_{i=1}^{N} \gamma_i + \max_i \lambda_M(P_i) \sum_{i=1}^{N} \sum_{j=1} \beta_{ij} \leq \frac{1}{2} \min_i \lambda_m(w_i) \qquad (6.9.33)$$

which completes the proof.

We note that we can guarantee asymptotic stability for the system if the sufficient condition is satisfied. Moreover, the condition only uses subsystem information and not global matrices. That makes it particularly useful for large scale systems. Also, having incorporated the interconnection matrix E_{ij} explicitly into the analysis, we see that asymptotic stability under structural perturbations is also guaranteed. Thus, the design is a robust one.

6.9.2 The ship example

Next we illustrate this powerful approach on the ship example.

Hassan and Singh [16] suggested a way of choosing A_{z_i}. They chose some elements in the diagonal blocks of the A matrix in order to construct A_{z_i}. This provides a good model for the case where the elements chosen are the dominant ones in the rows of A and this is clearly the case for their river example. For other examples, their choice of A_{z_i} may mean a longer correction period on-line.

Proposed choice of A_{z_i}

It should be recognised that there is no simple infallible way of choosing the crude model by picking out certain entries from the A matrix. Another way of choosing A_{z_i} may be to concentrate on the eigenvalues. However, in the multivariable situation, the relationship between the modes and the states is not a clear one.

If we recall the basic notion of the model follower for centralised control

[17], we start with a certain model and then we minimise the errors between the plant output and the model output. Thus, the choice of the model is largely governed by the response that we desire from the plant.

A simple way of specifying this response is in terms of assignment of the poles of the model in the left half plane. Thus, we could choose the matrices A_{z_i} to be diagonal with negative elements. We will see the effect of this when we study the ship model.

Let us consider once again the block diagram of Fig. 6.18. First of all, no matter how A_{z_i} is chosen, the errors will be corrected to a large extent by the feedback channel. Of course, the accuracy will partly depend on the numbers used in A_{z_i}. Here $\hat{z}_i$ which is produced by the minor loop I acts as a reference input. It is independent of any controls. The same notion is given by (6.9.6). Here the left side of $\tilde{A}_{z_i}$ is zero, and the lower part of $\tilde{B}$ is also zero. This means that no controls or other states can affect $\underline{z}_i$, i.e. the pair $(\tilde{A}_i, \tilde{B}_i)$ is not completely controllable. For stabilising the minor loop I, we should choose real negative diagonal elements in A_{z_i}. If one does not choose a diagonal matrix, at least one must ensure that the eigenvalues of A_{z_i} have negative real parts; and they match the required responses.

The effect of interacting controls

The other respect in which the problem of the ship, that we will study, is nonstandard in that here both the controls and states are interacting so that A and B are full matrices. Our choice of the decomposition of A still leaves B as a full matrix albeit with at least one control for each subsystem. In fact the full B matrix poses no real problem - the off-diagonal blocks of A and B combine to form the interactions. Thus as far as the subsystem models are concerned, we neglect these off-diagonal blocks. Their influence only appears through in the input $C_i \underline{z}_i$ which forces the subsystems and this effect is compensated for by the model follower loops.

Let us now consider the ship example.

We recall that the system under consideration is of 20th order and it splits into 3 interacting subsystems of order:

Subsystem 1 : 10 states and 2 controls

Subsystem 2 : 4 states and 2 controls

Subsystem 3 : 6 states and 2 controls

The global A and B matrices are given in tables 5.1 and 5.2 respectively in Chapter 5.

Using the approach of Hassan and Singh [16], the interaction models would be:

$$A_{z_1} = \begin{bmatrix} -0.106 & 0 & 0 \\ 0 & -0.04 & 0 \\ 0 & 0 & -0.0761 \end{bmatrix} \tag{6.9.34}$$

$$A_{z_2} = -0.114$$

$$A_{z_3} = -0.233$$

Here, roughly speaking, loop I in Fig. 6.18 has a time constant of 250 seconds approximately. Simulation results showed that $\hat{z}_1$ went to zero in approximately 1000 seconds. This shows that although it was possible for the states $\underline{x}_i$ to go to their steady state values in approximately 200 seconds, they oscillated over a further 800 seconds before settling down. Thus the model that is chosen is not a good one since it does not give a desirable response.

In fact, this problem is such that it is possible to calculate the global optimal controls off-line. On analysing the modes of the global optimal control we found that these were spread over a range of (-0.049) to (-31.2). Thus, we changed A_{z_i} to

$$A_{z_1} = \begin{bmatrix} -0.106 & 0 & 0 \\ 0 & -0.1 & 0 \\ 0 & 0 & -0.1 \end{bmatrix}$$

while keeping A_{z_2} and A_{z_3} as above. Based on this choice of A_{z_i} and for $S_i = I$ (the identity matrix) we found that although the responses were better, they still took too long to settle down to their desired values. However, the designer can also manipulate H_i and S_i. From equation (6.9.17) we see that the effect of S_i is to penalise the error vector $\hat{\underline{z}}_i$. Thus, to reduce this error, we need to increase S_i on using $S_i = 100\ I$, the responses given in Figs. 6.19, 6.20 and 6.21 were obtained. We see that these are much more reasonable. If we compare them with globally optimal responses as given in Figs. 6.22, 6.23 and 6.24 we see that the overshoot and settling time are similar.

Finally, in this chapter, we outline the powerful concept of decentralised controller design using overlapping information sets. This notion was first developed by Siljak and his co-workers [18].

6.10 SUBOPTIMAL DECENTRALISED CONTROL OF SYSTEMS WITH OVERLAPPING INFORMATION SETS [18]

The controllers that we have considered so far were designed on the basis that they have disjoint information sets, i.e. there is no sharing of information amongst the individual controllers. However, Siljak argues that there are many systems, particularly in traffic regulation economics, power

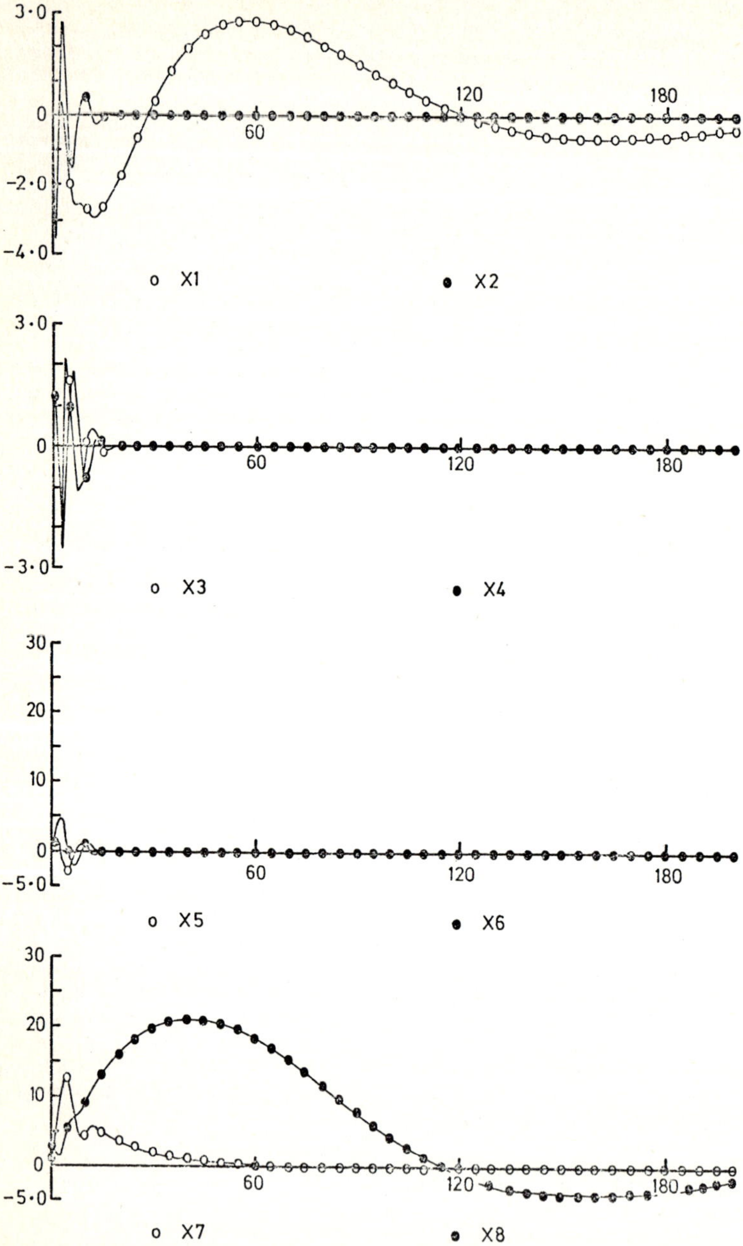
3·0
0
-2·0
-4·0
60
120
180
o X1
• X2
3·0
0
-3·0
60
120
180
o X3
• X4
30
20
10
0
-5·0
60
120
180
o X5
• X6
30
20
10
0
-5·0
60
120
180
o X7
• X8

Figure 6.19

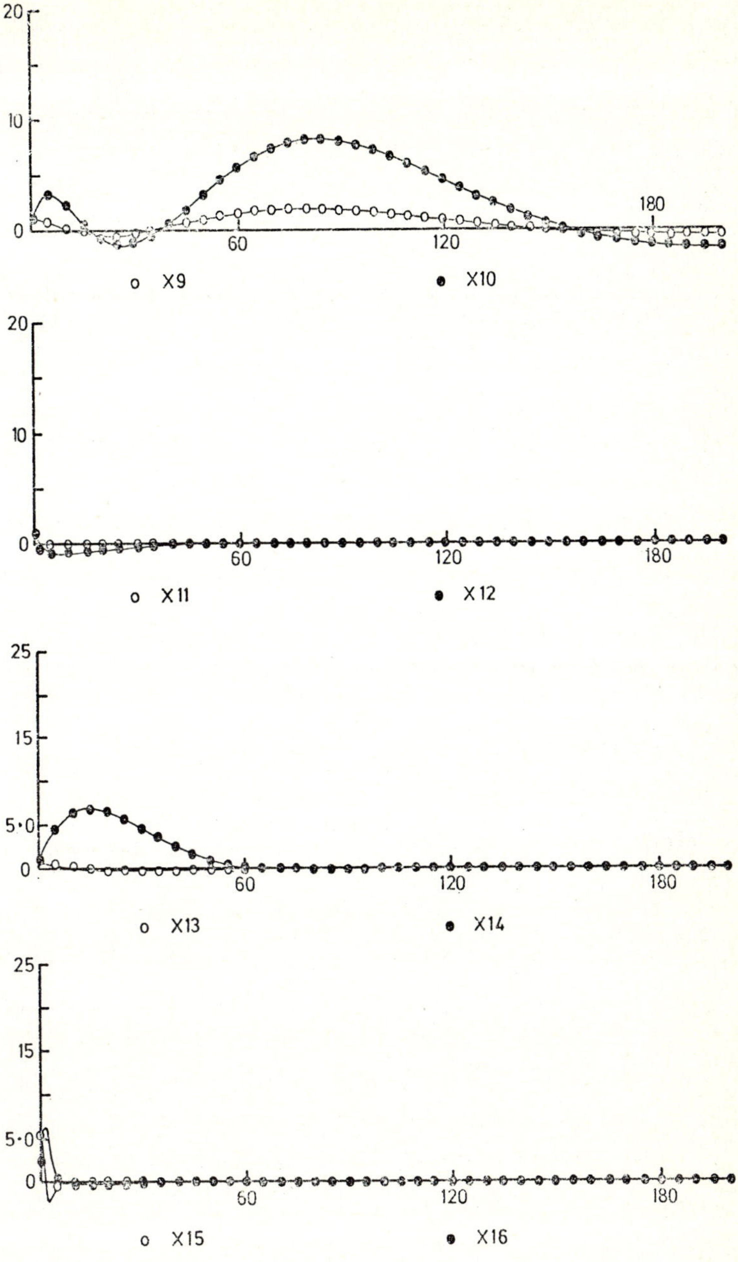
20
10
0
60
120
180
o X9
• X10
20
10
0
60
120
180
o X11
• X12
25
15
5·0
0
60
120
180
o X13
• X14
25
15
5·0
0
60
120
180
o X15
• X16

Figure 6.20

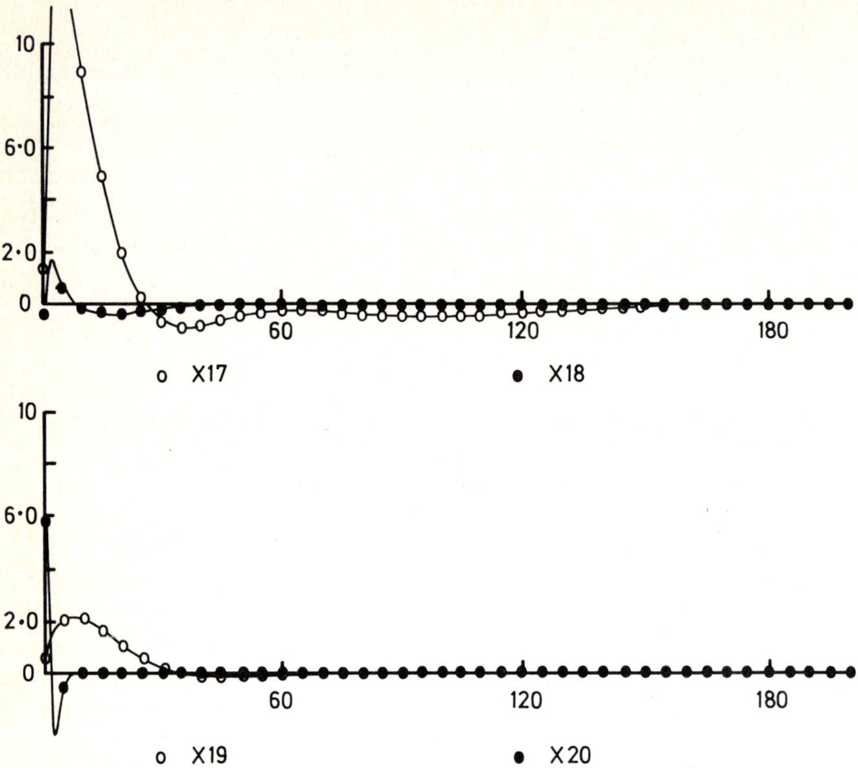

Figure 6.21

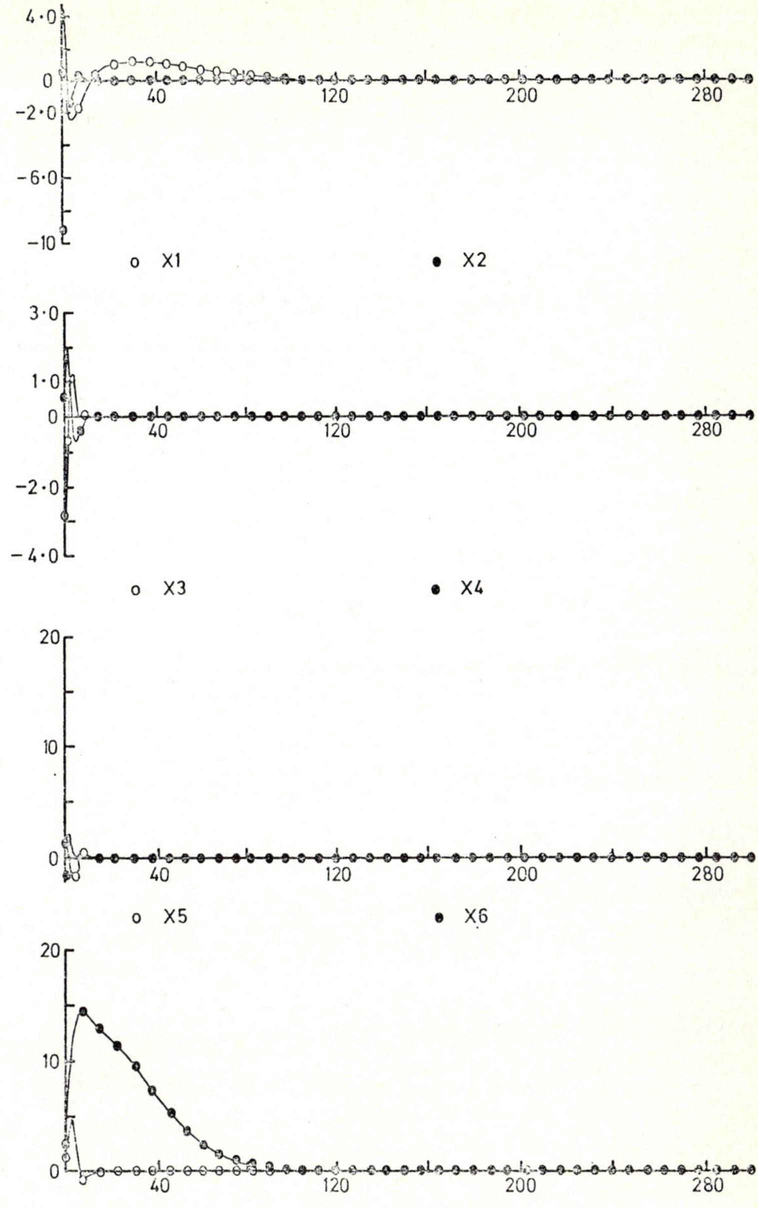
4·0
0
-2·0
-6·0
-10
40
120
200
280
o X1
• X2
3·0
1·0
0
-2·0
-4·0
40
120
200
280
o X3
• X4
20
10
0
40
120
200
280
o X5
• X6
20
10
0
40
120
200
280
o X7
• X8

Figure 6.22

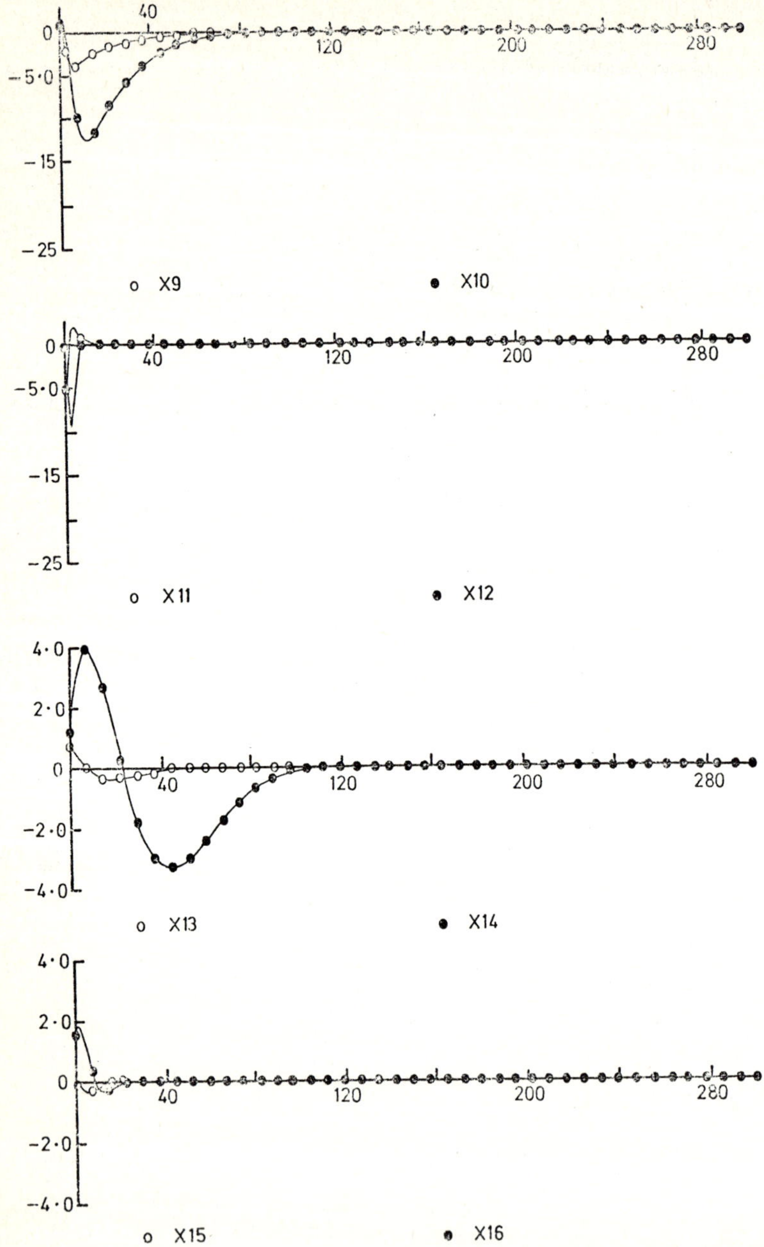
40
0
120
200
280
-5·0
-15
-25
o X9
• X10
0
40
120
200
280
-5·0
-15
-25
o X11
• X12
4·0
2·0
0
40
120
200
280
-2·0
-4·0
o X13
• X14
4·0
2·0
0
40
120
200
280
-2·0
-4·0
o X15
• X16

Figure 6.23

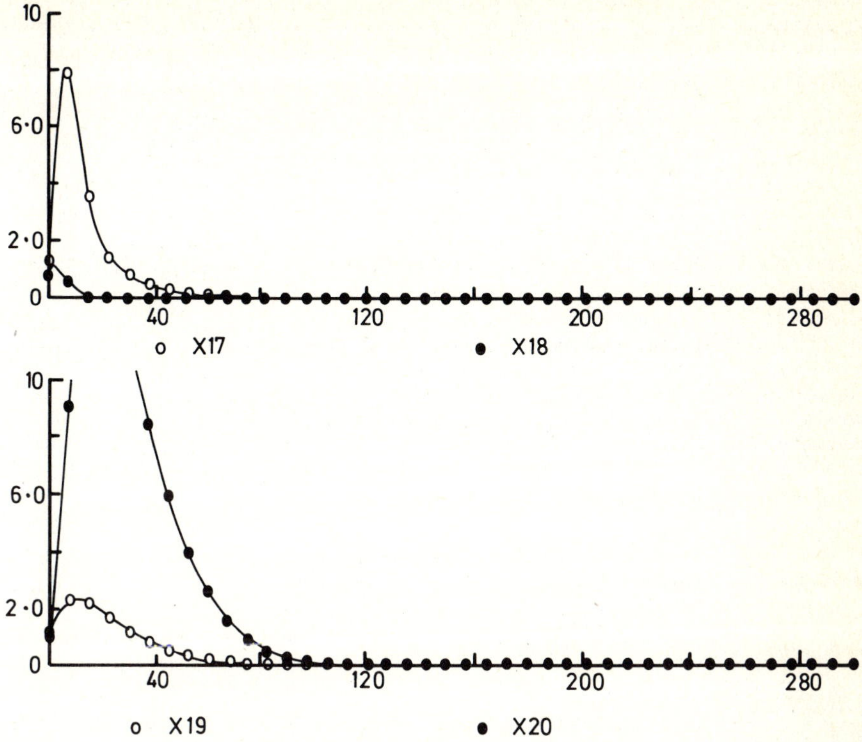
10
6·0
2·0
0
40
120
200
280
o X17
• X18
10
6·0
2·0
0
40
120
200
280
o X19
• X20

Figure 6.24

systems, etc. where reliability can be improving using shared information. The design of these overlapping controllers is carried out by expansions and contractions.

6.10.1 Expansions and contractions of performance indices

Consider the linear system S defined by

$$\dot{x} = Ax + Bu \tag{6.10.1}$$

where $x \in R^n$ and $u \in R^m$. Let the performance index for the system S be given by:

$$J(x_o,u) = \int_0^\infty (x^TQx + u^TRu)dt \tag{6.10.2}$$

where x_o is the initial state of S and as usual $Q \geq 0$, $R \geq 0$. Let us associate with the pair (S,J) another pair $(\tilde{S},\tilde{J})$ where the system S is defined by:

$$\dot{\tilde{x}} = \tilde{A}\tilde{x} + \tilde{B}u \tag{6.10.3}$$

where $\tilde{x} \in R^{\tilde{n}}$. The performance index $\tilde{J}$ is defined by

$$\tilde{J}(\tilde{x}_o,u) = \int_0^\infty (\tilde{x}^T\tilde{Q}\tilde{x} + u^T\tilde{R}u)dt \tag{6.10.4}$$

where $\tilde{x}_o$ is the initial state and $\tilde{Q} \geq 0$ and $\tilde{R} > 0$. We assume that $\tilde{n} \geq n$. Let us define a linear transformation i.e.

$$\tilde{x} = Tx \tag{6.10.5}$$

where T is an $\tilde{n}$xn constant matrix with full column rank. We denote by $x(t;x_o,u)$ and $\tilde{x}(t,\tilde{x}_o,u)$ the motions of the systems S and $\tilde{S}$ starting respectively from the initial states x_o and $\tilde{x}_o$ and using a fixed input u. Siljak [18] uses the transformation (6.10.4) to relate the pairs (S,J) and $(\tilde{S},\tilde{J})$ in the context of inclusion as

Definition

The pair $(\tilde{S},\tilde{J})$ includes the pair (S,J) if there exists a matrix T such that for any initial state x_o of S, the choice

$$\tilde{x}_o = Tx_o \tag{6.10.6}$$

of the initial state $\tilde{x}_o$ of $\tilde{S}$ implies that

$$x(t;x_o,u) = T^I\tilde{x}(t;\tilde{x}_o,u) \;\forall\; t \geq 0 \tag{6.10.7}$$

and $$J(x_o,u) = \tilde{J}(\tilde{x}_o,u) \tag{6.10.8}$$

for any fixed input u(t). Here T^I is the generalised inverse of T.

If the pair $(\tilde{S},\tilde{J})$ includes the pair (S,J) then $(\tilde{S},\tilde{J})$ is said to be an <u>expansion of</u> (S,J) and (S,J) is said to be a <u>contraction</u> of $(\tilde{S},\tilde{J})$. We note that if $(\tilde{S},\tilde{J})$ includes (S,J) then from (6.10.7) and (6.10.8), the optimal control problem corresponding to the expansion $(\tilde{S},\tilde{J})$ is equivalent to that of (S,J) provided (6.10.6) holds.

Next let us see under what conditions $(\tilde{S},\tilde{J})$ includes (S,J).

Under the transformation (6.10.5),

$$\tilde{A} = TAT^{I} + M; \quad \tilde{B} = TB + N$$

$$\tilde{Q} = (T^{I})^{T}QT^{I} + M_{Q}, \quad \tilde{R} = R + N_{R}$$

where $M, N, M_Q\ N_R$ are constant complimentary matrices of proper dimensions. The only restrictions imposed on the complimentary matrices is that $(\tilde{S},\tilde{J})$ includes (S,J). These restrictions are given by the following theorem:

<u>Theorem</u>

The pair $(\tilde{S},\tilde{J})$ includes the pair (S,J) if either

a) $$MT = 0, \; N = 0, \; T^{T}M_{Q}T = 0, \; N_{R} = 0 \tag{6.10.9}$$

or

b) $$T^{I}M^{i}T = 0, \; T^{I}N^{i-1}N = 0, \; M_{Q}M^{i-1}T = 0$$

$$M_{Q}M^{i-1}N = 0, \; N_{R} = 0, \qquad i=1,2,\ldots\bar{n} \tag{6.10.10}$$

Next let us see under what conditions we can contract a control law $u = -\tilde{K}\tilde{x}$ to $u = -Kx$ for implementation on S.

<u>Definition</u>

The control law $-\tilde{K}\tilde{x}$ for the expansion $\tilde{S}$ is contractable to the control law $-Kx$ for the original system S if $\tilde{x}_o = Tx_o$ implies that

$$Kx(t;x_o,u) = \tilde{K}\tilde{x}(t;\tilde{x}_o,u) \;\forall\; t \geq 0$$

for any fixed u(t)

It is easy to show that if

$$MT = 0 \text{ and } N = 0 \tag{6.10.11}$$

then any control law $-\tilde{K}\tilde{x}$ is contractable to the control law $-Kx$ and K is given by

$$K = \tilde{K}T \tag{6.10.12}$$

We note that the prime motivation for this development is that when we consider the optimisation of large scale systems with overlapping subsystems, we can carry out the design in the expanded space where the overlapping subsystems appear disjoint and thus standard techniques can be applied. These are then implemented on the original system.

Next, we consider overlapping decompositions.

6.10.2 Overlapping decompositions

Let us consider once again the linear constant system S given by

$$\dot{x} = Ax + Bu \tag{6.10.13}$$

Let us assume that the state x of S is composed of 3 components x_1, x_2, x_3 which are respectively of dimension n_1, n_2 and n_3, i.e.

$$x = (X_1^T, X_2^T, x_3^T)^T \text{ and } n = n_1 + n_2 + n_3$$

The input u is given by

$$u = (u_1^T, u_2^T)^T \text{ where } u_1 \in R^{m_1}, u_2 \in R^{m_2} \text{ and } m = m_1 + n_2$$

Thus S can be rewritten as:

$$\begin{pmatrix} \dot{x}_1 \\ \dot{x}_2 \\ \dot{x}_3 \end{pmatrix} = \begin{pmatrix} A_{11} & A_{12} & A_{13} \\ A_{21} & A_{22} & A_{33} \\ A_{31} & A_{32} & A_{33} \end{pmatrix} + \begin{pmatrix} B_{11} & B_{12} \\ B_{21} & B_{22} \\ B_{31} & B_{32} \end{pmatrix} \begin{pmatrix} u_1 \\ u_2 \end{pmatrix} \tag{6.10.14}$$

For simplicity, we only consider the overlapping decompositions for the case of 2 subsystems in (6.10.14). The generalisation to an arbitrary number of subsystems is quite straightforward.

Let us decompose the state x of S into 2 overlapping components $\tilde{x}_1 = (X_1^T, x_2^T)^T$, $\tilde{x}_2 = (x_2^T, x_3^T)^T$ and let us form a new vector $\tilde{x} = (\tilde{x}_1^T, \tilde{x}_2^T)^T$. $\tilde{x}$ is related to x by

$$\tilde{x} = Tx \tag{6.10.15}$$

where T is an nxn non-square matrix with $n = n_1 + 2n_2 + n_3$ and it is defined as:

$$T = \begin{pmatrix} I_1 & 0 & 0 \\ 0 & I_2 & 0 \\ 0 & I_2 & 0 \\ 0 & 0 & I_3 \end{pmatrix} \tag{6.10.16}$$

where I_1, I_2, I_3 are appropriate identity matrix.

As we have seen before, the linear transformation (6.10.15) defines the expansion

$$\tilde{S} : \dot{\tilde{x}} = \tilde{A}\tilde{x} + \tilde{B}u \tag{6.10.17}$$

where $\tilde{A} = TAT^I + M; \; \tilde{B} = TB + N$ (6.10.18)

As pointed out by Siljak [18], in (6.10.18), the choice

$$T^I = \begin{bmatrix} I_1 & 0 & 0 & 0 \\ 0 & \tfrac{1}{2}I_2 & \tfrac{1}{2}I_2 & 0 \\ 0 & 0 & 0 & I_3 \end{bmatrix}, \quad M = \begin{bmatrix} 0 & \tfrac{1}{2}A_{12} & -\tfrac{1}{2}A_{12} & 0 \\ 0 & \tfrac{1}{2}A_{12} & -\tfrac{1}{2}A_{22} & 0 \\ 0 & -\tfrac{1}{2}A_{22} & -\tfrac{1}{2}A_{22} & 0 \\ 0 & -\tfrac{1}{2}A_{32} & -\tfrac{1}{2}A_{32} & 0 \end{bmatrix} \tag{6.10.19}$$

$$N = 0$$

satisfies the conditions given by part (a) of the theorem in section 6.10.1. However, this choice is by no means unique. With (6.10.16) and (6.10.19), the expansion $\tilde{S}$ can be described by the equations

$$\begin{bmatrix} \tilde{x}_1 \\ \tilde{x}_2 \end{bmatrix} \quad \left[\begin{array}{cc|cc} A_{11} & A_{12} & 0 & A_{13} \\ A_{21} & A_{22} & 0 & A_{23} \\ \hline A_{21} & 0 & A_{22} & A_{23} \\ A_{31} & 0 & A_{32} & A_{33} \end{array}\right] \begin{bmatrix} \tilde{x}_1 \\ \tilde{x}_2 \end{bmatrix}$$

$$+ \left[\begin{array}{c|c} B_{11} & B_{12} \\ B_{21} & B_{22} \\ \hline B_{21} & B_{22} \\ B_{31} & B_{32} \end{array}\right] \left[\begin{array}{c} u_1 \\ \hline u_2 \end{array}\right] \tag{6.10.20}$$

on computing (6.10.14) with (6.10.20), we see that the overlapping decomposition of the system S has become a disjoint decomposition of the expansion $\tilde{S}$ of S. Of course, standard optimisation techniques can now be applied independently to each subsystem of $\tilde{S}$. Let these subsystems be given by

$$\left.\begin{aligned}\tilde{S}_1 : \dot{\tilde{x}}_1 &= \tilde{A}_1 \tilde{x}_1 + \tilde{B}_1 u_1 + \tilde{A}_{12} \tilde{x}_2 + \tilde{B}_{12} u_2 \\ \tilde{S}_2 ; \dot{\tilde{x}}_2 &= \tilde{A}_2 \tilde{x}_2 + \tilde{B}_2 u_2 + \tilde{A}_{21} \tilde{x}_1 + \tilde{B}_{21} u_1\end{aligned}\right\} \tag{6.10.21}$$

where

$$\tilde{A}_1 = \begin{pmatrix} A_{11} & A_{12} \\ A_{21} & A_{22} \end{pmatrix}, \tilde{B}_1 = \begin{pmatrix} B_{11} \\ B_{21} \end{pmatrix}, \tilde{A}_2 = \begin{pmatrix} A_{22} & A_{23} \\ A_{32} & A_{33} \end{pmatrix},$$

$$\tilde{B}_2 = \begin{pmatrix} B_{22} \\ B_{32} \end{pmatrix}$$

are the matrices corresponding to the two decoupled subsystems

$$\begin{aligned}\tilde{S}_1^D : \dot{\tilde{x}}_1 &= \tilde{A}_1 \tilde{x}_1 + \tilde{B}_1 u_1 \\ \tilde{S}_2^D : \dot{\tilde{x}}_2 &= \tilde{A}_2 \tilde{x}_2 + \tilde{B}_2 u_2\end{aligned} \tag{6.10.22}$$

and

$$\tilde{A}_{12} = \begin{pmatrix} 0 & A_{13} \\ 0 & A_{23} \end{pmatrix}, \tilde{A}_{21} = \begin{pmatrix} A_{21} & 0 \\ A_{31} & 0 \end{pmatrix},$$

$$\tilde{B}_{12} = \begin{pmatrix} B_{12} \\ B_{22} \end{pmatrix}, \tilde{B}_{21} = \begin{pmatrix} B_{21} \\ B_{31} \end{pmatrix} \tag{6.10.23}$$

are the interconnection matrices. Let us associate the following cost functions with each decoupled subsystem of (6.10.22).

$$\tilde{J}_1 (\tilde{x}_{10}, u_1) = \int_0^\infty (\tilde{x}_1^T \tilde{Q}_1 \tilde{x}_1 + u_1^T \tilde{R}_1 u_1) dt$$

and

$$\tilde{J}_2 (\tilde{x}_{20}, u_2) = \int_0^\infty (\tilde{x}_2^T \tilde{Q}_2 \tilde{x}_2 + u_2^T \tilde{R}_2 u_2) dt \tag{6.10.24}$$

where $\tilde{x}_{10}$, $\tilde{x}_{20}$ are the initial states of $\tilde{S}_1^D$ and $\tilde{S}_2^D$ and $\tilde{Q}_1$, $\tilde{R}_1$, $\tilde{Q}_2$, $\tilde{R}_2$ are appropriate matrices. The joint cost function $\tilde{J}(\tilde{x}_0, u)$ for the expansion $\tilde{S}$ is:

$$\tilde{J}(\tilde{x}_0, u) = \int_0^\infty (\tilde{x}^T \tilde{Q} \tilde{x} + u^T \tilde{R} u) dt \tag{6.10.25}$$

where $\tilde{Q} = \text{diag}(\tilde{Q}_1,\tilde{Q}_2)$, $\tilde{R} = \text{diag}(\tilde{R}_1,\tilde{R}_2)$.

By part (a) of the theorem in section (6.10.1), J of (6.10.25) is an expansion of

$$J(x_0,u) = \int_0^\infty (x^TQx + u^TRu)dt \tag{6.10.26}$$

with $Q = T^T\tilde{Q}T$ and $R = \tilde{R}$ which is the performance index of the original system S.

We can thus calculate the decentralised feedback control laws

$$u_1 = -\tilde{K}_1x_1 \text{ and } u_2 = -\tilde{K}_2x_2 \tag{6.10.27}$$

which are calculated so as to minimise $\tilde{J}_1$ and $\tilde{J}_2$ subject to the dynamical constraints of $\tilde{S}_1^D$ and $\tilde{S}_2^D$. The control law

$$u = -Kx$$

which needs to be implemented on the original system is given by

$$K = \left(\begin{array}{c|c|c} K_{11} & K_{12} & 0 \\ \hline 0 & K_{23} & K_{24} \end{array}\right)$$

Siljak *[18]* has also developed a suboptimality index when such a control law is implemented.

6.11 CONCLUSIONS

In this final chapter, we have described the main currently available design techniques of decentralised control. This is, however, an active field and many new methods will no doubt be developed in the future.

6.12 REFERENCES

1. Wang S. and Davison E.J. "On the stabilisation of decentralised control systems" IEEE Trans AC 18, pp. 473-478, 1973.

2. Geromel J. and Bernussou J. "An algorithm for optimal decentralised regulation of linear quadratic interconnected systems" Automatica, 15, pp. 489-491, 1979.

3. Bellman R. "Introduction to matrix analysis" Academic Press, New York.

4. Singh M.G. and Titli A. "Systems: decomposition, optimisation and control" Pergamon Press, Oxford, 1978.

5. Levine W. and Athans M. "On the determination of optimal output feedback gains for linear systems" IEEE Trans AC 15, 44, 1970.

6. Hassan M.F. and Singh M.G. "A hierarchical structure for computing near optimal decentralised control" IEEE Trans SMC8, 7, pp. 575-579, 1978.

7. Hassan M.G. and Singh M.G. "Hierarchical successive approximation algorithms for non-linear systems" LSS, 1, 3, 1980.

8. Singh M.G. and Hassan M.F. "A closed loop hierarchical solution for the continuous time river pollution control problem" Automatica, 12, pp. 359-363, 1976.

9. Siljak D. "On stability of large scale systems under structural perturbations" IEEE Trans SMC, pp. 415-417, 1975.

10. Anderson B. and Moore J. "Linear optimal control" Prentice Hall, 1971.

11. McBrinn D. and Roy R. "Stabilisation of linear multivariable systems by output feedback" IEEE Trans AC, pp. 243-245, 1972.

12. Singh M.G., Hassan M.F. and Titli A. "A feedback solution for large interconnected dynamical systems using the prediction principle" IEEE Trans SMC6, pp. 223-229, 1976.

13. Johnson C.D. "Further study of the linear regulator with disturbances: the case of vector disturbances satisfying a linear differential equation"

14. Hassan M.F., Singh M.G. and Titli A. "near optimal decentralised control with a fixed degree of stability" Automatica, 15, pp. 483-488, 1979.

15. Hassan M.F. and Singh M.G. "robust decentralised controller for linear interconnected dynamical systems" Proc. IEE, Vol. 125, 5, pp. 429-432, 1978.

16. Hassan M.F. and Singh M.G. "decentralised controller with on-line interaction trajectory improvement" Proc. IEE, Vol. 127, pt D3, pp. 142-148, 1980.

17. Markland C. "optimal model following control system synthesis technique" Proc. IEE, Vol. 117, 3, March 1970.

18. Ikeda M., Siljak D. and White C. "Suboptimal decentralised control of systems with overlapping information sets" JOTA, 1980.

19. Armentano V. and Singh M.G. "A new approach to the decentralised controller initialisation problem" CSC Report 485, Manchester, 1980.

20. McBrinn D.E., Roy R.J. "Stabilisation of Linear Multivariable Systems by Output Feedback", IEEE Transactions on Automatic Control, pp. 243-245, April, 1972.

21. Lancaster P., "On Eigenvalues of Matrices Dependent on a Parameter", Numerische Mathematik 6, pp. 377-387, 1964.

22. Davison E.J., Wang S.H. "Properties of linear time-invariant, multivariable system subject to arbitrary output and state feedback",

IEEE Transactions on Automatic Control, Vol. AC-18, pp. 24-32, February, 1973.

23. Singh M.G., Hassan M.F. and Titli A. "A feedback solution for large interconnected dynamical systems using the prediction principle", IEEE Trans. SMC, 6, pp. 223-239, 1976.

AUTHOR INDEX

SUBJECT INDEX